TECHNIQUES OF CHEMISTRY

VOLUME XI

CONTEMPORARY LIQUID CHROMATOGRAPHY

BY

R. P. W. SCOTT
Chemical Research Department
Hoffmann-La Roche Inc.
Nutley, New Jersey

A WILEY-INTERSCIENCE PUBLICATION

JOHN WILEY & SONS

New York · London · Sydney · Toronto

Copyright © 1976 by John Wiley & Sons, Inc.

All rights reserved. Published simultaneously in Canada.

No part of this book may be reproduced by any means, nor transmitted, nor translated into a machine language without the written permission of the publisher.

Library of Congress Cataloging in Publication Data:

Scott, Raymond Peter William, 1924-
  Contemporary liquid chromatography.

  (Techniques of chemistry; v. 11)
  "A Wiley-Interscience publication"
  Includes bibliographical references.
  1. Liquid chromatography. I. Title.

QD61.T4 vol. 11 [QD79.C454] 542'.08s [543'.08] 76-15553
ISBN 0-471-92900-X

Printed in the United States of America

10 9 8 7 6 5 4 3 2 1

VOLUME IX
> In Preparation

VOLUME X
> APPLICATIONS OF BIOCHEMICAL SYSTEMS IN ORGANIC CHEMISTRY, in Two Parts
> *Edited by J. Bryan Jones, Charles J. Sih, and D. Perlman*

VOLUME XI
> CONTEMPORARY LIQUID CHROMATOGRAPHY
> *R. P. W. Scott*

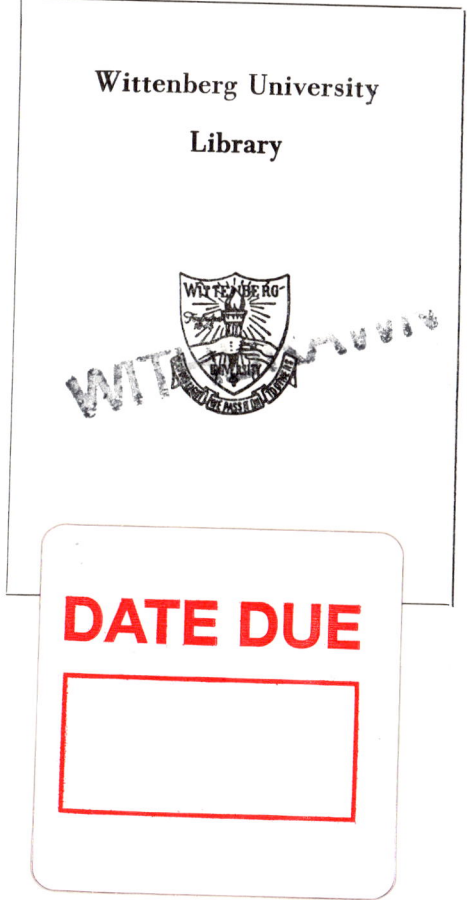

TECHNIQUES OF CHEMISTRY

ARNOLD WEISSBERGER, *Editor*

VOLUME XI

CONTEMPORARY LIQUID CHROMATOGRAPHY

INTRODUCTION TO THE SERIES

Techniques of Chemistry is the successor to the Technique of Organic Chemistry Series and its companion—Technique of Inorganic Chemistry. Because many of the methods are employed in all branches of chemical science, the division into techniques for organic and inorganic chemistry has become increasingly artificial. Accordingly, the new series reflects the wider application of techniques, and the component volumes for the most part provide complete treatments of the methods covered. Volumes in which limited areas of application are discussed can be easily recognized by their titles.
Like its predecessors, the series is devoted to a comprehensive presentation of the respective techniques. The authors give the theoretical background for an understanding of the various methods and operations and describe the techniques and tools, their modifications, their merits and limitations and their handling. It is hoped that the series will contribute to a better understanding and a more rational and effective application of the respective techniques.
Techniques used in the *separation and purification* of materials have been included in several volumes of *Techniques of Organic Chemistry*. Volume III (1950) describes Thermal Diffusion of Organic Liquids; Barrier Separations; Dialysis and Electrodialysis, Zone Electrophoresis; Laboratory Extraction and Countercurrent Distribution; Crystallization and Recrystallization; Centrifuging; Filtration; Solvent Removal, Evaporation and Drying. Volume V (1951) treats Adsorption and Chromatography. Volume X (1957) deals with the fundamentals of Chromatography.
Volume IV, Distillation, published in 1951, was revised into the second edition in 1965. At that time Dr. E. S. Perry joined the series editor, Dr. A. Weissberger, to assist in the editorial responsibilities for volumes on *Techniques for Separation and Purification*. Volume XII, Thin Layer Chromatography, and Volume XIII, Gas Chromatography, were published under this joint editorship.
With Volume VII of Techniques of Chemistry, Membranes in Separations by Hwang and Kammermeyer, this collaboration was transferred to the new series. It continues with the present volume on Contemporary Liquid Chromatography

by Raymond P. W. Scott and with volumes on other *Techniques of Separation and Purification* now in preparation.

Authors and editors hope that readers will find the volumes in this series useful and will communicate to them any criticisms and suggestions for improvements.

EDMOND S. PERRY
ARNOLD WEISSBERGER

Research Laboratories
Eastman Kodak Company
Rochester, New York

PREFACE

In the early 1960s a renaissance took place in the field of liquid chromatography. This renaissance resulted from both the successes and failures of gas chromatography, a technique that had developed with amazing rapidity over the previous decade. On the one hand, the development of gas chromatography provided a real theoretical basis to explain the processes involved in chromatographic separations, and this knowledge could be applied with appropriate modification to improve the separations obtainable by liquid chromatography. On the other hand, the failures of gas chromatography to cope with highly polar and involatile substances provided an impetus for workers to develop the liquid chromatograph in order to solve the difficult and sometimes long-standing separation problems associated with such materials. As a result, modern high-performance liquid chromatography was born accompanied by the introduction of first the pellicular type packings; these in turn gave way to the microparticulate packings, new and improved detectors, and columns operating at pressures of many thousands of pounds per square inch.

The industrial and academic scientist, in his everyday problems, is faced with a far greater proportion of highly polar and involatile materials to separate than volatile substances amenable to separation by gas chromatography, and it follows that liquid chromatography has an even greater potential use in the laboratory then gas chromatography. It is likely that over three-quarters of the laboratory separations carried out in the future will be by some form of liquid chromatography.

This book has been written to provide the reader with a basic understanding of chromatographic principles, a knowledge of liquid chromatographic apparatus and how to use it, and finally to acquaint him with the more recent developments in the technique. An extensive theoretical treatment of the subject has been given to provide a real appreciation of the fundamentals of liquid chromatography, but the mathematics involved have been kept as simple as possible. In dealing with the experimental aspects of the technique, much has been taken from practical experience, both of my own and that of my colleagues.

I would like to express my grateful thanks to my coworkers Dr. C. G. Scott,

Mr. P. Kucera, and Mr. C. Reese for their many helpful suggestions in the preparation of this book, to Professor Eli Grushka and Dr. James Edwards for reading the manuscript and their very useful comments, and to Mrs. C. Caso for typing the manuscript.

R. P. W. SCOTT

Nutley, New Jersey
February 1976

CONTENTS

Chapter I
Principles, Classification, and Nomenclature. 1

Chapter II
The Theory of Chromatography 23

Chapter III
Liquid Chromatography Apparatus 101

Chapter IV
Stationary and Mobile Phases
for Liquid Chromatography 193

Chapter V
Chromatographic Procedures. 245

Chapter VI
The Combination of Liquid Chromatography
with Spectroscopic Techniques 273

Chapter VII
Preparative Liquid Chromatography. 297

Index 323

Chapter I

PRINCIPLES, CLASSIFICATION, AND NOMENCLATURE

1 Brief History of Chromatography   1
2 Definition of Chromatography   4
3 Classification of Different Forms of Chromatography   6
4 The Scope and Application of the Different Forms of Chromatography   7
   Gas-Solid Chromatography   7
   Gas-Liquid Chromatography   8
   Critical-State Chromatography   8
   Thin-Layer Chromatography   8
   Paper Chromatography   8
   Liquid-Solid Chromatography   8
   Liquid-Liquid Chromatography   9
5 The Basic Liquid Chromatograph   9
   Mobile Phase Supply   9
   Injection System   10
   Column and Column Oven   10
   Detectors   10
   Detectors Electronics   10
   Recorder   10
6 Methods of Development in Liquid Chromatography   11
   Frontal Analysis   11
   Elution Development   12
   Displacement Development   15
7 The Chromatogram, Terms and Nomenclature   16

1 BRIEF HISTORY OF CHROMATOGRAPHY

The first scientist to recognize chromatography as an efficient method of separation was the Russian botanist Tswett, who used a primative form of liquid-solid chromatography to isolate various plant pigments. In his paper published in 1903 he described the results he had obtained using over 100 different adsorbents. Little further work was carried out until 1931 when Kuhn and Lederer repeated some of Tswett's experiments, again using chromatography to separate plant pigments. Kuhn and Lederer used alumina and calcium carbonate as adsorbents, and their results encouraged other workers to investigate chromatography as a new method of separation.

Elution chromatography was first effectively employed by Reichstein in the late 1930's and in the early 1940's Martin and Synge introduced liquid-liquid partition chromatography using silica gel loaded with water as a chromatographic stationary phase. The separations carried out by Martin and Synge were, in fact, a mixture of adsorption and partition chromatography. About the same time, Tiselius developed frontal analysis and displacement analysis. In a paper by Martin and Synge published in 1941 the possibilities of gas chromatography were suggested; Martin stated that if the liquid mobile phase were replaced by a gas, the increase in the transfer rate of the solute in the mobile phase should result in highly efficient separations.

However, during this time, Martin and his co-workers were developing liquid-liquid partition chromatography using paper as the support material, and it is surprising that no other workers took up Martin's suggestion for gas chromatography. In 1939 Brown suggested that an adsorbent held between two glass plates, through which the mobile phase was allowed to percolate, could be an effective chromatographic system, and in 1946 Williams developed the idea into a practical separation system. Many workers elaborated on this new idea, but it was not until 1949, when Maclean and Hall introduced the starch binder into an alumina absorbent, that the first effective form of thin-layer chromatography (TLC) made its appearance.

It was found that TLC had advantages over paper chromatography because of the larger forces involved between solute molecules and adsorbent relative to solute molecules and a liquid.

TLC was extensively developed and, with the introduction of efficient TLC spreaders by Stahl in 1956, it became an extremely effective separation technique with a very wide field of application. Up to this time, development in liquid column chromatography had been relatively slow and those developments that had occurred were largely confined to the introduction of new phase systems and some crude detectors. The work in this field was also eclipsed by the indroduction of a new and very powerful chromatographic process, gas chromatography.

In the early 1950s, James and Martin turned their attention to the possibility of separating substances by the distribution of solutes between gas and liquid, and in 1952 they published their paper describing the first gas chromatograph. The results were startling and attracted a great deal of attention. Every major oil company and university interested in volatile substances commenced development of the technique. In a few years GLC had completely replaced analytical low-temperature distillation and was being employed by the solvent industry, the essential oil industry, and in the field of biochemistry. The rate of development of this technique was unique in the history of analytical instrumentation. It attracted all types of scientists, ranging from well-established, international research teams to imaginative gadgeteers. There were several reasons for this. First, the basic chromatograph was simple and inexpensive to make; providing the operator was prepared to plot the elution curves by hand, it could be constructed for less than $100. Second, the problem of detection was relatively simple, since the presence of a solute vapor in a permanent gas significantly modifies the physical characteristics of that gas, and the measurement of thermal conductivity, density, and calorific value became effective but simple methods for detection. Finally, even for the inexperienced, coating a support and packing a column was a very simple matter and resulted in impressive separations for that time. Between 1954 and 1962 the development of GLC was almost meteroric but after 1962 development slowed down and was largely confined to the association of the gas chromatograph with acillary techniques such as mass spectroscopy, infrared spectroscopy, and pyrolysis.

During the development of the GLC, liquid chromatography became the Cinderella of the chromatographic techniques; although a few new phase systems were introduced, no significant improvements were described. However, despite the wide range of applications of GLC, a large number of separation problems remained, particularly in the high molecular weight field and where thermally labile and highly polar materials had to be separated. Thus attention was turned to the development of liquid column chromatography, but the problems associated with liquid chromatography were found to be far greater than those met with in gas chromatography. Detection systems based on the bulk properties of the column eluents were insensitive because a solute did not significantly modify the overall properties of the mobile phase and where high sensitivity detectors were devised (e.g., UV absorption) they were highly selective and limited the choice of mobile phase. Furthermore, the mobile phase was a liquid and whereas gases could be obtained directly from cylinders in a state of high purity, liquids had to be purified before they were satisfactory for use. Probably the most difficult problem has resulted from the fact that whereas in gas chromatography the distribution coefficients are directly a function of the forces between the solutes and the absorbents since there is no interaction in the gas phase, in liquid chromatography distribution coefficients depend on the relative difference between forces exerted on the solute molecule in one liquid and another or between a liquid and a solid surface. Where significant distribution coefficients could be obtained between liquids, the absolute solubility of the solute in the mobile phase was so small that detection became difficult at the present levels of sensitivity. Nevertheless, slow but steady progress has been made in column chromatography. The theory of the band spreading in a column is fairly well understood. The sensitivities of detectors have been increased, although they are still many orders of magnitude lower than gas chromatography detectors, and the liquid chromatographs can now have a performance approaching that of the early gas chromatographs. Much remains to be developed before the techniques can match the sophistication and performance of modern gas chromatography. Furthermore, liquid column chromatography will not develop with the same rapidity because the instrumental problems encountered are far more difficult to overcome. However, significant contributions have been made over the past few years and there are now a number of commercial liquid chromatographs available.

2 DEFINITION OF CHROMATOGRAPHY

Chromatography has been classically defined as a separation method that is achieved by the distribution of substances between two phases, a mobile phase and a stationary phase. Those substances distributed preferentially in the mobile phase will move through the system more rapidly than those preferentially distributed in the stationary phase.

Although scientifically correct, this definition is rather vague; while it introduces the essential concept of a mobile and a stationary phase, it obscures the basic phenomena by which a separation is effected in the term "distribution." The distribution of a solute between two phases results from the balance of forces between solute molecules and the molecules of each phase. These forces can be polar in nature, arising from permanent or induced electric fields associated with both solute and solvent molecules, or be due to London's dispersion forces (van der Waals forces), which depend on the relative masses of the solute and solvent molecules. In ion-exchange chromatography, the forces on the solute molecules will be substantially ionic in nature but will include the so-called polar and nonpolar forces as well. Separations in chromatography are, therefore, achieved by exploiting the different molecular forces that can occur between each solute and the two phases.

The role played by the stationary phase now becomes apparent. If it is required to separate substances of different polarity, then polar forces may be used to effect the separation. Thus a polar stationary phase should be chosen and, at the same time, to ensure that the selectivity is maintained predominantly in the stationary phase, a relatively nonpolar mobile phase should be employed. Conversely, if it is required to separate substances of roughly the same polarity but of different molecular weights, then a nonpolar stationary phase will effect selectivity by London's dispersion forces and the mobile phase would now be polar or semipolar. It is essential to appreciate that the whole basis of chromatography (like many other separation systems) depends on exploiting different intermolecular forces. The magnitude of the difference that is necessary to achieve a separation of two substances will depend on the efficiency of the chromatographic apparatus, but however efficient the apparatus may be, if the forces exerted by the molecules of the mobile and stationary phases on the two solute types are identical, then separation can never be effected.

It follows that any change in the environment of a chromatographic system that can effect these intermolecular forces could be used to change or enhance a separation. Consider the effect of temperature on the intermolecular forces. Strictly, all distribution systems are dynamic and should be treated statistically. However, for simplicity, one can consider that the net force holding a molecule in one phase is the difference between the static forces between molecules and the dynamic forces (due to molecular vibration) that tend to break the molecular association. Raising the temperature of the system will increase the kinetic energy of each molecule and reduce the net force holding it in a particular phase. This effect will occur between the solute molecules and those of both phases, but if the net effect of a change in temperature on one type of solute differs from that on another then an enhancement of the separation can result. From this it can be seen that temperature must be provided as an operating parameter of any chromatographic apparatus.

In a similar way it would appear that operating a chromatographic system in an intense electric field would increase the polar forces between molecules and affect the separation obtained. Although such a procedure is theoretically possible, unfortunately the electric fields close to the molecule are so large that it would be impossible, practically, to produce a sufficiently high external field to have any significant effect without electrical breakdown of the medium.

If the mobile phase is a gas, and thus compressible, then the intermolecular forces between the solute and gas molecules can be increased by raising the absolute pressure. At normal pressures, molecular interactions in the gas phase are small but at high pressures they become significant and can affect the separation obtained [1]. The recent work [2] using volatile liquids above their critical temperature as the mobile phase effects an interesting compromise between gas and liquid chromatography. Under these conditions intermolecular forces in the mobile phase become large enough to give a solute selectivity and increase absolute solubility of normally involatile substances while maintaining some of the desirable transfer characteristics of a gas.

Chromatography may, therefore, also be defined as a separation process that is achieved by exploiting the different intermolecular forces that are exerted on solutes when distributed between a mobile and stationary phase. Those substances that are held more strongly in the mobile phase pass through or from the system more rapidly than those that are held more strongly in the stationary phase. Thus, the individual substances will move through or from the system in order of the increasing forces that hold them in the stationary phase.

3 CLASSIFICATION OF DIFFERENT FORMS OF CHROMATOGRAPHY

The various forms of chromatography can be deduced by permutating any two from the three basic phases, gas, liquid, and solid, with the proviso that the two chosen phases be immisible and one of the pair be mobile. This suggests two of the three basis forms of chromatography—gas chromatography and liquid chromatography, where the mobile phases are gas and liquid, respectively. Each type of mobile phase can be paired with either a liquid or solid stationary phase. Thus, each of the above basic forms of chromatography gives rise to two secondary forms: gas chromatography gives rise to gas-liquid and gas-solid chromatography and liquid chromatography to liquid-liquid and liquid-solid chromatography. Where the mobile phase is a liquid operating close to its critical temperature and pressure, then the mobile phase cannot be termed a liquid or a gas and

Table 1.1. Classification of Chromatography

GC Gas Chromatography Mobile Phase: Gas	
GSC Gas-solid Chromatography Stationary phase: solid	GLC Gas-liquid chromatography Stationary phase: liquid
CSC Critical-State Chromatography Mobile Phase: Liquid in Critical State	
CSSC Cirtical-state solid chromatography Stationary phase: solid	CSLC Critical-state liquid chromatography Stationary phase: liquid
LC Liquid Chromatography Mobile Phase: Liquid	
LSC Liquid-solid chromatography Stationary phase: solid	LLC Liquid-liquid chromatography Stationary phase: liquid

the process has been given the name critical-state chromatography. It is better to consider critical-state chromatography as a third basic chromatographic system; it then gives rise to critical-state liquid-solid chromatography and critical-state liquid-liquid chromatography.

Table 1.1 show how the various forms of chromatography can be classified. The table is self-explanatory, but some comments should be made with reference to LSC and LLC. Both these forms of chromatography have been developed utilizing distribution systems that have different geometric forms. If the distribution occurs on a lamina, with a solid absorbent (LSC), this is called thin-layer chromatography (TLC), and if it occurs on liquid absorbed on paper (LLC) then it is called paper chromatography (PC). If the separations are carried out in a column system, then the two forms of liquid chromatography are liquid-liquid column chromatography [LLC(C)] or liquid-solid column chromatography [LSC(C)]. The only apparent anomaly is the classification of paper chromatography as liquid-liquid chromatography. This is because the paper for the most part only plays a supporting role for the liquid (usually water) that is absorbed on it and acts as a stationary liquid phase. Any separations involving ion-exchange resin will also be classed as liquid-solid chromatography. Gel permeation, however, cannot be classed as a form of chromatography since it does not affect separations on the basis of molecular interaction but on a basis of molecular size and is, in fact, a filtration or sieving procedure. It follows that the term gel permeation is a misnomer and the original term gel filtration gives a more correct description of the separation technique. Since gel permeation is not a chromatographic procedure it will not be discussed in this book.

4 THE SCOPE AND APPLICATION OF THE DIFFERENT FORMS OF CHROMOTOGRAPHY

Gas-Solid Chromatography

Gas-solid chromatography, using stationary phases such as silica gel, alumina, or molecular sieves, is employed almost exclusively for the analysis of permanent gases and low-boiling hydrocarbons. The reason for this is that when high molecular weight substances, or substances that have even slight polarity, are distributed between a gas and a solid adsorbent, the forces between the solute and stationary phase are excessively high and do not permit the movement of the substances through the chromatographic system in reasonable time. Attemps have been made to reduce the activity of adsorbents such as alumina, which has permitted the separation of normal hydrocarbons to carbon numbers of 40 or 50, when operating at 300-400°C. However, the deactivation is not sufficiently effective to reduce the retention of polar substances and the use of gas-solid chromatography for applications other than those mentioned above is still very limited. The most successful attempts to chromatograph polar compounds using solid adsorbents has been with benzene di-vinyl styrene copolymers; in this case the materials have high nonpolar surface activity and, while they strongly retain hydrocarbons, they virtually repel polar compounds such as water, acetic acid, and alcohol.

Gas-Liquid Chromatography

GLC is used to separate all types of volatile materials provided they are stable and provided that they are not excessively polar. The inability of GLC to separate volatile, highly polar materials is largely due to the adsorbtion of the support material. In GLC the stationary liquid phase has to be held on a support, usually an inert diatomaceous earth, such as Celite. Although relatively inert, Celite still has some residual absorption that holds the polar materials in the system and can render the separations impossible, or a least very poor in quality. However, some success has been achieved in the separation of highly polar materials by forming derivatives of lower polarity and in some instances by the use of special adsorbents.

Critical-State Chromatography

The technique of critical-state chromatography is very much in its infancy; in fact, relatively few papers have been published on the subject. So far its field of application is not clearly defined. It has been used for the successful separation of polynuclear aromatic hydrocarbons and of chlorinated hydrocarbons.

Thin-Layer Chromatography

TLC is now widely employed, particularly in the biochemical field, and covers diverse applications ranging from the separation of lipids, fat-soluble dyes, and polynuclear aromatic hydrocarbons to peptides and amino acids, where it has now replaced paper chromatography.

Paper Chromatography

The use of paper chromatography is now confined almost exclusively to the separation of sugars, nucleotides, polypeptides, and inorganic ions. It is likely that some of these applications may eventually be taken over by TLC, and in due course, by column chromatography.

Liquid-Solid Chromatography

Liquid-solid chromatography has been used for some time, employing adsorbents such as silica gel, alumina, Sephadex, and ion-exchange beads as stationary phases in the separation of amino acids, polynuclear hydrocarbons, and a wide range of both high molecular weight substances and substances of high polarity. The full application of liquid-solid chromatography has not yet been realized, because the techniques and equipment are still being developed.

Liquid-Liquid Chromatography

Liquid-liquid chromatography has been used in the separation of many substances of biological origin, including free fatty acids, but it has not been as widely employed as liquid-solid chromatography. One of the difficulties encountered in this system is that of obtaining a permanent coating of stationary phase on the support medium. If this is not achieved, then the presence of the stationary phase in the mobile phase may affect the performance of the detector. This problem has been to some extent overcome by the new "bonded phases" where substances that exhibit the properties of liquid phases are chemically bonded to the surface of suitable supports. A bonded phase is, however, not strictly a liquid stationary phase but a hybrid between solid and liquid. Chromatography carried out using such materials as stationary phases therefore exhibit, in some degree, the properties of both a liquid-solid system and a liquid-liquid system.

5 THE BASIC LIQUID CHROMATOGRAPH

A diagram of a basic liquid chromatograph is shown in Fig. 1.1. The various

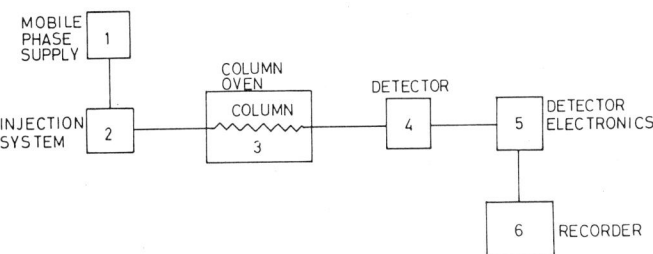

Fig. 1.1 The basic liquid chromatograph.

parts of the system will be discussed in detail in Chapter III. However, for the sake of continuity, some brief notes on the individual components of the apparatus are given below.

Mobile Phase Supply

This system supplies mobile phase to the chromatographic column at suitable flow rates. Depending on the pressure drop across the column the mobile phase driving force can range from gravity feed devices to a high-pressure pump. In some instances pressures up to 5000 lb/in.^{2} are necessary to cope with either very small diameter columns, very long columns or, in some cases, column packing consisting of particles of small diameter.

Injection System

This is a device that allows the introduction of a sample of the mixture to be separated into the stream of mobile phase, with or without arresting the column flow so that the sample is carried as a sharp band on to the column. One common form of injection system employs a suitably supported plastic septum and hypodermic syringe. Suitably constructed, such a system can operate satisfactorily at very high inlet pressures.

Column and Column Oven

A column can be a tube of virtually any length and diameter, and can be constructed from any suitable material, ranging from glass to metal. For reproducible results, the column must be thermostated and its temperature must be maintained constant at a chosen value over the range $-40°C$ to $+200°C$. In some instances, temperature programming may be advisable, in which case the oven-column system must have relatively low thermal inertia. Some detecting systems require the column eluent to enter the detector at ambient or near-ambient temperature. Under such circumstances a suitable heat exchanger will have to be employed to bring the temperature of the eluent back to ambient, subsequent to having passed through the column.

Detectors

The detector can work on the principle of measuring some bulk property of the eluent such as dielectric constant or refractive index, in which case the detector usually takes the form of a flow-through cell. If the detector is designed to operate *solely* on the physical properties of the solute, then the detector can embody a transporting system such as a wire to continuously take a sample from the eluent for subsequent detection. The solvent is then removed by evaporation and the solute subsequently detected on the transporting medium, for example, by pyrolysis.

Detector Electronics

The electronics associated with the detector will depend on the detecting system employed. The transport detector may incorporate a flame ionization detector amplifier. A UV absorption detector will include a photoelectric cell and associated equipment. The deteeter electronics will also provide some means of sensitivity control and may incorporate integration facilities.

Recorder

The recorder presents a concentration profile of the solutes leaving the column as an analog curve. The full-scale deflexion (FSD) of the recorder will depend on the output from the detector electronics and may range from 1 mV to 1 V. To date, the demands on the recorder speed of response by liquid chromatography are not as severe as those of gas chromatography. A balancing time of about 1 sec is usually more than adequate and chart speeds varying between 3 cm/hr and 3 cm/min will cope with most separations.

It is hard to predict the form that the liquid chromatograph will take in the distant future, but the basic system described will be applicable to the great majority of liquid chromatographs that will be produced within the next decade. There will no doubt be new types of detectors developed that may require special ancillary equipment and it is probable that many liquid chromatographs will be fitted with automatic fraction collectors. However, the latter can hardly be considered part of the basic liquid chromatograph. In the future, eluents from the column will be passed automatically to suitable spectroscopic equipment to produce absorption spectra, and this may well be achieved by means of an interrupted elution procedure already employed in gas chromatography.

6 METHODS OF DEVELOPENT IN LIQUID CHROMATOGRAPHY

The use of the term development in chromatography may have historical connotations. The colored bands originally produced by Tswett and the method of displaying solute bands on paper and thin-layer plates by reaction with suitable reagents, possibly suggested a similarity to a photographic procedure. Like the term chromatography itself, which now has no general connection with color, the use of the word development has persisted to describe the process of effecting a chromatographic separation.

There are three basic systems of chromatographic development: displacement development, frontal analysis, and elution development, although displacement analysis might be considered a special case of elution chromatography. Of the three methods, elution chromatography is by far the most used method of development, because the separations can be complete and discrete bands of each solute obtained from the chromatographic system.

Frontal Analysis

There are two important ways in which frontal analysis differs from elution development. First, in frontal analysis the sample is continuously fed onto the column during development, either as a pure sample or as a solution in the mobile phase, whereas in elution development a discrete sample is placed on the column and the chromatogram subsequently developed. Second, elution development can, under the right circumstances, be made to completely separate all the components of a mixture, whereas in frontal analysis only part of the first compound eluted is in a relatively pure state, each subsequent component being mixed with those previously eluted.

Consider a three-component mixture, containing equal quantities of each component, fed onto a column; because of the forces between solute and stationary phase, each solute will be retained to a different extent as it saturates the stationary phase while passing through the column. The first component to elute will be that which is held least strongly in the stationary phase, then the second component will elute but in conjunction with the first component, and finally, the most strongly held of the three will elute in conjunction with the first and second components. Subsequently there will be no change in concentration of solute in the mobile phase and the concentration of the respective solutes will be the same as the feed mixture. The concentration profile resulting from such a frontal analysis, when operated under ideal conditions, is shown in Fig. 1.2a. The continuous curve shows the total concentration of solute in the eluent, plotted against volume of mobile phase passed through the column, and the dotted curves represent a similar concentration profile but for the individual components, not as a mixture. The ideal nature of this separation is represented by the vertical steps as each component passes from the column. In practice, sharp fronts are not obtained from concentration profiles due to the various spreading processes that occur in a column (Fig. 1.2b). These processes will be dealt with in detail in Chapter II, but at this stage the result of these spreading processes is a diffuse front to each solute profile. The effect of this is shown in the continuous curve in Fig. 1.2b, and the dotted curves again represent diffuse fronts of each component run separately. It can also be seen from the diagram that even the first component to be eluted contains traces of the other two components due to the diffuse fronts.

Frontal analysis was employed as a development procedure in the early stages of chromatography and before detection procedures were fully effective; it is not often used today. The reason for this is that no individual component is completely separated from the others in the mixture, except possibly the first component and even this will contain significant quantities of the other materials. If, after the complete profile is obtained, the sample feed is replaced by pure mobile phase, then the frontal analysis becomes elution development of a very large sample, the elution curve is the reverse of the frontal analysis curve, and the final step will give a relatively pure sample of the most strongly retained component. This situation is depicted in Fig. 1.2c. Today, the use of frontal analysis is restricted to specific separations for samples where, because of the nature of the components, there is no alternative procedure.

Elution Development

Elution development can be best described as a series of extraction-absorption processes that operate continuously during the passage of a solute band down a chromatographic column. The rate of absorption and rate of extraction will depend upon the relative forces between the solute molecule in the stationary phase and the mobile phase, respectively. If the forces on the solute in the mobile phase are greater than those in the stationary phase then the solute mole-

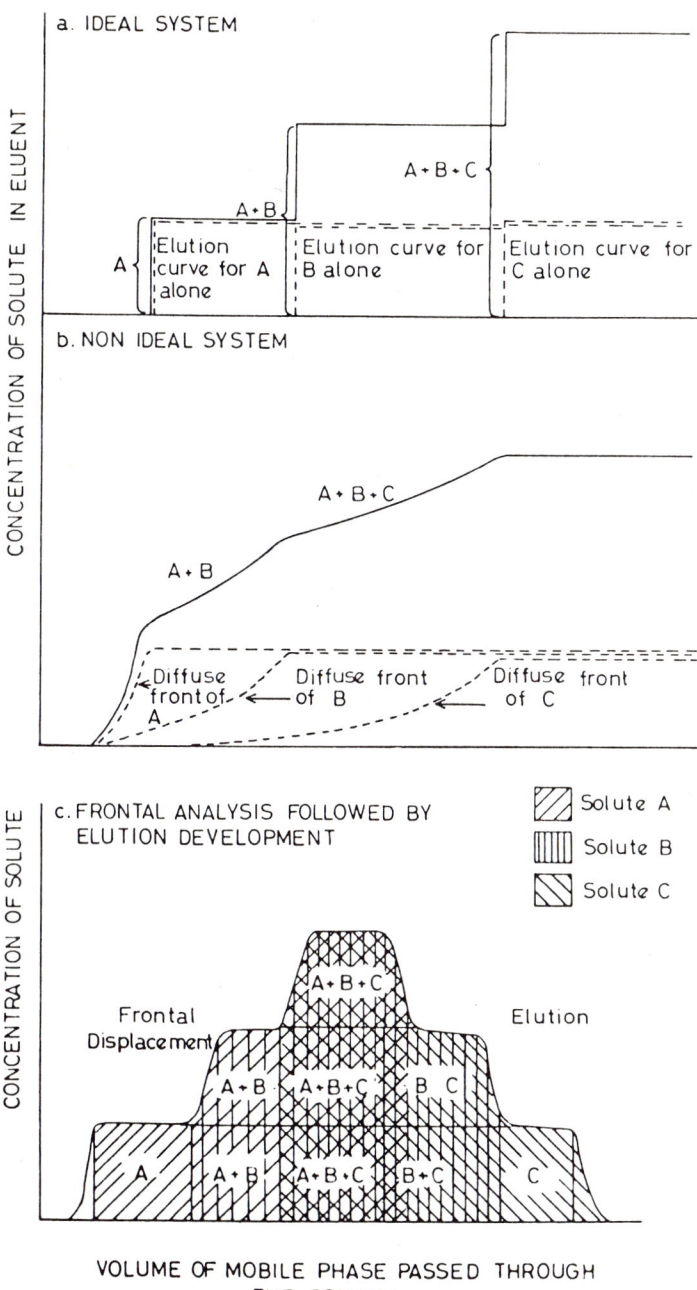

Fig. 1.2. Concentration profile for the frontal displacement of a mixture of three components.

cule on the average will spend more time in the moving phase and thus be transported more rapidly through the column. If, however, the forces on the solute in the stationary phase are greater than those in the mobile phase, then the molecules will spend more of their time in the stationary phase and thus travel slowly through the column system. The absorption and extraction process is continuous from the instant of injection to elution. The solute is first absorbed into the stationary phase and when fresh solute-free mobile phase passes over the point of absorption the solute is extracted back into the mobile phase. It will then be reabsorbed into the stationary phase at a point further along the column and the process repeated. For ideal elution development, where no band spreading processes occur in the column, the concentration profile is depicted in Fig. 1.3a. It is seen that each solute has been separated completely from its

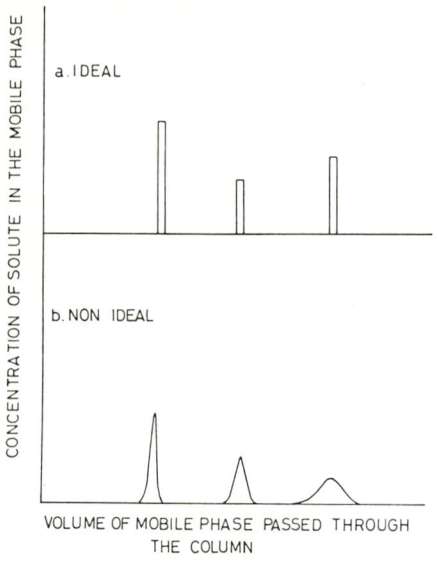

Fig. 1.3. Elution development of a three-component mixture.

neighbors. All the bands are the same width and identical with the band width of the injected sample. However, in practice, the exchange processes between mobile and stationary phase are not instantaneous; this, together with longitudinal diffusion, causes the bands to spread. Both these factors result in a chromatogram being formed having the shape shown in Fig. 1.3b. It is seen that a typical Gaussian elution curve is produced for each solute and that the extent of band-spreading increases with the extent to which the solute is retained in the column.

Elution development is by far the most effective method of development used in chromatography. Although other methods of development will be described, and some examples given, this book will deal largely with liquid chromatography by elution development.

Displacement Development

Displacement development depends on the competition between solutes for the active sites of the absorbent and is only really effective in separating very strongly absorbed materials. In displacement development all the substances in the sample will be held on the stationary phase so strongly that they cannot be eluted by the mobile phase; they can, however, be displaced by substances being held on the surface by stronger forces. However, there will be competition between individual solutes and when the sample is placed on the column, all the immediately available active sites of the adsorbent will be occupied by the most strongly held component. As the band of sample is moved down the column, the next available sites will be occupied by the next most strongly retained component. Thus, all the components array themselves along the column in order of their adsorption strength or, in effect, in order of the forces between them and the absorbent. To develop the chromatogram, another substance called the displacer is introduced into the mobile phase stream; the displacer has an even higher affinity for the absorbent than any of the components to be separated. Thus, on coming into contact with the sites occupied by the most strongly absorbed component, it will displace this into the mobile phase, which will then move it onto the next group of sites occupied by the next component and will displace this. Thus, the displacer forces the absorbed components progressively along the solumn, each component displacing the one in front, until they are eluted in the same order in which they were absorbed on the column, the least polar being eluted first. The concentration profile monitored at the end of the column from displacement development will take the form shown in Fig. 1.4.

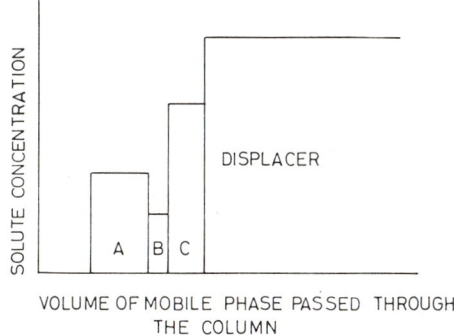

Fig. 1.4. Displacement development of three-component mixture.

16   PRINCIPLES, CLASSIFICATION, AND NOMENCLATURE

The order in which the solutes emerge will characterize the individual components and the length of the band, not the height, will be proportional to the concentration of any individual component. Because the capacity of the absorbent may vary for each type of molecule absorbed, the proportionality constant may not be the same for each substance. If the concentration-sensitive device used to monitor the concentration of solutes in the eluent has been made to have a response that is different for the individual components, then the height of the steps would also give a guide to the identity of the solute. For example, if a UV absorption system was used to detect the solutes then high steps would indicate the presence of substances having a high absorption coefficient in the UV.

Displacement chromatography has very limited applications as a separation technique. It depends on the substances to be separated having different response characteristic to the monitoring device, and furthermore, does not produce discrete local concentrations of individual components.

7 THE CHROMATOGRAM, TERMS AND NOMENCLATURE

The normal chromatogram is an analog curve relating concentration of solute in the eluent from the column with respect to time. If a constant column flow is maintained during the development of the chromatogram, the time axis will be directly proportional to the volume of eluent passed through the column. A typical chromatogram for a simple component eluted from the column is shown as the normal curve in Fig. 1.5. It will be seen later that provided the adsorption isotherm is linear, the elution curve is in the form of a error function curve. Because of this it would be worthwhile to consider the characteristics of error function curves and their derivatives. The basic error function is described by the equation

$$X = \frac{X_0}{\sqrt{2\pi n}} e^{-W^2/2n} ,$$

where X is the dependent variable and corresponds to the solute concentration.

W is the independent variable and corresponds to the column flow and X_0 and n are constants. This form of the error function is chosen because it can be directly derived as the equation for the elution curve from the plate theory of chromatography.

The first derivative of this is given by the equation

$$\frac{dX}{dW} = \frac{-X_0}{\sqrt{2\pi n}} \frac{W}{n} e^{-W^2/2n}$$

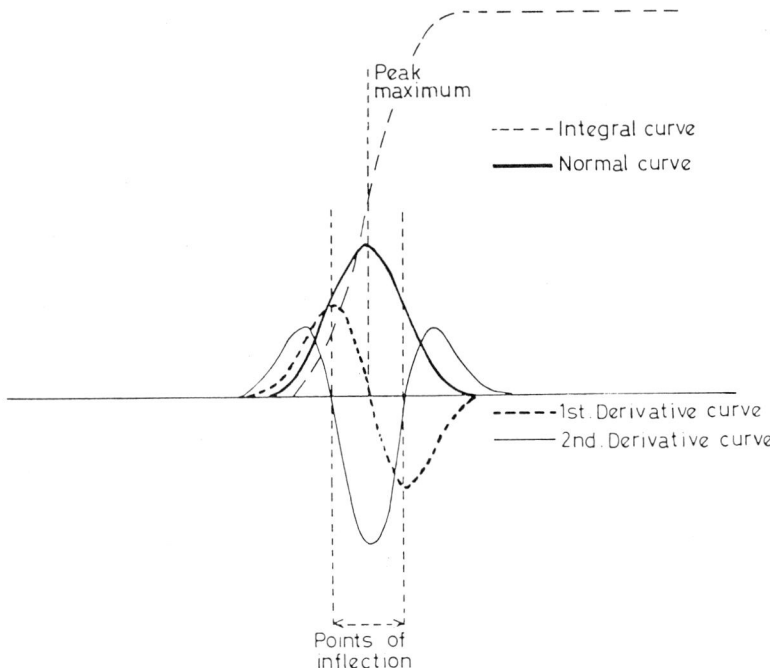

Fig. 1.5. The normal, integral and first- and second-derivative curves of the error function.

and the second derivative

$$\frac{d^2X}{dW^2} = \frac{X_0}{\sqrt{2\pi n}} \, e^{-W^2/2n} \, \frac{W^2-n}{n^2}$$

The curves represented by the equation are shown together with the integral curve in Fig. 1.5. The height of the integral curve is proportional to the area under the normal error function curve; this is useful in quantative analyses for measuring the total peak area on the chromatogram. The first derivative curve has also useful characteristics for measurement of peak areas.

The first derivative can identify the position of the peak maximum of a peak as, at that point, the value of this function goes from positive through zero to negative. An electrical analog of the normal curve that has been differentiated electronically can be made to actuate a time-counting print-out unit as the drivative signal becomes negative, and thus identify the retention time.

It is seen that at the point of inflection of the normal curve, the second derivative passes through zero and becomes negative. At the second point of inflection the second derivative passes through zero and becomes positive. The distance between these two points is equal to twice the standard deviation of the peak, that is, the peak width at 0.6065 of the peak height. Thus again if the chromatogram is produced as an electrical analog, then by differentiating the signal twice, the peak width can be determined by identifying the points where there is a change in sign of the second derivative. It should be pointed out, however, that if any noise is present, differentiating the signal will decrease the signal-to-noise ratio and thus decrease the sensitivity of the system.

In chromatography there are names given to these elution curves: the *normal chromatogram,* which is the curve relating concentration to time; the *integral chromatogram,* which relates the area under the normal curve of the eluted component to time; and the *first* and *second derivative chromatograms,* which are self-descriptive.

The majority of work is carried out using the normal chromatogram, and Fig. 1.6 depicts a chromatogram showing two resolved substances. When an injection is made a mark is usually made on the chromatogram; this is usually called the *injection point.* With any sample that is chromatographed, a substance that is not retained in the column is usually added to the mixture to aid in identification. The point at which this unretained substance is eluted is usually

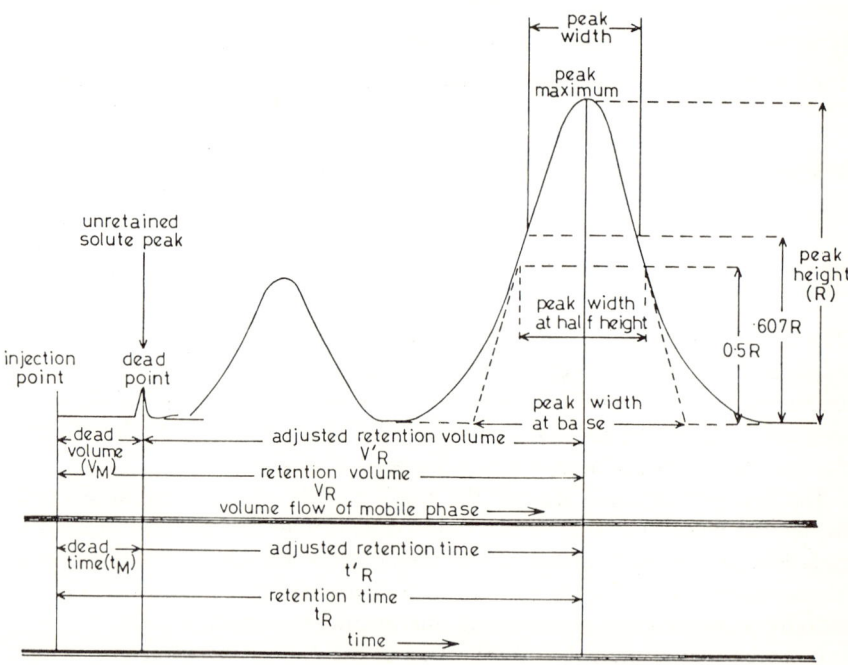

Fig. 1.6. The nomenclature of a chromatogram.

called the *dead point*. The volume of mobile phase that has passed through the column between the injection point and the dead point is called the *dead volume* of the column and in fact includes all the volume of the system between the injection point and the detector that is not occupied by stationary phase or support. This assumes that the two phases are in equilibrium and that there is no mobile phase isolated from the bulk mobile phase by the stationary phase. The *dead volume* is obtained by multiplying the dead time by the flow rate. The *baseline* is that portion of a chromatogram recorded when only mobile phase is emerging from the column. The point at the maximum concentration of any peak eluted is the peak *maximum* and the distance between the peak maximum and a line joining the base of the peak by extrapolation of the baseline is the *peak height*. There are various measurements used for the peak width. The width that has a theoretical significance and which should be used wherever possible is twice the standard deviation of the elution curve; it is measured at 0.6065 of the peak height. The product of that width and the peak height will always give 79.8% of the total peak area for a true error function curve.

For convenience, in quantitative analyses the peak area is often taken as the product of the peak height and the peak width at half the peak height. This parameter is given the term *peak width at half-height*. Another measure of the peak width, which is sometimes used, involves constructing the tangents to the points of inflection of the normal curve and measuring the distance between their intersection with the baseline produced. This is called the *peak width at the base* and is, in fact, equivalent to twice the peak width at 0.607 of the peak height.

Those characteristics of a chromatogram that have a theoretical significance will be discussed later, but the common measurements that are made on a chromatogram are given here. The time between the injection point and the peak maximum of any peak is called the *retention time,* and the time between the dead point and the peak maximum is called the *adjusted retention time.* If the retention time is multiplied by the flow rate, one obtains the *retention volume,* and similarly multiplying the adjusted retention time by the flow rate gives the *adjusted retention volume*.

In order to obtain some rationalization between gas and liquid chromatography the symbols used for the above parameters of the chromatogram are taken from those recommended for gas chromatography by the Gas Chromatography Discussion Group [3,4] and are as follows:

dead time, t_M
dead volume, V_M
retention time, t_R
retention volume, V_R
adjusted retention time, t_R'
adjusted retention volume, V_R'

Since the physical processes involved in gas chromatography are identical with those in liquid chromatography, the use of the same symbols for both techniques is permissible and will produce no confusion.

One further parameter ought to be mentioned, which to some extent is arbitrary but which is a useful measure of the column performance, and that is *resolution*. The significance of resolution will also be discussed in Chapter II. The value determined from the chromatogram, which is normally taken as a measure of resolution, is the ratio of the distance between the peak maxima of the two respective peaks to the sum of their peak widths at 0.6065 of the peak height.

Unfortunately, pure error function curves are not always produced in chromatographic separations. The reasons for this will also be discussed in due course but the shape that the peaks take will be given terms. Figure 1.7a shows a

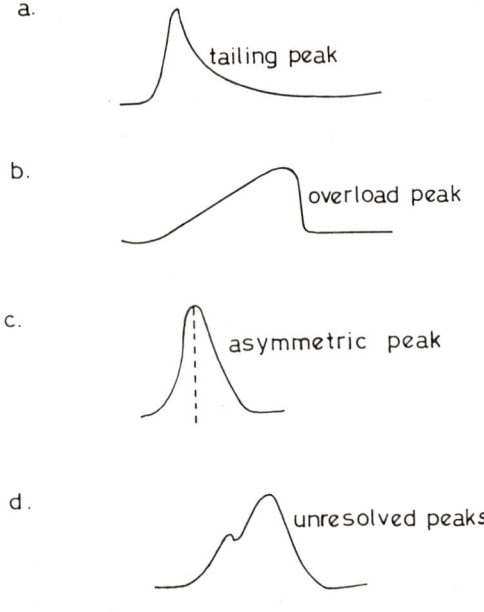

Fig. 1.7. Asymmetric peaks.

peak with a sharp front and a long tail. This is called a *tailing peak*. Figure 1.7b shows a peak with a sloping front and a sharp tail; this is called an *overloaded peak*. Figure 7.1c shows a curve that cannot be related directly to any one specific process and is called merely an *asymmetric peak*. Figure 1.7d shows two peaks where separation is incomplete; these are called *unresolved peaks*.

References

1. D. H. Desty, H. Goldup, G. R. Luckhurst, and W. R. Swanton, in *Gas Chromatography,*
   matography, M. van Swaay, Ed., Butterworths, London, 1962, p. 67-81.
2. J. C. Giddings, H. N. Myers, and J. W. King, *Advances in Chromatography,* A. Zlatkis, Ed., Preston Technical Abstracts Co., 1969, p. 346.
3. R. G. Primavesi et al., *J. Inst. P.,* **53**, No. 527, 367-381 (1967).
4. *Pure and Appl. Chem.,* **1**, No. 1, 177-186 (1960).

Chapter II

THE THEORY OF CHROMATOGRAPHY

1 **The Plate Theory** 25
   Retention Volume 29
   Peak Asymmetry 31
   Column Efficiency 33
   Position of the Points of Inflexion 35
   Resolving Power of a Column 36
   Effective Plate Number 40
   Gaussian Form of the Elution Equation 43
   Maximum Sample Volume 45

2 **The Rate Theory** 47
   Multipath Process 49
   Longitudinal Diffusion 49
   Dispersion due to Resistance to Mass Transfer in the Mobile Phase 50
   Dispersion due to Resistance to Mass Transfer in the Stationary Phase 51
   Resistance to Mass Transfer in Liquid-Liquid Chromatography 56
   The Transfer Coefficient 56
   Resistance to Mass Transfer in Liquid-Solid Chromatography 60
   Experimental Support of the Rate Theory 66
   Empirical Approach to the Relationship Between h and u 71

3 **Radial Dispersion** 72

4 **Dispersion in Capillary Connecting Tubes** 76

5 **The Peak Capacity of a Chromatographic System** 81
   Dependence of k' on Sample Concentration and Detector Sensitivity 86
   Dependence of k' on the Absorptive Capacity of the Stationary Phase 89
   Liquid-Solid System 89
   Liquid-Liquid Systems 93
   Injection of a Sample as a Solution in the Stationary Phase 94

The detailed study of the theory of any phenomenon or process permits a better understanding of the process and provides information on how it will be affected by the conditions under which it is carried out. Furthermore, theory can predict the optimum conditions that will ensure that the process operates at its maximum efficiency.

Since liquid chromatography has such a large number of operating variables and such a wide range of application, a clear understanding of the theory of the process is more essential for its effecient operation than with many other techniques. The theory of chromatography is now fairly well understood and can, indeed, show how solute characteristics and operating conditions affect the speed, resolution, and loading capacity that can be obtained from a given column system. It can also indicate how the system may be optimized to give any one of these attributes in excess. To do this the theory provides the basis for choosing a suitable distribution system, gives the necessary details for column design, and predicts the operating conditions necessary to effect a given separation.

There are two approaches to the theory of chromatography, the plate theory and the rate theory; the former was introduced by Martin and Synge [1] and is applicable to all forms of partition chromatography. The rate theory was developed for gas chromatography by van Deemter, Zuiderweg, and Klinkenberg [2], and was further extended for gas chromatography by Golay [3] for capillary columns and by Giddings [4] for packed columns. The extension of the rate theory to liquid chromatography columns has been carried out by Huber [5], Giddings [6], Horvath et al. [7] Kennedy and Knox [8], and others.

The approach to the plate theory is more straightforward and involves relatively simple mathematics, and for this reason will be considered first. The equations derived are quite general and can apply to either gas chromatography or liquid chormatography. Before discussing the plate theory in detail it would be useful to consider more specifically what kind of knowledge the theory should be expected to provide. Consider the chromatogram shown in Fig. 2.1. It would be

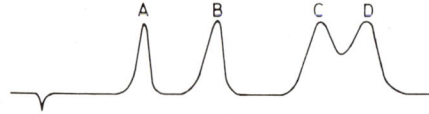

Fig. 2.1. Chromatogram of partially resolved mixture.

useful to know what factors affect the order of elution of the components A, B, C, and D and whether any of the factors are related to the chemical characteristics of the solutes and so enable their identification from retention time and retention volume measurements. The determination of those factors that affect retention times will have other important ramifications since they will also condition the separation or analysis time Figure 2.1 depicts components A and B completely resolved whereas C and D are only partially separated. It is clear that C and D would be resolved if either the peaks were moved further apart or their widths were reduced. It would therefore be of value to know those factors that affect the width of a peak, in order to modify the chromatographic system and to improve the resolution. The theory can therfore be expected to provide the following information:

1. An equation that shows those parameters that govern the retention volume and retention time of a solute.
2. An equation that describes the degree of separation between two peaks.
3. An equation that shows those parameters that determine the width of a peak.

1 THE PLATE THEORY

Primarily the plate theory provides an equation for the elution curve of a solute relating the concentration of solute at any point in the column to the volume of mobile phase that has passed it. It is from the elution curve equation that the various characteristics of a chromatographic system can be determined using the data provided by the chromatogram.

Let the chromatographic column be considered to be divided into a number of cells or plates such that the solute can be assumed to be in equilibrium with the two phases in each plate. Thus in each plate,

$$X_s = KX_m \tag{2.1}$$

where X_m and X_s are the concentrations of the solute in the mobile and stationary phases, respectively, and K is the distribution coefficient of the solute between the two phases. (This only approximates to the thermodynamic distribution coefficient except for ideal solutions where activity coefficients approximate to unity). K is a dimensionless constant and thus in liquid-liquid systems X_s and X_m are conveniently measured as mass of solute per unit volume of phase; in liquid-solid systems a convenient method of measurement would be mass of solute per unit mass of phase. An excellent discussion on the meaning of K is given by Karger et al. [8a], and those wishing to pursue the meaning of distribution coefficient further are recommended to refer to this book.

Equation (2.1) merely states that the general distribution law applies to the system and that the absorption isotherm is linear. At the concentrations normally used in liquid chromatography this will be true for all liquid-liquid systems and for most liquid-solid systems. It will be shown later that the absorption isotherm must be very close to linear if the system is to be practically useful, since nonlinear isotherms produce asymmetrical peaks.

Differentiating (2.1),

$$dX_s = KdX_m. \tag{2.2}$$

26   THE THEORY OF CHROMATOGRAPHY

Consider three consecutive plates in the column, $p-1, p$, and $p+1$, and let there be a total of n plates in the column. The three plates are depicted in Fig. 2.2.

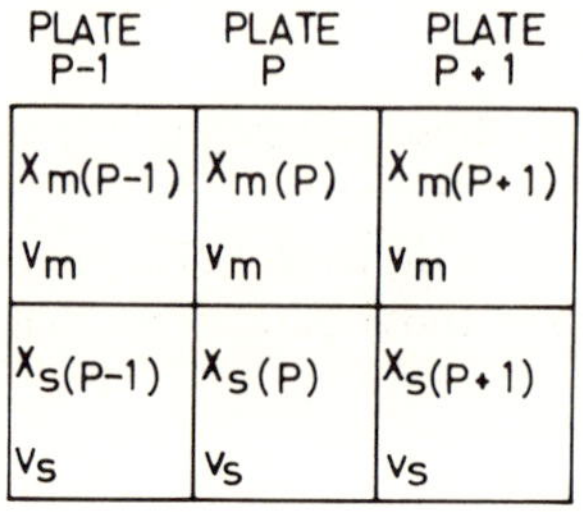

PLATE P−1	PLATE P	PLATE P+1
$X_{m(P-1)}$	$X_{m(P)}$	$X_{m(P+1)}$
v_m	v_m	v_m
$X_{s(P-1)}$	$X_{s(P)}$	$X_{s(P+1)}$
v_s	v_s	v_s

Fig. 2.2. Theoretical diagram of three consecutive plates in a chromatographic column.

Let the volumes of the mobile phase and stationary phase in each plate be v_m and v_s, respectively. Let the concentrations of solute in the mobile phase and stationary phase in each of the three plates be $X_{m(p-1)}$, $X_{s(p-1)}$, $X_{m(p)}$, $X_{s(p)}$, $X_{m(p+1)}$, and $X_{s(p+1)}$, respectively. Let a volume of mobile phase, dV, pass from plate $p-1$ into plate p at the same time displacing the same volume of mobile phase from plate p to plate $p+1$. In doing this there will be a change of mass of solute in plate p that will be numerically equal to the difference between the mass of solute entering plate p from plate $p-1$ and the mass of solute leaving plate p and entering plate $p+1$. Thus the change of mass of solute in plate p is

$$dm = (X_{m(p-1)} - X_{m(p)})dV. \quad (2.3)$$

Now if equilibrium is maintained in plate p, this mass dm will distribute itself between the two phases, which will result in a change of solute concentration in the mobile phase of $dX_{m(p)}$ and in the stationary phase of $dX_{s(p)}$. Thus

$$dm = v_s dX_{s(p)} + v_m dX_{m(p)}. \quad (2.4)$$

Substituting for $dX_{s(p)}$ from (2.2), (2.4) becomes

$$dm = (v_m + Kv_s)dX_{m(p)}. \quad (2.5)$$

Equating (2.3) and (2.5) and rearranging,

$$\frac{dX_{m(p)}}{dV} = \frac{X_{m(p-1)} - X_{m(p)}}{v_m + Kv_s}. \quad (2.6)$$

Now to aid in algebraic manipulation it is necessary to effect a change of variable. The volume flow of mobile phase will be measured in units of $v_m + Kv_s$ instead of milliliters. Thus a new variable v can be defined where

$$v = \frac{V}{v_m + Kv_s} \tag{2.7}$$

The function $v_m + Kv_s$ is given the name plate volume and thus the flow of mobile phase will, for the present, be measured in plate volumes instead of milliliters. The plate volume is that volume of mobile phase that can contain all the solutes at the equilibrium concentration in the mobile phse. Differentiating (2.7),

$$dv = \frac{dV}{v_m + Kv_s} \tag{2.8}$$

Substituting for dV from (2.8) in (2.6),

$$\frac{dX_{m(p)}}{dv} = X_{m(p-1)} - X_{m(p)} \tag{2.9}$$

Equation (2.9) is the basic differential equation that describes the rate of change of concentration of solute in any one plate with the volume flow of mobile phase through it, and thus the integration of this equation will provide the equation of the elution curve for any plate in the column. There are several ways of solving this equation but the one that involves the simplest mathematics will be used. First consider the conditions for the above equation when an initial charge of concentration X_m^0 has been placed on the column but development has not been commenced, that is, $v=0$. Then

$$X_{m(p)} = X_m^0 \quad \text{when } p = 0 \text{ (i.e., the first plate)}$$

and

$$X_{m(p)} = 0 \text{ when } p > 0 \text{ (i.e., any plate other than the first).}$$

The first condition merely states that before the chromatographic development commences the concentration on plate $p = 0$ is that resulting from the injection of the sample on the column. Incidentally, this plate only holds the initial charge and is not part of the chromatogrpahic system. The second condition states that the remainder of the column is free of solute before development Thus for the plate $p = 0$,

$$\frac{dX_{m(0)}}{dv} = -X_{m(0)} \quad \text{there being no plate } (p-1)$$

$$\frac{dX_{m(0)}}{X_{m(0)}} = -dv$$

Integrating,

$$\log_e X_{m(0)} = -v + \text{constant.}$$

When $v = 0$, $X_{m(0)} = X_m^0$. Thus, constant $= \log_e X_m^0$. Hence,

$$\log_e X_{m(0)} = -v + \log_e X_m^0$$

or

$$X_{m(0)} = X_m^0 \, e^{-v}. \tag{2.10}$$

For plate 1,

$$\frac{dX_{m(1)}}{dv} = X_{m(0)} - X_{m(1)}. \tag{2.11}$$

Substituting for $X_{m(0)}$ from (2.10) in (2.11),

$$\frac{dX_{m(1)}}{dv} = X_m^0 \, e^{-v} - X_{m(1)}$$

or

$$\frac{dX_{m(1)}}{dv} + X_{m(1)} = X_m^0 \, e^{-v}.$$

Multiplying throughout by e^v,

$$e^v \frac{dX_{m(1)}}{dv} + X_{m(1)} e^v = X_m^0.$$

New this equation can be recognized as the differential of a product. Hence

$$\frac{d(X_{m(1)} e^v)}{dv} = X_m^0$$

Integrating,

$$X_{m(1)}e^v = X_m^0 v + \text{constant.}$$

When $v = 0$, $X_{m(1)} = 0$; thus constant = 0. Furthermore,

$$X_{m(1)} = X_m^0 e^{-v} v. \qquad (2.12)$$

in a similar way it can be shown that

$$\text{for plate 2} \quad X_{m(2)} = X_m^0 \frac{e^{-v} v^2}{1 \cdot 2}$$

$$\text{for plate 3} \quad X_{m(3)} = X_m^0 \frac{e^{-v} v^3}{1 \cdot 2 \cdot 3}$$

Thus for the nth plate,

$$X_{m(n)} = X_m^0 \frac{e^{-v} v^n}{n!} \qquad (2.13)$$

Equation (2.13) is the equation of the elution curve and as the detector on a column measures the concentration leaving the nth plate, the chromatogram is a voltage-time analog curve of this equation. Equation (2.13) is a Poisson function, but it will be shown later that if n is large the function approximates closely to the normal error function or Gaussian function. Thus since n is always greater than 100, it would be expected that for the conditions assumed for this equation all chromatographic peaks are Gaussian or nearly Gaussian in shape.

Retention Volume

A chromatogram containing two peaks is shown in Fig. 2.3 with the axis labeled according the the elution curve equation. The dead point on the chromatogram is the position of the maxima for a completely unretained solute. The volume of mobile phase passed through the column between the injection point and the dead point will be equivalent to the dead volume of the column. The retention volume of any substance is the volume of mobile phase that has passed through the column up to the elution of the peak maximum. The equation for the retention volume can, therefore, be obtained by differentiating the elution curve equation and equating to zero to obtain the position of the maximum:

$$X_{m(n)} = X_m^0 \frac{e^{-v} v^n}{n!}$$

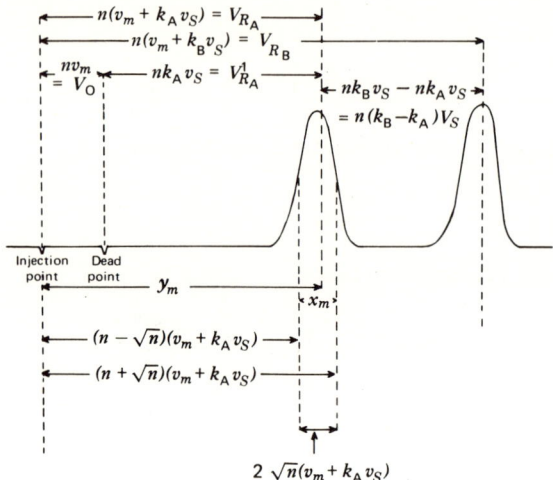

Fig. 2.3. The characteristics of a chromatogram.

Differentiating,

$$\frac{dX_{m(n)}}{dv} = \frac{X_m^0}{n!}(-e^{-v}v^n + e^{-v}nv^{(n-1)})$$

$$= \frac{X_m^0 e^{-v} v^{(n-1)}}{n!}(n-v). \qquad (2.14)$$

At the peak maximum, (2.14) must be zero; thus

$$v = n. \qquad (2.15)$$

As v is measured in plate volumes then, the peak maximum, n plate volumes of mobile phase have passed through the column. Thus, from (2.7), the retention V_R in milliliters of mobile phase is given by

$$V_R = n(v_m + Kv_s). \qquad (2.15a)$$

Now if there are n plates in the column and v_m and v_s are the volumes of mobile phase and stationary phase, respectively, in each plate, then nv_m and nv_s are the total volumes of mobile phase and stationary phase in the column. Thus

$$V_R = V_m + KV_s \qquad (2.16)$$

where V_m is the total volume of mobile phase in the column and V_s is the total volume stationary phase in the column. It should be remembered in practice that measured values of V_m include the total free volume of the column system and will include all the mobile phase between the point of injection and the detector. Volumes associated with detector connections will be included in V_m. If the solute is completely unretained then $K = 0$, and thus the dead volume of the column is given by

$$V_0 = V_m. \qquad (2.17)$$

Since the adjusted retention volume $V'_R = V_R - V_0 = V_R - V_m$, therefore

$$V'_R = V_m + KV_s - V_m = KV_s. \qquad (2.18)$$

Since K will be a characteristic of the substance eluted, for a column carrying a given quantity of stationary phase V_s, the adjusted retention volume $V'_R = KV_s$ could act as a parameter by which the solute could be identified.

Now since V'_R is a function of K, and K is a function of temperature, change in temperature will modify V'_R in the same way that it modifies K and thus measurement of the corrected retention volume over a range of temperatures can provide useful thremodynamic data on a given solute-solvent system.

Peak Asymmetry

By dividing (2.16) by the flow rate of the mobile phase Q, an equation for the retention time is obtained:

$$\frac{V_R}{Q} = t_R = \frac{V_m + KV_s}{Q}.$$

Now the velocity of the band along the column Z is given by the ratio of the column length l to the retention time t_R:

$$Z = \frac{l}{t_R} = \frac{lQ}{V_m + KV_s}.$$

Thus the band velocity Z is inversely proportional to $(V_m + KV_s)$ and for a significantly retained solute, $V_m \ll KV_s$; thus for such a column system,

$$Z \propto \frac{l}{V'_R} \propto \frac{l}{K}. \qquad (2.19)$$

Consider the isotherms shown in Fig. 2.4. Each curve represents a different

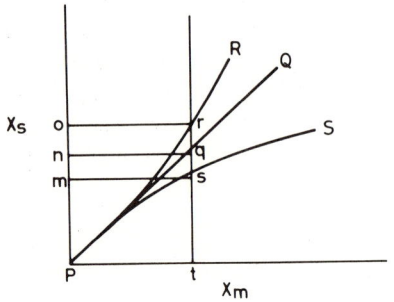

Fig. 2.4. Adsorption isotherms.

isotherm relating the concentration of the solute in the mobile phase X_m to the concentration of the solute in the stateionary phase X_s.

PqQ represents a linear isotherm, and for this line the distribution coefficient is given by

$$\frac{X_s}{X_m} = \frac{tq}{qn} = K_1 .$$

Since PqQ is a linear isotherm, K_1 is constant for all values of X_s and thus, from (2.19), all concentrations in the band will travel at the same velocity and a symmetrical elution curve will be produced as depicted by Fig. 2.5a. This, of course, is expected from the plate theory.

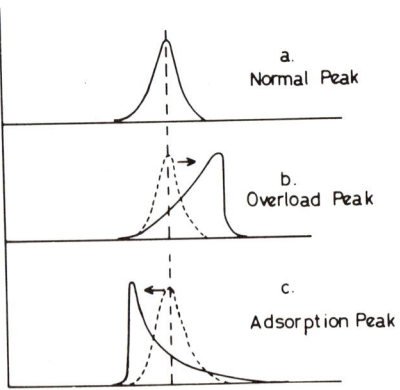

Fig. 2.5. Peak asymmetry.

Now consider the situation where a large charge is placed on a liquid-liquid column and the solute molecules, when in the stationary phase, are no longer surrounded solely by solvent molecules but the solute *and* solvent molecules. Under these circumstances the forces holding the solute molecules in the stationary phase are increased because the solute-*solute* forces are usually greater than the solute-*solvent* forces. In effect the average molecular weight of the stationary phase has been changed. Thus K will increase as X_s increases and the result is an isotherm of the form shown by curve *PrR* in Fig. 2.4. The distribution coefficient at the higher concentration will now be $tr/ro = K_2$ and because $tr > tq$ and $ro = qn$, then $K_2 > K_1$. From (2.19) it is seen that those parts of the solute band containing the higher concentrations of solute will move more slowly through the column than the regions of lower solute concentration. This will result in an elution curve of the form shown in Fig. 5*b*, an asymmetric peak that, for the above reason, has been given the name overloaded peak.

Conversely, in some liquid-solid systems high concentrations of solute will cover the adsorbent surfaces with multilayers of solute molecules. Because of the screening effect of these solute multilayers, the forces holding further molecules on the surface will be significantly reduced since they will only experience long-range forces from the solid surface. The effect of this will be the opposite to that of overload and the adsorption isotherm will take the form of the curve *PsS* in Fig. 2.4. Thus at high concentrations, $K_3 = ts/sm$. Now since $ts < tq$ and $sm = qn$, then $K_3 < K_1$. Thus again from (2.19) it follows that those parts of the solute band containing the higher concentrations of solute will move more rapidly through the column than the regions of lower solute concentrations. The resulting elution curve will take the form shown in Fig. 2.5*c*. This form of peak asymmetry is called an adsorption peak; the term comes from gas chromatography and is not really an appropriate term for use in liquid chromatography.

Column Efficiency

In order to effect the separation of two substances, the bands of the two solutes must be moved apart during their passage through the chromatographic column. However, having moved the two bands apart their peak widths must remain sufficiently narrow to permit the discrete elution of each peak. Thus the peak width of a substance eluted from a column could give a measure of the separating power of the column. Because the peak will be Gaussian in form, the peak widths at the points of inflexion of the elution curve (which will correspond to twice the standard deviation of the curve) will be determined. At the points of inflexion,

$$\frac{d_2(X_m^0 e^{-v} v^n/n!)}{dv^2} = 0.$$

Now

$$\frac{d^2(X_m^0 e^{-v} v^n/n!)}{dv^2} = \frac{X_m^0(e^{-v}v^n - e^{-v}nv^{(n-1)} - e^{-v}nv^{(n-1)} + e^{-v}n(n-1)v^{(n-2)})}{n!}$$

$$= \frac{X_m^0 e^{-v} v^{(n-2)}(v^2 - 2nv + n(n-1))}{n!}.$$

Thus, at the points of inflexion

$$v^2 - 2nv + n(n-1) = 0.$$

Hence

$$v = \frac{2n \pm (4n^2 - 4n(n-1))^{\frac{1}{2}}}{2}$$

$$= \frac{2n \pm (4n)^{\frac{1}{2}}}{2}$$

$$= n \pm n^{\frac{1}{2}}.$$

The points of inflexion occur after $n-n^{\frac{1}{2}}$ and $n+n^{\frac{1}{2}}$ plate volumes of mobile phase have passed through the column. Thus the volume of mobile phase that passes through the column between the points of inflexion will be $(n+n^{\frac{1}{2}}) - (n-n^{\frac{1}{2}})$ plate volumes and so the peak width in milliliters of mobile phase will be given by (2.20):

$$\text{Peak width} = 2n^{\frac{1}{2}}(v_m + Kv_s). \tag{2.20}$$

Because the peak width at the points of inflexion of the curve is twice the standard deviation measured in plate volumes, the variance of the curve (the square of the standard deviation) will be n. Thus the variance is numerically equal to the number of theoretical plates in the column and is a measure of the column separating power. The variance has been given the name column effeciency, which is expressed as a number of theoretical plates. The variance of the elution curve has been chosen as a measure of the separating power of a column as opposed to the apparently more logical standard deviation of peak width for a very good reason. The overall spread of a solute peak is due to a number of individual, unrelated band-spreading processes that together contribute to the final peak width. Each process is random in nature; therefore, if any one were the sole process that produced band spreading, then the resultant elution curve would be narrower but still Gaussian in shape. If the individual spreading

1 THE PLATE THEORY

processes are determined, they must be added together to account for the final bandwidth. From the theory of error functions it can be shown that it is not possible to add the standard deviations resulting from each process to provide the final peak width. It *is* possible, however, to add the variance from each process to give the final band variance. For this reason the column efficiency is measured as n, the number of theoretical plates in the column, which, as has been shown from the plate theory, is a measure of the band variance.

It is therefore important to be able to measure the efficiency of any column system to determine its separating power. Let the distance between the injection point and the peak maximum of a given solute band, measured on the chromatogram, be y cm and the peak width at the points of inflexion be x cm (see Fig. 2.3). From (2.15a) and (2.20),

$$\frac{\text{Ret. distance}}{\text{Peak width}} = \frac{y}{x} = \frac{n}{2n^{1/2}} \frac{v_m + Kv_s}{v_m + Kv_s} = \frac{n^{1/2}}{2}.$$

Hence

$$n = 4\left(\frac{y}{x}\right)^2 \qquad (2.21)$$

Equation (2.21) allows the efficiency of any solute peak from any column to be calculated from measurements made directly on the chromatogram. The various characteristics of a chromatogram that have so far been derived from the plate theory are shown in Fig. 2.3.

The Position of the Points of Inflexion

Since the measurement of efficiency is important for evaluating the quality of a column it is necessary to know the position of the points of inflexion in order to measure the peak width x used in (2.21). The inflexion points are ill defined on the chromatogram, and to determine them it is necessary to know at what fraction of the peak height they occur. Thus if their height above the base line is f and the peak height is h it is necessary to know $\frac{f}{h}$. From (2.13) replacing v by n and $n-\sqrt{n}$

$$\frac{f}{h} = \frac{X_{m(n-\sqrt{n})}}{X_{m(n)}} = \frac{X_0 e^{-(n-\sqrt{n})}(n-\sqrt{n})^n}{n!} \bigg/ \frac{X_0 e^{-n} n^n}{n!}$$

$$= \frac{e^{-(n-\sqrt{n})}(n-\sqrt{n})^n}{e^{-n} n^n}$$

THE THEORY OF CHROMATOGRAPHY

$$= e^{\sqrt{n}} \left(\frac{n-\sqrt{n}}{n}\right)^n = e^{\sqrt{n}} \left(1 - \frac{1}{\sqrt{n}}\right)^n$$

Thus

$$\log_e \frac{f}{h} = \sqrt{n} + n \log_e \left(1 - \frac{1}{\sqrt{n}}\right) \qquad (2.22)$$

Now if $x \ll 1$,

$$\log_e (1-x) = -x - \frac{-x^2}{2} - \frac{x^3}{3} - \frac{x^4}{4} \cdots .$$

Thus since n will always be significantly greater than 100,

$$n \log_e \left(1 - \frac{1}{\sqrt{n}}\right) = n \left[\frac{-1}{\sqrt{n}} \frac{-1}{2n} \frac{-1}{3n\sqrt{n}} \frac{-1}{4n^2}\right] \cdots$$

$$= -\sqrt{n} - \frac{1}{2} - \frac{1}{3\sqrt{n}} - \frac{1}{4n}. \qquad (2.23)$$

Substituting (2.23) in (2.22)

$$\log_e \frac{f}{h} = \sqrt{n} - \sqrt{n} - \frac{1}{2} - \frac{1}{3\sqrt{n}} - \frac{1}{4n} \cdots .$$

Because $\frac{1}{3\sqrt{n}} \ll \frac{1}{2}$, hence $\log_e \frac{f}{h} = -\frac{1}{2}$. Thus $\frac{f}{h} = e^{-\frac{1}{2}} = \frac{1}{e^{\frac{1}{2}}} = 0.6065$; that is

$$f = 0.6065h.$$

The Resolving Power of a Column

It has already been stated that for two peaks to be resolved they must be moved apart in the column and at the same time be maintained sufficiently narrow to permit them to be eluted as discrete peaks. The criterion for two peaks to be resolved will be arbitrary but, if the areas are to be measured to give reasonable quantitative accuracy, the peak maxima of the two solutes must be at least two peak widths apart. It should be pointed out that two adjacent peaks from solutes of different chemical types will not necessarily have precisely the same peak widths. However, the difference will not be significant and in the following argument will be assumed to be negligible.

1 THE PLATE THEORY

The difference between the peaks of two solutes A and B, measured in volume flow of mobile phase, will be

$$n(v_m + K_B v_s) - n(v_m + K_A v_s) = n(K_B - K_A)v_s. \quad (2.25)$$

Now because it has been assumed that the widths of the two peaks are the same, then the peak width in volume flow of mobile phase will be

$$2n^{1/2}(v_m + K_A v_s). \quad (2.26)$$

Taking as the criterion of resolution that the distance between the peak maxima must be twice the peak width, then from (2.25) and (2.26),

$$4n^{1/2}(v_m + K_A v_s) = n(K_B - K_A)v_s.$$

Thus rearranging,

$$n^{1/2} = \frac{4(v_m + K_A v_s)}{(K_B - K_A)v_s}.$$

Dividing by v_s and defining $a = \dfrac{v_m}{v_s}$ as the *phase ratio* of the column, then

$$n^{1/2} = \frac{4(a + K_A)}{K_B - K_A}.$$

Hence the efficiency required to separate solutes A and B is

$$n = \frac{16(a + K_A)^2}{(K_B - K_A)^2}. \quad (2.27)$$

Equation (2.27) permits the calculation of the efficiency required to separate two substances of known distribution coefficients on a column of known phase ratio.

In practice, (2.27) is of limited use since the distribution coefficients of the substances to be separated are not usually known and the phase ratio is difficult to measure accurately.

Developing the equation in a different way and again using the same criterion for resolution,

$$4n^{1/2}(v_m + K_A v_s) = n(v_m + K_B v_s) - n(v_m + K_A v_s).$$

Thus

$$4n^{1/2} = \frac{n(v_m + K_B v_s)}{v_m + K_A v_s} - n = n\left(\frac{v_m + K_B v_s}{v_m + K_A v_s}\right) - 1.$$

Now referring to Fig. 2.3 it can be seen that $\dfrac{(v_m + K_B v_s)}{(v_m + K_A v_s)}$ is the ratio of the retention volume of A and B, that is, $\dfrac{V_{R(B)}}{V_{R(A)}}$, which will be written in the form V_{BA}; hence

$$4n^{1/2} = n(V_{BA} - 1).$$

Rearranging, the efficiency required to effect the separation is given by:

$$n = \frac{16}{(V_{BA} - 1)^2}. \tag{2.28}$$

Equation (2.28) allows the operator to assess the potential of the column for a particular separation but it can only be applied to columns with the same phase system, having the same phase ratio, and operating at the same temperature. Thus its application will be limited to data obtained solely from the operator's own equipment. At present there is only a limited amount of retention data available on liquid chromatography, but that which is available and the data that will become available in the future will be in the form of retention ratios. These are calculated from the ratio of the adjusted retention volumes and will be quoted for particular phase systems at particular temperatures. For the example taken the retention ratio will be

$$V'_{BA} = \frac{nK_B v_s}{nK_A v_s} = \frac{K_B}{K_A}. \tag{2.29}$$

It is therefore desirable to develop an equation that will provide a value for the efficiency required to separate components A and B using retention ratio values (V'_{BA}) and some property of the chromatogram that will take into account the phase ratio of the column used.

Restating the conditions for resolution,

$$4n^{1/2}(v_m + K_A v_s) = n(v_m + K_B v_s) - n(v_m + K_A v_s)$$
$$= n v_s (K_B - K_A)$$

1 THE PLATE THEORY

$$= nv_s K_A (V'_{BA} - 1).$$

Dividing throughout by $n^{1/2} v_s$ and substituting $\dfrac{v_m}{v_s}$ by a, then

$$4(a + K_A) = n^{1/2} K_A (V'_{BA} - 1).$$

Dividing throughout by K_A, rearranging and putting $\dfrac{a}{K_A} = V_{aA}$, V_{aA} is the ratio of dead volume of the column to the adjusted retention volume of solute A. From (2.29) this can be seen to be $\dfrac{nv_m}{nK_A v_s}$, which takes into account the capacity ratio of the column with respect to the solute being chromatographed. To maintain the rational development of the theory the notations V'_{BA} and V_{aA} are employed, but it should be noted and will be shown later that $V'_{BA} = \alpha$, the separation ratio, and $V_{aA} = \dfrac{1}{k'_A}$. Hence the efficiency required to effect the separation is given by

$$n = \frac{16(V_{aA} + 1)^2}{(V'_{BA} - 1)^2} = \frac{16(1 + k'_A)^2}{k'^2_A (\alpha - 1)^2}. \tag{2.30}$$

Thus the efficiency required to effect the seaparation can be calculated from published retention ratios and data obtained by chromatographing one of the solutes under chosen conditions.

Taking a range of practical values for $V_{\alpha A}$ and V'_{BA} and substituting them in (2.30) the efficiency n required to effect each separation can be calculated. Curves relating n and V_{aA} for different values of V'_{BA} are shown in Fig. 2.6. As would be expected, the curves show that as V'_{BA} increases, the number of theoretical plates required to effect the separation decreases. However, it can also be seen that when V_{aA} is greater than 1, the efficiency required for separation rapidly increases. For example, to separate a pair of substances that have a V'_{BA} value of 1.10 requires 7000 theoretical plates if V_{aA} equals 1 and 70,000 theoretical plates if V_{aA} equals 10. Now V_{aA} is the ratio of the dead volume of the column to the adjusted retention volume, that is,

$$V_{aA} = \frac{v_m}{K_A V_S}.$$

Thus to achieve separations with the minimum number of plates, either K_A or v_s must be made as large as possible such that $\dfrac{v_m}{K_A V_S} < 1$. Thus in practice

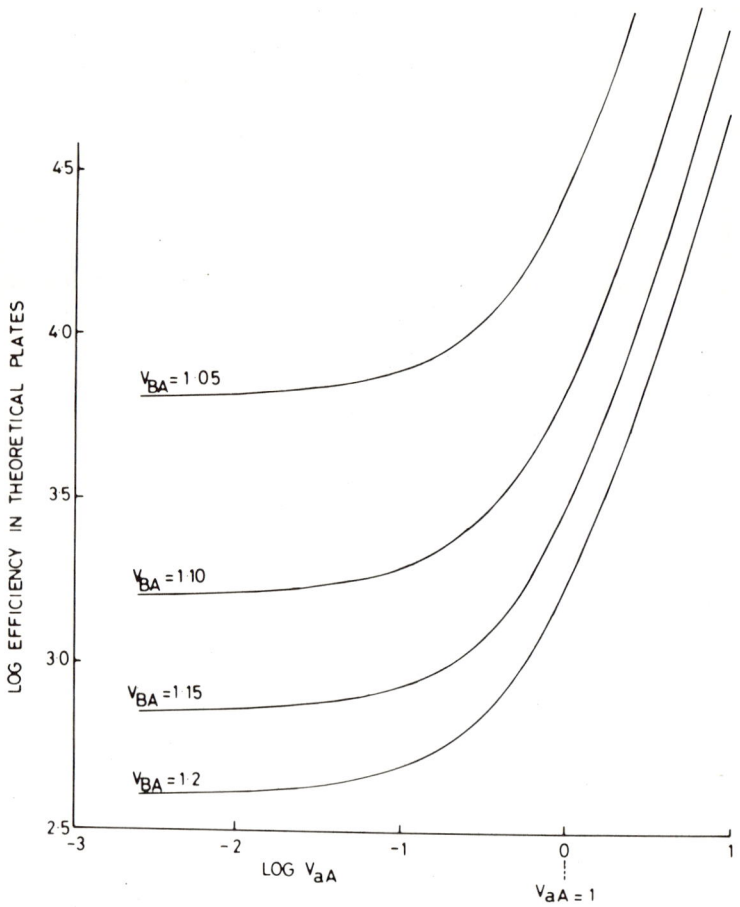

Fig. 2.6. Graphs showing the efficiency required to effect a given separation for different values of V'_{BA} and V_{aA}.

the column loading of stationary phase and the distribution system K_A must be chosen such that the first eluted component of the given pair has a retention time that is greater than twice the column dead time.

The Effective Plate Number

The concept of effective plate number was introduced and employed in the late 1950s by Purnell, Desty, and others [9, 10]. Its introduction resulted directly from the development of capillary columns which, if employed with the correct operating conditions, could provide efficiencies of up to a million theoretical plates [11]. It was noted, however, that these high efficiencies were only realized for solutes eluted close to the column dead volume or at low k' values. Furthermore, they in no way reflected the increase in resolving power that would be expected from such high efficiencies on the basis of the performance of packed columns. This can be shown theoretically to result from the high phase ratio of capillary columns, that is, the ratio of mobile phase to stationary phase in the column. Because of the high phase ratio a significant proportion of the solutes is eluted between k' values of zero and unity where, as it has been shown previously, very high efficiencies are needed to provide reasonable resolution. To compensate for such misleading efficiency values the effective plate number was introduced that, in fact, utilized the corrected retention distance as opposed to the to the total retention distance in the formula used for calculating normal efficiency values. In this way the effective plate number was significantly smaller than the true efficiency for solutes eluted at low k' values and more nearly corresponded to the column resolving power. Modern liquid chromatographic columns using microparticle absorbents resemble packed gas chromatographic columns with respect to the phase ratio of such columns. However, whereas the efficiency fo packed gas chromatographic columns increases with the k' value of the eluted solute, the packed liquid chromatographic columns resemble the capillary column in that, because of the nature of the solute transfer process in the column, efficiencies fall with the increasing k' value of the solute. Because liquid chromatographic columns are similar to packed gas chromatographic columns with respect to phase ratio but similar to capillary columns with respect to their efficiency-k' relationship, it is of interest to determine whether the use of the effective plate number in liquid chromatography would be more useful than the normal plate number. Furthermore, both real and effective plates relate both to retention and bandwidth, but neither relates directly to the resolving power of the columns. The practicing chromatographer would like a simple number that would relate directly to the capability of his column to separate substances and it would therefore also be of interest to determine whether a simpler function of either real plates or effective plates can provide him with such a parameter. From (2.2),

$$n = \frac{4y^2}{x^2}$$

where n is the number of theoretical plates in the column, y is the retention distance, and x is the bandwidth at the point of inflexion of the elution curve. The number of effective plates N is given by

$$N = 4\left(\frac{y-y_0}{x^2}\right)^2$$

where y_0 is the retention distance of a nonretained peak. Now from the plate theory,

$$\frac{y}{x} = \frac{n(v_m + Kv_s)}{2\sqrt{n}\,(v_m + Kv_s)}.$$

Thus

$$\frac{y - y_0}{x} = \frac{n(v_m + Kv_s) - nv_m}{2\sqrt{n}\,(v_m + Kv_s)}$$

$$= \frac{\sqrt{n}}{2} \frac{Kv_s}{v_m + Kv_s} = \frac{\sqrt{n}}{2} \frac{k'}{1+k'}.$$

Noting that

$$\frac{kv_s}{v_m} = k',$$

thus

$$4\left(\frac{y - y_0}{x}\right)^2 = n\left(\frac{k'}{1+k'}\right)^2 = N.$$

The above equation describes the relationship between n and N.

Giddings [12] suggested a function for resolution of $\frac{k'}{\Delta k'}$ as an analog to the function $\frac{\lambda}{\Delta \lambda}$ used by spectroscopists to measure the resolving power of spectrometric devices.

where $\Delta k'$ is the bandwidth at the base of peak eluted at k' and hence $\Delta k'$ was the equivalent to 4 standard deviations of the eluted peak.

Thus from the plate theory

$$R = \frac{k'}{\Delta k'} = \frac{nKv_s}{4\sqrt{n}\,(v_m + Kv_s)} = \frac{\sqrt{n}}{4} \frac{k'}{1+k'} = \frac{\sqrt{N}}{4}$$

where R is the resolving power of the column. Thus the resolving power of the column will be directly proportional to the square root of the effective plates. Therefore, $\sqrt{N}$ can be used by the chromatographer to directly compare columns of any size packed with any type of stationary phase on the basis of their capacity for separation. This assumed that resolution is acceptable when the two peaks of interest are separated by the equivalent to the average peak width at the base. Furthermore, the value of $\sqrt{N}$ will vary with the k' of the eluted solute and must, therefore, be determined over a range of k' values.

However, the practicing chromatographer would also like to know the minimum separation ratio α of the two peaks that he can separate. In fact this has been suggested (13) as a basis for comparing the resolving power of different columns. The disadvantage of this measure is that it becomes smaller as the resolving power of the column increases. However, the minimum α value is important in practice and it is of interest to see if it can be related to the effective plate number.

Now the minimum α value will be given by

$$\alpha = \frac{nKv_s + 4\sqrt{n}(v_m + Kv_s)}{nKv_s}$$

$$= 1 + \frac{4}{\sqrt{n}} \frac{1+k'}{k'}.$$

Thus from (2.1),

$$\alpha = 1 + \frac{4}{\sqrt{N}} = 1 + \frac{1}{R}.$$

Thus the chromatographer can also arrive at the minimum α value of a pair of solutes he can separate on his column from a simpler function of the effective plate number.

The Gaussian Form of the Elution Equation

In order to change the Poisson form of the elution equation into a Gaussian form it is necessary to effect a change of origin. Consider the elution curve shown in Fig. 2.7. The origin for the Poisson form of the elution equation is at the point of injection whereas the origin for the Gaussian equation will be at the peak maximum, n plate volumes from the injection point. Thus a point A, v plate volumes from the injection point O will be $v-n = w$ plate volumes from the peak maximum. Now

$$X_{m(n)} = \frac{X^0 e^{-v} v^n}{n!}.$$

Substituting $w + n$ for v,

THE THEORY OF CHROMATOGRAPHY

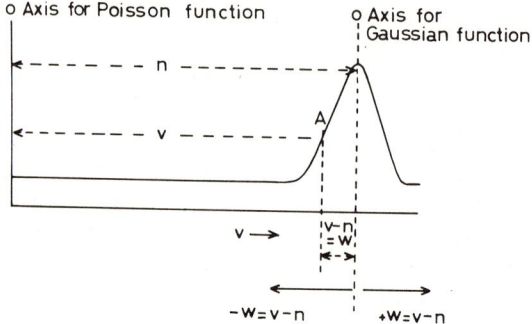

Fig. 2.7. Relationship between Poisson and Gaussian functions of the elution curve.

$$X_{m(n)} = \frac{X^0 e^{-(w+n)}(w+n)^n}{n!}.$$

Now from Sitrling's theorem, $n! = e^{-n}n^n\sqrt{2\pi n}$. Therefore,

$$X_{m(n)} = \frac{X^0 e^{-(n+w)}(n+w)^n}{e^{-n}n^n\sqrt{2\pi n}}$$

$$= \frac{X^0}{\sqrt{2\pi n}} e^{-w} \left(1 + \frac{w}{n}\right)^n.$$

Thus

$$\log_e X_{m(n)} = \log_e \frac{X^0}{\sqrt{2\pi n}} - w + n \log_e \left(1 + \frac{w}{n}\right)$$

$$= \log_e \frac{X^0}{\sqrt{2\pi n}} - w + n \left(\frac{w}{n} - \frac{w^2}{2n^2} + \frac{w^3}{3n^3} \cdots\right)$$

$$= \log_e \frac{X^0}{\sqrt{2\pi n}} - w + w - \frac{w^2}{2n} + \frac{w^3}{3n^2} \cdots$$

Now if n is large and w is always less than n, $\frac{w^2}{2n}$ will always be very much greater than $\frac{w^3}{3n^2}$ and thus $\frac{w^3}{3n^3}$ and higher terms can be ignored with respect to $\frac{w^2}{2n}$. Thus

$$\log_e X_{m(n)} = \log_e \frac{X^0 e^{-w^2/2n}}{\sqrt{2\pi n}}$$

or

$$X_{m(n)} = \frac{X^0 e^{-w^2/2n}}{\sqrt{2\pi n}}. \qquad (2.31)$$

Equation (2.31) is the well-known Gaussian form of the elution curve having variance n.

The Maximum Sample Volume

In order to produce an effective separation it has been shown that the solutes must be significantly retained in the column, that is, $V_{aA} \ll 1$. However, this will mean that the distribution of the solutes is strongly in favor of the stationary phase, and this can often result in the absolute solubility of the solutes in the mobile phase being relatively small. Therefore, in liquid chromatography in order to place sufficient sample on the column to meet the sensitivity requirements of the detector, it may be necessary to inject the sample contained in the largest volume of sample solution that can be placed on the column without seriously impairing the column efficiency. The increase in bandwidth due to sample volume that can be tolerated is obviously a matter of choice. However, the generally accepted maximum increase in bandwidth owing to sample volume is 5%, which will result in approximately 10% reduction in column efficiency.

Consider a volume of sample V_i injected on to the column forming a rectangular distribution at the front of the column. The variance of the final peak will equal the sum of the variance of the sample volume plus the normal variance of the peak for a small sample. Now the variance of the rectangular distribution of a sample of volume V_i at the beginning of the column is $\frac{V_i^2}{12}$ and assuming the peak width is increased by 5% due to the sample volume and using (2.20),

$$\frac{V_i^2}{12} + [\sqrt{n}(v_m + Kv_s)]^2 = [1.05\sqrt{n}(v_m + Kv_s)]^2.$$

Thus

$$\frac{V_i^2}{12} = n(v_m + Kv_s)^2 (1.05^2 - 1)$$

$$= n(v_m + Kv_s)^2 \, 0.102.$$

Thus

$$V_i^2 = n(v_m + Kv_s)^2 \quad 1.23$$

or

$$V_i = \sqrt{n}(v_m + Kv_s) \, 1.1 \, . \quad (2.32)$$

Equation (2.32) gives an expression for V_i, the maximum permitted sample volume, in terms of the efficiency n, the volume of mobile phase and stationary phase per plate v_m and v_s, respectively, and the distribution coefficient of the solute K. However, most of these variables are not usually known and a more useful expression for V_i can be derived from (2.32) as follows:

The retention volume V_R given by (2.15a) is

$$V_R = n(v_m + Kv_s)$$

or

$$\frac{V_R}{\sqrt{n}} = \sqrt{n}(v_m + Kv_s).$$

Substituting for $\sqrt{n}(v_m + Kv_s)$ in (2.32),

$$V_i = \frac{1.1 V_R}{\sqrt{n}}. \quad (2.33)$$

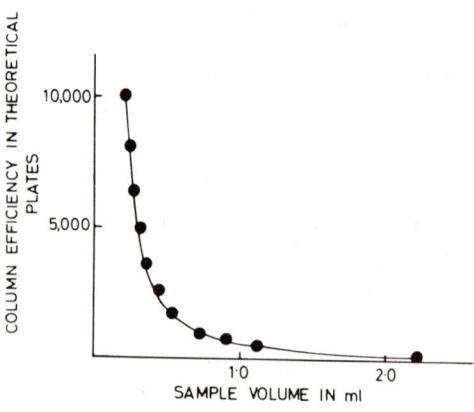

Fig. 2.8. Graph detailing the maximum sample volume against column efficiency for a solute having a retention volume of 20 ml.

Now V_R and n can be easily determined for any solute on a given column system and thus V_i can be obtained. In Fig. 2.8 the maximum permitted sample volume has been calculated for a solute having a retention volume of 20 ml chromatographed on columns with efficeincies ranging from 100 to 10,000 theoretical plates. It is seen that even with an efficiency of 10,000 plates a maximum sample volume of 0.22 ml can be tolerated.

2 THE RATE THEORY

An elution curve or chromatogram can be expressed using parameters other than the volume flow of mobile phase as the independent variable. Instead of using milliliters of carrier gas, time or distance traveled by the solute band along the column can be plotted against solute concentration in the mobile phase and proportionally the same chromatogram would be obtained, as shown in Fig. 2.9.

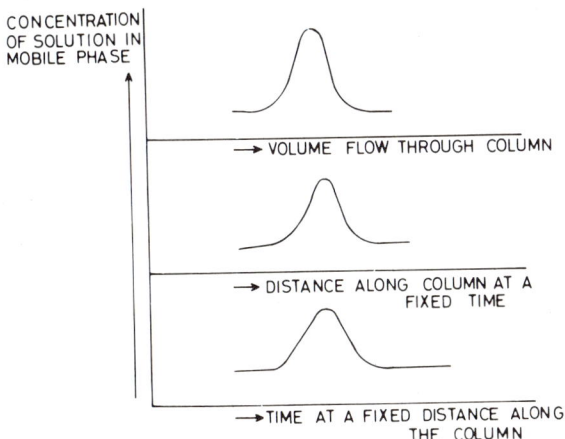

Fig. 2.9. An elution curve plotted on different axes.

Because the curves are describing the same chromatogram, by proportion the ratio of the variance to the square of the retention distance, in the respective units in which the independent variables are defined, will all be equal. Thus

$$\frac{\sigma v^2}{V_R^2} = \frac{\sigma x^2}{l^2} = \frac{\sigma t^2}{t_R^2} \qquad (2.34)$$

where σv, σx, and σt, respectively, are the standard deviations of the elution curves when related to the volume flow of mobile phase, distance traveled by the solute along the column, and time. Similarly V_R, l, and t_R refer to the retention volume, column length, and retention time, respectively.

Now from (2.15a) and (2.20) of the plate theory it has been shown that

$$\sigma v = \sqrt{n}(v_m + Kv_s) \text{ and } V_R = n(v_m + Kv_s).$$

Thus (2.34) becomes

$$\frac{n(v_m + Kv_s)^2}{n^2(v_m + Kv_s)^2} = \frac{\sigma x^2}{l^2}.$$

Thus

$$\frac{1}{n} = \frac{\sigma x^2}{l^2}$$

or

$$\frac{l}{n} = \frac{\sigma x^2}{l} \tag{2.35}$$

The ratio l/n, the column length divided by the number of theoretical plates in the column, has for obvious reasons been given the term height equivalent to a theoretical plate (HETP) and the symbol h. However, h is numerically equal to $\frac{\sigma x^2}{l}$ which has been termed the variance per unit length of the column. When concerned with the rate theory and the band-spreading processes that occur in a column, it is more useful to think of h as the variance per unit length than the height equivalent to a theoretical plate, since the latter, in this context, has little significance. The variance per unit length gives a value related to the total band spread that occurs in a unit length of column. It is, therefore, independent of the total column length and thus allows the "goodness" of the column construction to be compared for columns of any length.

The basic assumption of the rate theory is that the overall band spread per unit length of column is due to a number of individual, noninteracting, random spreading processes that can be added together to account for the final spreading effect. The validity of this assumption depends on the *noninteraction* of any of the individual spreading processes. That is to say that the extent to which one spreading process develops does not affect the rate or extent of the band spreading due to any other process. Whether there is any interaction between the individual random spreading processes is still to some extent a matter of opinion, but whatever interaction may be present is sufficiently small to allow the rate theory to predict fairly accurately the overall band spreading that will occur in a given column.

Assuming there are n noninteracting random processes occurring in the column, then any process p acting alone will produce a Gaussian elution curve having a variance σ_p^2. Thus by the principle of the summation of variances,

$$\sigma_1^2 + \sigma_2^2 + \sigma_3^2 + \cdots + \sigma_n^2 = \sigma^2 = hl.$$

The rate theory considers each individual process that is likely to occur in the column, determines its contribution to the variance per unit length of the column, and then sums each to provide the final value for h.

The procedure was first carried out by van Deemter et al. [2] to evaluate h in a gas chromatographic column; and their equation probably still represents the most generally effective form of the HETP equation for h. Van Deemter's treatment will be given briefly and the final equation for h developed. The resulting equation for h for a gas chromatographic column will then be logically modified to meet the requirements of liquid chromatography and finally substitution will be made in the light of the more recent equations for liquid chromatography developed by Huber [5].

Van Deemter and his colleagues considered that there were four basic processes that contributed to the band variance in a gas chromatographic column.

The Multipath Process

In a packed column the solute molecules describe a tortuous path between the interstices of the support. It is fairly obvious that some molecules will randomly travel shorter paths than others. Those that on an average pass along the shorter path will move ahead of the maximum of the concentration profile while those molecules that pass along paths of greater length will lag behind the maximum of the concentration profile. This will result in a spreading of the band and its variance contribution to h was deduced by van Deemter to be proportional to the particle diameter of the support and a constant depending on the physical nature of the packing. The variance per unit length due to the multipath effect as a contribution to h is given as

$$\gamma d_p$$

where γ is the packing constant and d_p the particle diameter.

Longitudinal Diffusion Process

If a local solute concentration is placed at the midpoint of a long tube filled with either a liquid or a gas, the solute will slowly diffuse to either end of the tube. It will first produce a Gaussian distribution with the maximum concentration at the center and finally, when the solute vapor reached the end of the tube, end effects will occur and the solute will diffuse until there is a constant concentration of solute throughout the whole tube. The latter effect is never realized in chromatography, but the initial spreading process does occur in the mobile phase of a column. The degree of spreading will obviously be proportional to some function of the time that the solute exists in the mobile phase and thus, if the mobile phase is flowing through the column at a linear velocity of u, then the extent of the spreading process will be some function of $1/u$. The variance due to this spreading will also be directly dependent upon the difffusivity of the solute in the mobile phase and the geometry of the spaces occupied by the mobile phase. Van Deemter et al. showed that the variance due to this process in the packed columns used for gas chromatography was $\dfrac{2\lambda D_m}{u}$ where λ is a constant depending upon the column packing and D_m is the diffusivity of the solute in the mobile phase.

Dispersion due to the Resistance to Mass Transfer in the Mobile Phase

During the movement of a solute band along a column, the solute molecules are continually transferring from the mobile phase into the stationary phase and back from the stationary phase into the mobile phase. This transfer process is not instaneous, because a finite time is required for the molecules to traverse the mobile phase in order to enter the stationary phase. Thus those molecules close to the stationary phase will enter it almost immediately, whereas those molecules some distance away from the stationary phase will find their way to it a significant time interval later. However, as the mobile phase is traveling at a given velocity along the column, the solute molecules will move a finite distance along the column during this time interval. Thus they will be absorbed into the stationary phase further along the column than those that were originally in close proximity to the stationary phase. The result of this delayed transfer of some solute molecules in the mobile phase results in band spreading and is termed the band dispersion due to the resistance to mass transfer in the mobile phase. This dispersion effect is further amplified by the parabolic velocity profile of the mobile phase flowing between the particals that result from normal viscous flow. Thus the layer of mobile phase close to the stationary phase surface is static and solute molecules that diffuse from this static layer into the moving fluid move along the column at a rate proportional to the distance they diffuse into the bulk of the moving mobile phase. Because of the parabolic nature of the velocity profile of the mobile phase between the particle those molecules diffusing the greatest distance into the moving fluid move away from those molecules that diffuse only a small distance from the stationary phase surface thus causing band spreading or dispersion. Van Deemter showed that the variance of this dispersion was a function of number of molecules that were present in the mobile phase at any time (i.e., a function of the distribution coefficient K), the phase ratio a, the particle diameter d_p, the diffusivity of the solute in the mobile phase D_m, and the mobile phase linear velocity u. The contribution that this dispersion effect made to h, the variance per unit length, was deduced to be

$$\frac{K^2}{\psi a(a+K)^2} \frac{d_p^2 u}{D_m}$$

where ψ is a constant. It should be pointed out, however, that the precise form of this function is still a matter of some discussion.

Dispersion due to the Resistance to Mass Transfer in the Stationary Phase

The dispersion resulting from the resistance to mass transfer in the stationary phase is exactly analagous to that in the mobile phase. Those solute molecules close to the surface of the stationary phase will leave the surface and enter the mobile phase before those that have diffused further into the stationary phase. This again will produce band spreading and contribute to the total variance per unit length. The two forms of dispersion resulting from resistance to mass transfer are shown diagramatically in Fig. 2.10. Because the six molecules in the stationary phase of bandwidth xy take a finite time to leave the stationary phase, and the mobile phase is moving, the resultant bandwidth when all the molecules have left the stationary phase is XY. Similarly, the six molecules in the mobile phase of bandwidth pq will take a finite time to transfer to the stationary phase and thus, by the time they have all enetered the stationary phase, because of the effect of the movement of the mobile phase, the band has spread to the length PQ. The contribution to h from the dispersion due to the resistance to mass transfer in the stationary phase was deduced by van Deemter to be

$$\frac{2}{3}\left[\frac{aK}{(a+K)^2}\right]\frac{d_f^2}{D_s}u$$

where d_f is the film thickness of stationary phase and D_s is the diffusivity of the solute in the stationary phase. The factor $\frac{2}{3}$ is correct for a thin uniform film on a capillary tube, but it is possible that for packed columns this factor may take another value close to but not necessarily equal to $\frac{2}{3}$.

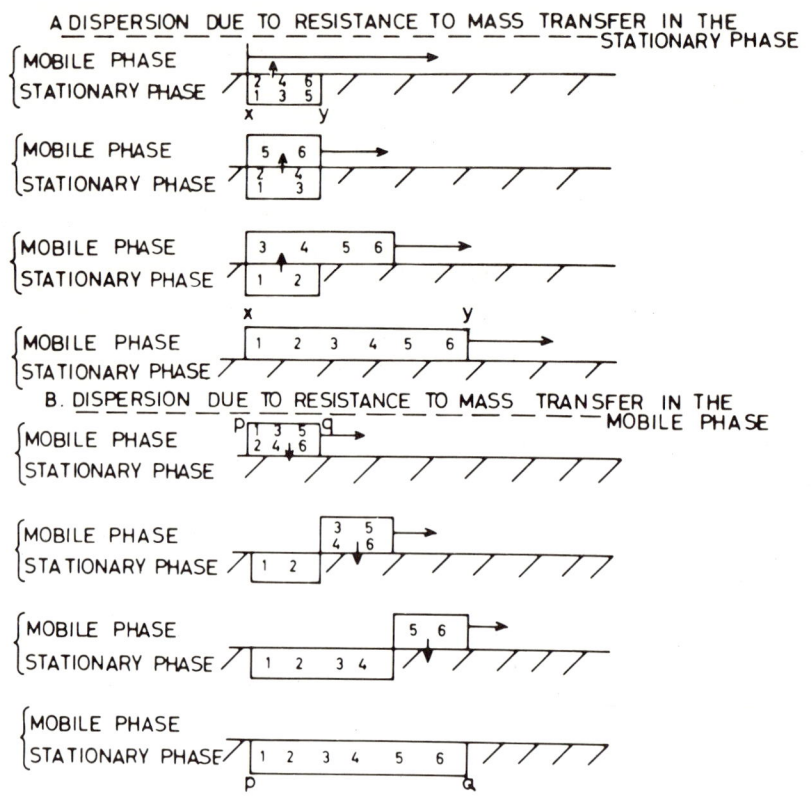

Fig. 2.10. Band dispersion resulting from resistance to mass transfer.

Combining the variances due to the four band spreading processes to give the final variance per unit length will result in the following equation:

$$h = \gamma d_p + \lambda \frac{D_m}{u} + \frac{K^2}{\psi a(a+K)^2} \frac{d_p^2 u}{D_m} + \frac{2}{3}\left[\frac{aK}{(a+K)^2}\right]\frac{d_f^2}{D_s} u. \quad (2.36)$$

| Multipath Effect | Longitudinal Diffusion | Dispersion due to Resistance to Mass Transfer in the Mobile Phase | Dispersion due to Resistance to Mass Transfer in the Stationary Phase |

In gas chromatography the curve obtained by plotting h/u is shown in Fig. 2.11 and it is seen that the curve exhibits a minimum value of h for a particular value of u called the optimum gas velocity. This type of curve is often called the HETP curve.

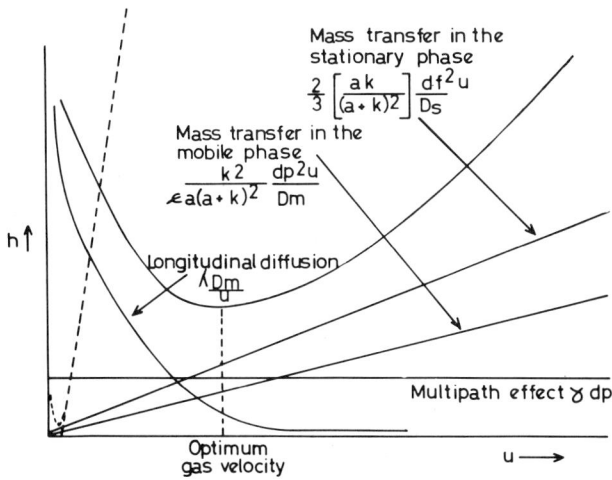

Fig. 2.11. Typical HETP curve for a gas chromatographic column.

If (2.36) is now considered from the point of view of liquid chromatography, the major difference will result in the change of the value of D_m (the diffusivity of the solute in the mobile phase) for the two systems. For the solutes normally separated by gas chromatography, D_m will be between 10^{-1} and 10^{-2} cm²/sec whereas the values of D_m for solutes in a liquid will lie between 10^{-5} and 10^{-6} cm²/sec. Thus D_m will be more than four orders smaller in liquid chromatography than in gas chromatography. This difference in D_m will cause the longitudinal diffusion term D_m to become insignificant and the dispersion due to the resistance to mass transfer in the mobile phase to become inordinately large. Such a situation should produce a curve of the form shown by the broken line in Fig. 2.11. As a result, to obtain narrow bands and thus reasonable resolution, the mobile phase velocity would have to be very small, which would result in impossibly long analysis times. If this situation did in fact exist then liquid chromatography could not be considered a practical separation technique. However, in liquid chromatography interparticle eddies occur as a result of the continuous change in direction of flow of the mobile phase as it winds its way between the interstices of the support. This does not constitute turbulance although it might be considered that the eddies can form the source of turbulent movement at high velocities. Turbulence does not occur in packed beds until the mobile phase velocities reach values of 10-100 cm/sec, and such velocities are seldom present in normal liquid chromatography. The eddies formed between the particles aid the mixing in the mobile phase and, as will be seen later, decrease the resistance to mass transfer in the mobile phase. The mixing of the solute in the mobile phase therefore results from two processes, the normal diffusion process that is brought about by concentration gradients, and convective mixing, which is brought about by interparticle eddies. Hiby [14] considered that the diffusion coefficient should be replaced by a dispersion coefficient that include both these mixing effects. The equation he suggested is as follows:

$$D_e = \frac{D_m}{T_m} + \frac{\lambda_2 2 d_p u}{1 + \lambda_1 (D_m/u d_p)^{\frac{1}{2}}} \qquad (2.37)$$

where D_e is the dispersion coefficient, T_m is the interparticle tortuosity factor, and λ_1 and λ_2 are constants.

The ratio D_m/T_m is the contribution due to diffusion and T_m corrects the diffusion coefficient D_m for the varying size and direction of the interstitial pores. The second term is the contribution due to convective mixing, and the different parts of the function can be broadly assigned to the two different convective processes. $2\lambda_2 d_p u$ describes the multipath effect that is reduced by the eddy effects given by the function $\lambda_1 (D_m/u d_p)^{\frac{1}{2}}$.

It is now necessary to return to the basic equation h derived by van Deemter where substitution can be made for D_e from (2.37). However, before doing so some new terms will be defined. In liquid chromatography the pores of the support will be filled with stationary phase together with some trapped "mobile phase." This mobile phase trapped in the pores will not be flowing with the main bulk and thus will be referred to as "static" mobile phase to differentiate it from flowing mobile phase. The two chromatographic phases will now therefore be referred to as phase α and phase β where phase β is stationary and phase α is mobile including "static" mobile phase. The fractions of the column volume that are occupied by phase α and phase β will be termed ϵ_α and ϵ_β, respectively. The fractions of the column volume that are occupied by flowing fluid and static fluid will be termed ϵ_m and ϵ_s, respectively. It should be pointed out that

$$\epsilon_\alpha \neq \epsilon_m$$
$$\epsilon_\beta \neq \epsilon_s$$
$$\epsilon_s = \epsilon_\beta + (\epsilon_\alpha - \epsilon_m).$$

Furthermore, it can be stated that if the total fraction of the column occupied by the two phases is ϵ, then $\epsilon = \epsilon_\alpha + \epsilon_\beta = \epsilon_m + \epsilon_s$.

Two further fractions should be defined: one is the fraction of the column that is occupied by the static phases plus the inert support material; this will obviously be $1 - \epsilon_m$. The other is the volume fraction of the inert support alone, $1 - \epsilon$.

2 THE RATE THEORY

Having completed these definitions, the basic equation of van Deemter can be dealt with. The derivation of the following equation will not be given, owing to its complexity, but it may be obtained from the original paper [2]:

$$h = \underbrace{\frac{2D_e}{u}}_{\substack{\text{Contribution to} \\ h \text{ due to mixing}}} + \underbrace{\frac{2}{6}\left(\frac{k'}{k'+1}\right)^2 \frac{\epsilon_m d_0}{1-\epsilon_m}\frac{u}{R}}_{\substack{\text{Contribution to } h \text{ due to resistance to} \\ \text{mass transfer at the interface}}} \qquad (2.38)$$

where R is the mass transfer coefficient at the interface, D_e is the dispersion coefficient, and k' is the capacity ratio of the column, defined as the ratio of the mass of solute in the stationary phase to that in the mobile phase at any point in the column. Numerically this will be given by

$$k' = \frac{Kv_s}{v_m}$$

It should be noted that R will be made up of two parts, resistance to mass transfer in the mobile phase and resistance to mass transfer in the stationary phase.

The basic equation of van Deemter therefore contains a term that describes a contribution to h due to resistance to mass transfer of solute from one phase to the other. Concerning ourselves for the moment solely with the mixing terms, substituting for D_e from (2.37) and (2.38),

$$h = \underbrace{\frac{2D_m}{T_m u}}_{\substack{\text{Longitudinal} \\ \text{Diffusion}}} + \underbrace{\frac{2\lambda_2 d_p}{1+\lambda_1(D_m/ud_p)^{1/2}}}_{\substack{\text{Convective} \\ \text{Mixing}}} + \underbrace{\frac{2}{6}\left(\frac{k'}{k'+1}\right)^2 \frac{\epsilon_m d_p}{1-\epsilon_m}\frac{u}{R}}_{\text{Resistance to Mass Transfer}} \qquad (2.39)$$

In gas chromatography van Deemter considered that $ud_p \gg D_m$ and thus (2.39) becomes

$$h = \frac{2D_m}{T_m u} + 2\lambda_2 d_p + \frac{2}{6}\left(\frac{k'^2}{k'+1}\right)^2 \frac{\epsilon_m d_p}{(1-\epsilon_m)}\frac{u}{R}$$

It is seen that the first two terms are identical with the longitudinal diffusion term and the multipath term in (2.36) where $\lambda = 2/T_m$. However, in liquid chromatography u can be more than one order less than in gas chromatography and furthermore d_p may have a value of only a few microns. For this reason the convection term must remain as given in (2.39).

Rewriting (2.39),

$$h = \frac{2D_m}{T_m u} + \frac{2\lambda_2 d_p}{1 + \lambda_1 (D_m/u d_p)^{1/2}} + \frac{2}{6}\left(\frac{k'}{k'+1}\right)^2 \frac{\epsilon_m d_p}{(1-\epsilon_m)} \frac{u}{R} \qquad (2.39)$$

It is now necessary to develop the function for the resistance to mass transfer, and this must be carried out as two separate operations, one for liquid-liquid chromatography and one for liquid-solid chromatography. This separate treatment results from the fact that in liquid-solid chromatography there is a third resistance to mass transfer effect at the adsorbent surface.

Resistant to Mass Transfer in Liquid-Liquid Chromatography

The development of the mass transfer function in (2.39) requires the evaluation of R and k', the capacity ratio. In determining the latter it must be remembered that the static phase consists of the stationary phase plus any mobile phase also entrained in the particles. If this were not so, k' would be merely K/a as in gas chromatography, remembering that $a = \dfrac{V_m}{V_s}$ is the phase ratio. For a liquid-liquid system,

$$k' = \frac{(\epsilon_\alpha - \epsilon_m + K\epsilon_\beta)}{\epsilon_m} \qquad (2.40)$$

Thus

$$\left(\frac{k'}{k'+1}\right)^2 = \left(\frac{\epsilon_\alpha - \epsilon_m + K\epsilon_\beta}{\epsilon_m}\right) \Big/ \left(\frac{\epsilon_\alpha - \epsilon_m + K\epsilon_\beta + \epsilon_m}{\epsilon_m}\right)^2$$

$$= \left(\frac{\epsilon_\alpha - \epsilon_m + K\epsilon_\beta}{\epsilon_\alpha + K\epsilon_\beta}\right)^2 \qquad (2.41)$$

Substituting for $[k'/(k'+1)]^2$ in (2.39) from (2.41),

$$h = \frac{2D_m}{T_m u} + \frac{2\lambda_2 d_p}{1 + \lambda_1 (D_m/u d_p)^{1/2}}$$

$$+ \frac{2}{6}\left(\frac{\epsilon_\alpha - \epsilon_m + K\epsilon_\beta}{\epsilon_\alpha + K\epsilon_\beta}\right)^2 \frac{\epsilon_m d_p}{(1-\epsilon_m)R} u \qquad (2.42)$$

The Transfer Coefficient

It is now necessary to obtain an expression for the transfer coefficient R. By analogy with electrical circuits, R can be compared to the total conductivity of two conductors in series, the overall conductance of the circuit being $1/R$ and its overall resistance being R. The concentration gradient across the interface is the potential that draws a given mass of solute per unit time (current) against a resistance to mass transfer (the electrical resistance). If the two conductors of resistance $1/R_m$ and $1/R_s$ are considered to represent the transfer coefficients of the phases, the overall resistance can be considered as the sum of the two individual resistances. Thus

$$\frac{1}{R} = \frac{1}{R_m} + \frac{1}{K'R_s} \tag{2.43}$$

where R_m is the mass transfer coefficient of the mobile phase, R_s is the mass transfer coefficient of the stationary phase and K' is the ratio of the concentration of solute in the stationary bed to that in the moving fluid. The constant K' takes into account the quantity of solute in the stationary phase relative to the mobile phase. Assuming unit concentrations of solute in the mobile phase, the mass of solute in the stationary bed m is given by

$$m = \epsilon_\alpha - \epsilon_m + K\epsilon_\beta \tag{2.44}$$

Now the volume of the stationary bed will be $1 - \epsilon_m$; thus the concentration of solute in the stationary bed in terms of concentrations in the mobile phase will be

$$K' = \frac{m}{1 - \epsilon_m} = \frac{\epsilon_\alpha - \epsilon_m + K\epsilon_\beta}{1 - \epsilon_m} \tag{2.45}$$

Substituting for k' in (2.43) from (2.45) the mass transfer coefficient becomes

$$\frac{1}{R} = \frac{1}{R_m} + \frac{1 - \epsilon_m}{\epsilon_\alpha - \epsilon_m + K\epsilon_\beta} \frac{1}{R_s} \tag{2.46}$$

Huber and Quaadrgrass [15] derived the following expressions for $1/R_m$ and $1/R_s$:

$$\frac{1}{R_m} = \frac{1}{1.9} \frac{d_p^{1/2} v^{1/6}}{\epsilon_m^{1/2} D_m^{2/3} u^{1/2}} \tag{2.47}$$

$$\frac{1}{R_s} = \frac{d_p T_s (1-\epsilon_m)}{10 D_s \epsilon_s} \tag{2.48}$$

where v is the kinematic viscosity of the solute in the mobile phase, T_s is the tortuosity factor for the intraparticle pores, and D_s is the diffusivity of the solute in the stationary phase. Substituting (2.47) and (2.48) and (2.46), expressions for R are obtained:

$$\frac{1}{R} = \frac{1}{1.9} \frac{d_p^{1/2} v^{1/6}}{D_m^{2/3} \epsilon_m^{1/2} u^{1/2}} + \frac{1}{10} \frac{(1-\epsilon_m)^2 d_p T_s}{(\epsilon_\alpha - \epsilon_m + K\epsilon_\beta)\epsilon_s D_s} \quad (2.49)$$

Finally, substituting for $1/R$ from (2.49) in (2.42) the final equation for h is obtained as given by Huber [5]:

$$h = \frac{2D_m}{T_m u} + \frac{2\lambda_2 d_p}{1 + \lambda_1 (D_m/ud_p)^{1/2}} +$$

$$\frac{1}{5.7} \left(\frac{\epsilon_\alpha - \epsilon_m + K\epsilon_\beta}{\epsilon_\alpha + K\epsilon_\beta} \right)^2 \frac{\epsilon_m^{1/2} d_p^{3/2} v^{1/6} u^{1/2}}{(1-\epsilon_m)D_m^{2/3}} + \quad (2.50)$$

$$\frac{1}{30} \left(\frac{\epsilon_\alpha - \epsilon_m + K\epsilon_\beta}{(\epsilon_\alpha + K\epsilon_\beta)^2} \right) \frac{\epsilon_m(1-\epsilon_m)T_s d_p^2 u}{\epsilon_s D_s}$$

The equations for h appear extremely complex relative to (2.36) for h given by van Deemter. This is largely a result of taking into account the static "mobile" phase contained in each particle. Because the static "mobile" phase contributes significantly to h but at the same time does not contribute to selectivity in liquid-liquid chromatography, the support particle should be optimumly loaded with stationary phase. That is to say that the pores of the particles should be completely filled with stationary phase. Then

$$\epsilon_\alpha = \epsilon_m \quad \text{and} \quad \frac{\epsilon_m}{\epsilon_\beta} = \frac{V_m}{V_s} = a$$

If at the same time it is assumed that $D_m \ll ud_p$, then (2.50) simplifies to

$$h = \frac{2D_m}{T_m u} + 2\lambda_2 d_p + \frac{1}{5.7} \frac{K^2}{(a+K)^2} \frac{\epsilon_m^{1/2} d_p^{3/2} v^{1/6} u^{1/2}}{(1-\epsilon_m)D_m^{2/3}}$$

$$+ \frac{1}{30} \frac{aK}{(a+K)^2} \frac{(1-\epsilon_m)T_s}{\epsilon_s D_s} d_p^2 u. \quad (2.51)$$

Equation (2.51) for a liquid-liquid column optimumly loaded with stationary phase and derived from the basic equation by Huber (5) now bears a close resemblance to equation (2.36) for gas liquid columns by van Deemter.

The important difference between the equations of Huber and van Deemter lies in the resistance to mass transfer for the mobile phase, which van Deemter concludes is a linear function of u whereas Huber concludes that it is proportional to $u^{1/2}$.

However, owing to the significance of the static "mobile" phase contributions to h in liquid chromatography, (2.50) must remain the basic equation for h in liquid-liquid columns. Furthermore, there is considerable experimental support for the equation of Huber when applied to liquid-liquid chromatogrpahy; this support will be discussed in some detail later. Equation (2.50) is thus made up of four fractions, each describing a particular band spreading effect:

$$h = \frac{2D_m}{T_m u} \quad \text{Longitudinal diffusion}$$

$$+ \frac{2\lambda_2 d_p}{1 + \lambda_1 (D_m/u d_p)^{1/2}} \quad \text{Convective mixing}$$

$$+ \frac{1}{5.7} \left(\frac{\epsilon_\alpha - \epsilon_m + K\epsilon_\beta}{\epsilon_\alpha + K\epsilon_\beta} \right)^2 \frac{\epsilon_m^{1/2} d_p^{3/2} \nu^{1/6} u^{1/2}}{(1 - \epsilon_m) D_m^{2/3}} \quad \text{Resistance to mass transfer in the mobile phase}$$

$$+ \frac{1}{30} \frac{(\epsilon_\alpha - \epsilon_m + K\epsilon_\beta)}{(\epsilon_\alpha + K\epsilon_\beta)^2} \frac{\epsilon_m(1-\epsilon_m) T_s d_p^2 u}{\epsilon_s D_s} \quad \text{Resistance to mass transfer in the stationary phase.}$$

A typical h-versus-u curve is shown in Fig. 2.12; included are the curves for

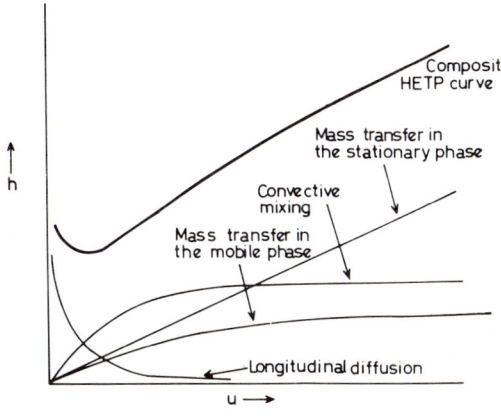

Fig. 2.12. Typical HETP curve for a liquid chromatographic column.

each individual function. It can be seen that at relatively high linear mobile phase velocity the resistance to mass transfer in the static phase has the predominant effect on h. At very low values of u the longitudinal diffusion effect is the major contribution to h. At intermediate velocities the convective mixing term together with the resistance to mass transfer in the mobile phase has the major effect. For reasonable analysis time it is desirable to operate at as high a linear mobile phase velocity as possible, and thus the column has to be designed to reduce the resistance to mass transfer in the static phase to a minimum.

Resistance to Mass Transfer in Liquid-Solid Chromatography

For liquid-solid chromatography, (2.50) is at least theoretically inadequate because it disregards any resistance to mass transfer at the surface of the adsorbent. A molecule held on the surface of the adsorbent will have a given potential energy P_e due to the intermolecular forces holding it there, and a kinetic energy K_e due to vibrational movement.

When K_e exceeds P_e the molecule will leave the surface. Now the molecules held on the surface will have a range of energies depending on their mass, shape, and temperature, and this range of energies will probably have a Boltzmann distribution. This means that in any group of molecules only a proportion will have a value of K_e that exceeds P_e and can therefore leave the surface. The remainder must await their statistical chance to achieve a kinetic energy K_e before they are desorbed. Now this means that there will be a finite time interval necessary between the time when the first of a group of molecules leaves the adsorbent and when the last of the molecules leaves the surface. As they are desorbed into a moving phase then, owing to this finite desorption time, the solute band will spread. In exactly the same way a molecule of solute in the moving phase will have an energy of K_e and on striking the adsorbent surface, if $K_e < P_e$ the molecule will be adsorbed; if $K_e > P_e$ then it will move back into the moving phase. It is probable that the distribution of kinetic energy of the molecules in the mobile phase is also Boltzmann in form and each molecule with $K_e > P_e$ will have to await its statistical chance to arrive when $K_e < P_e$ on striking the adsorbent surface before it can be adsorbed. Thus there will be a finite time interval for a group of molecules to *all* become adsorbed on the surface of the adsorbent, and as they are being adsorbed from the moving phase, this will result in a band spreading process. The situation is depicted in Fig. 2.13. In Fig. 2.13a the distribution curve represents the energy of the solute molecules on the surface of the adsorbent at any instant. Those exceeding P_e leave the surface; the rest remain until their statistical chance occurs when the vibrational energy exceeds P_e. In Fig. 2.13b, the distribution curve represents the energy of the solute moelcules in the mobile

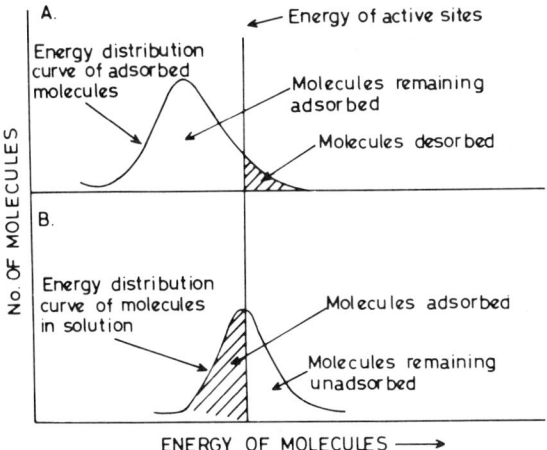

Fig. 2.13. Conditions for adsorption and desorption from solid to a liquid.

phase; those that have energies less than P_e are adsorbed on striking the surface while the rest await their statistical chance for their energies to fall below P_e when on striking the surface they will be adsorbed. At equilibrium the number of molecules adsorbed will equal the number of molecules desorbed; that is, the number of molecules in the shaded portions of the curves will be the same.

The above arguments assume, however, that all the sites have the same activity (P_e). In fact, the situation will be more complicated since the sites themselves will have a distribution of energies. If the adsorption isotherm is linear then the energy distribution curve will be symmetrical; if nonlinear, an asymmetrical energy distribution curve is likely. Thus the condition for adsorption and desorption will be more complex because it will depend not only on the probability of a molecule having less or more than a particular energy but also on its striking a respective site having a particular activity. Now because the distribution of energies for molecules in the liquid will be different from that on the solid since they have different degrees of freedom, the probabilities that a molecule will be desorbed will not be the same as the probability that a molecule will be adsorbed. Thus the resistance to mass transfer at the adsorbent liquid interface will be made up of two components, one associated with the static liquid in the adsorbent pore and the other associated with the adsorbent surface at the pore walls. This is analogous to resistance to mass transfer in the mobile phase and the resistance to mass transfer in the stationary phase at the interface between the two phases.

To develop an equation for h pertinent to liquid-solid chromatography, it is necessary to start again from (2.39), which is as follows:

$$h = \frac{2D_m}{T_m u} + \frac{2\lambda_2 d_p}{1 + \lambda_1 (D_m/u d_p)^{1/2}}$$

$$+ \frac{2}{6} \left(\frac{k'}{k'+1}\right)^2 \frac{\epsilon_m d_p}{(1-\epsilon_m)} \frac{u}{R} \qquad (2.39)$$

Before evaluating k' the distribution coefficient K_s for the liquid-solid system must be defined as follows:

$$K_s = \frac{C_s}{C_\alpha} \qquad (2.52)$$

where C_s is the concentration of solute on the surface in grams per square centimeter and C_α is the concentration of solute in the mobile phase in grams per milliliter. It should be noted that the dimensions of C_s and C_α are m/l^2 and m/l^{-3}, respectively; thus K_s has the dimensions of length (l).

Now k' is the ratio of the mass of solute in the stationary and static phases to that in the mobile phase. Thus

$$k' = \frac{\epsilon_\alpha - \epsilon_m + K_s \phi}{\epsilon_m} \qquad (2.53)$$

where ϕ is the ratio of the surface area of adsorbent to the column volume and has the dimensions of l^{-1} and, therefore, $K_s \phi$ is dimensionless as demanded by (2.39). It follows from (2.53) that

$$\frac{k'}{k'+1} = \frac{\epsilon_\alpha - \epsilon_m + K_s \phi}{\epsilon_\alpha + K_s \phi}$$

and therefore, substituting for $k'/k'+1$ in (2.39),

$$h = \frac{2D_m}{T_m u} + \frac{2\lambda_2 d_p}{1 + \lambda_1 (D_m/u d_p)^{1/2}}$$

$$+ \frac{2}{6} \left(\frac{\epsilon_\alpha - \epsilon_m + K_s \phi}{\epsilon_\alpha + K_s \phi}\right)^2 \frac{\epsilon_m d_p}{(1-\epsilon_m)} \frac{u}{R} \qquad (2.54)$$

It now remains to determine R for a liquid-solid system.

In a manner similar to liquid-liquid chromatography, R can be divded into two parts, one associated with the mobile phase and the other associated with the stationary phase. Thus restating (2.43), each term having the same definitions,

$$\frac{1}{R} = \frac{1}{R_m} + \frac{1}{KR_s} \qquad (2.43)$$

and K' for a liquid-solid system is given by

$$K' = \frac{\epsilon_\alpha - \epsilon_m + K_s\phi}{(1 - \epsilon_m)} \qquad (2.55)$$

It is seen that $K\epsilon_\beta$ in the function for K' in liquid-liquid chromatography given by (2.45) is replaced in liquid-solid chromatography by $K_s\phi$. Thus substituting for K' from (2.55) in (2.43),

$$\frac{1}{R} = \frac{1}{R_m} + \frac{(1 - \epsilon_m)}{\epsilon_\alpha - \epsilon_m + K_s\phi} \frac{1}{R_s}. \qquad (2.56)$$

The expressions for R_m derived by Huber and Quaadgrass [15] and given by (2.47) holds for liquid-solid chromatography as well as liquid-liquid chromatography. However, their expression for R_s given by (2.48) has to be modified to account for the desorption rate of the solute at the adsorbent surface.

Probably at the present state of the development of chromatographic theory, the best way of accounting for the rate of desorption of the solute from the surface is to substitute D_{AS} for D_s in (2.48) where D_{AS} is weighted combination of $D_s + D_A$ that takes into account the geometry of the pores, D_A is defined as the mass of solute transported across the interface per unit concentration difference.

Thus substituting D_{AS} for D_s in (2.48), and using (2.47) and (2.48) in (2.56)

$$\frac{1}{R} = \frac{1}{1.9} \frac{d_p^{1/2} v^{1/6}}{D_m^{2/3} \epsilon_m^{1/2} u^{1/2}} + \frac{1}{10} \frac{(1 - \epsilon_m)^2 d_p T_s}{(\epsilon_\alpha - \epsilon_m + K_s\phi)\epsilon_s D_{AS}} \qquad (2.57)$$

Substituting for $1/R$ from (2.57) in (2.54) we obtain an expression for h for liquid-solid chromatography:

$$h = \frac{2D_m}{T_m u} + \frac{2\lambda_2 d_p}{1 + \lambda_1 (D_m/ud_p)^{1/2}} +$$

$$\frac{1}{5.7} \left(\frac{\epsilon_\alpha - \epsilon_m + K\phi}{\epsilon_\alpha + K_s\phi} \right)^2 \frac{\epsilon_m^{1/2} d_p^{3/2} v^{1/6} u^{1/2}}{(1 - \epsilon_m) D_m^{2/3}} + \qquad (2.58)$$

$$\frac{1}{30} \frac{\epsilon_\alpha - \epsilon_m + K_s\phi}{(\epsilon_\alpha + K_s\phi)^2} \frac{\epsilon_m(1-\epsilon_m)T_s d_p^2}{\epsilon_s D_{AS}} u.$$

Both equation (2.50) for liquid-liquid chromatography and (2.58) for liquid-solid chromatography give the same form of h-versus-u curve as shown in Fig. 2.12. In order to reduce analysis time it is desirable to operate at high mobile phase velocities and thus the resistance to mass transfer in the static phase must be made as small as possible. The function for the resistance to mass transfer in the static and stationary phases in (2.50) and (2.58) shows that d_p must be made as small as possible and D_s or D_{AS} made as large as possible. Further, ϵ_m/ϵ_s, the ratio of mobile phase to static phase, should be reduced to a minimum and the volume of the support itself reduced in order to decrease $1 - \epsilon_m$.

Thus the theory has shown how the column must be packed to improve column efficiency, but some warning must be given with respect to (2.50) and (2.58) when they are used to give absolute values for h. Although the equations can predict how column systems may be improved and generally describe the relations of h to u, they contain many empirical constants. To indicate the difficulties that arise when trying to assign values to these constants consider the micrographs shown in Figs. 2.14a and 2.14b.

Figure 2.14a, a pair of stereo micrographs of a Celite praticle is shown, the

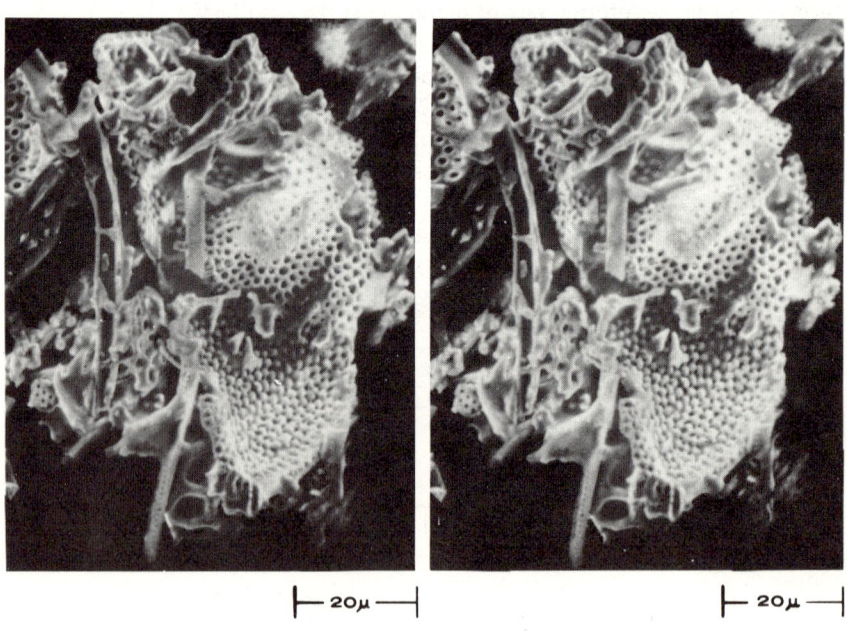

a

b

Fig. 2.14. (*a*) Electron micrographs of celite particles. (*b*). Electron micrograph of a surface of a silica gel particle.

holes having an average diameter of about 5000 Å. A support particle will consist of a conglomerate of thousands of these sieve plates randomly orientated. In liquid-liquid chromatography these particles are coated with stationary phase, and it can be seen how empirical and arbitrary any assumption of film thickness of stationary phase, tortuosity factor, or even paticle diameter of such a system must be.

Figure 2.14*b* is a micrograph of a section of a silica gel particle. It was obtained by embedding particles of silica gel, 10 μ in diameter, in a resin and cutting sections using a glass microtome. A carbon replica of the surface was then made and the micrograph obtained from this replica. The particle appears to be made up of small, almost spherical units 100 Å in diameter. The dark areas are the solid part of the matrix and the light areas represent holes through the surface. The holes vary in shape and their dimensions range from a few hundred Angstroms to a few thousand Angstroms. It is seen that any estimation of T_s, D_{AS}, or even d_p will be largely guesswork. Furthermore, in liquid-solid chromatography the structure of the adsorbent will vary from supplier to supplier and batch to batch, depending on the method of manufacture and the degree of activation. Equations such as (2.50) and (2.58) can only indicate the important parameters that affect band dispersion in liquid chromatography, but in doing so they show how the column or packing should be constructed to minimize peak dispersion.

Experimental Support of the Rate Theory

Rewriting (2.50) in a simplified form and assuming $K \gg 1$, which will be true for most practical purposes,

$$h = \frac{AD_m}{u} \qquad \text{Longitudinal diffusion}$$

$$+ \frac{2\lambda_2 d_p}{1 + \lambda_1 (D_m/ud_p)^{1/2}} \qquad \text{Convective mixing}$$

$$+ \frac{E d_p^{3/2} u^{1/2}}{D_m^{2/3}} \qquad \text{Resistance to mass transfer in the mobile phase}$$

$$+ \frac{F d_p^2 u}{KD_s} \qquad \text{Resistance to mass transfer in the stationary phase}$$

where $A = 2/T_m$ and

$$E = \frac{1}{5.7} \frac{\epsilon_m^{1/2} \, \nu^{1/6}}{(1-\epsilon_m)}$$

and it is assumed that $(\epsilon_\alpha - \epsilon_m + K\epsilon_\beta)/(\epsilon_\alpha + K\epsilon_\beta)$ tends to 1 when $K \gg 1$, and

$$F = \frac{1}{30} \frac{\epsilon_m(1-\epsilon_m)T_s}{\epsilon_s}$$

assuming that $(\epsilon_\alpha - \epsilon_m + K\epsilon_\beta)/(\epsilon_\alpha + K\epsilon_\beta)^2$ tends to $1/K$ when $K \gg 1$.

Scott, Blackburn, Wilkins [16] evaluated the effect of particle diameter on column efficiency using data obtained from two sets of experiments as follows. First, three columns were packed with 200-240-, 100-120-, and 85-100-mesh brickdust in water and h/u curves were obtained for solute bands of salt solution over a mobile phase velocity range of 0.02-0.2 cm/sec. The results are shown in Fig. 2.15. The second set of experiments involved four columns, two of which

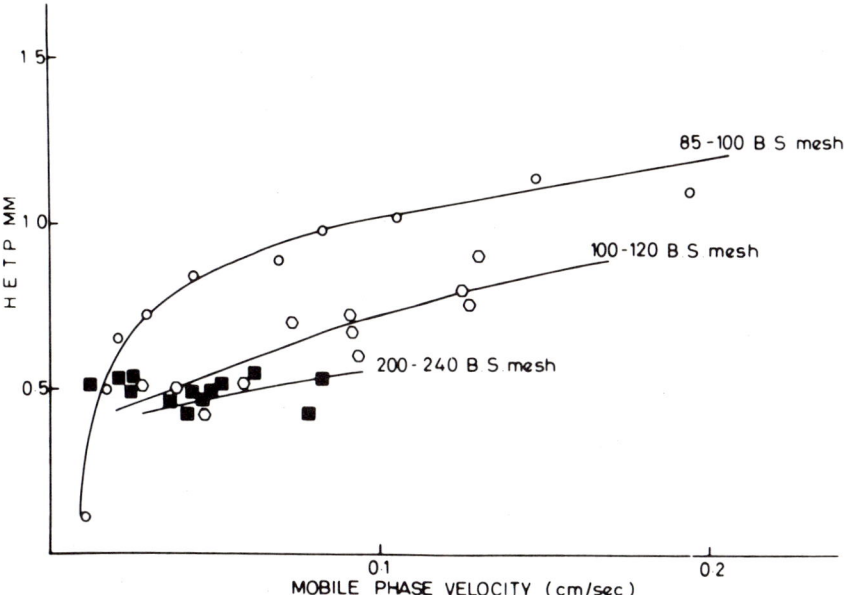

Fig. 2.15. Graph of HETP against mobile phase velocity for columns packed with different grades of brickdust.

were packed with 100-120 brickdust and 100-120 silica gel in water, and HETP curves for bands of salt solution were obtained over mobile phase velocities of 0.02-0.5 cm/sec. The second two columns were packed with 100-120 brickdust and 100-120 silica gel in petroleum ether. Solute bands of squalane were used to obtain HETP curves over a similar range of mobile phase velocities but employing the moving wire detecting system for monitoring the elution curve instead of the conductiometric detection system. The results obtained are shown in Fig. 2.16.

68   THE THEORY OF CHROMATOGRAPHY

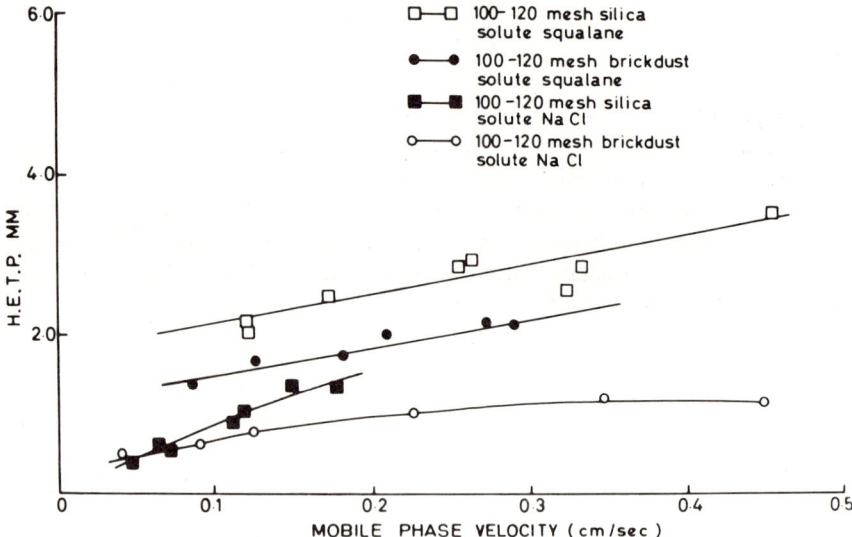

Fig. 2.16. Graph of HEPT against mobile phase velocity for columns with different packing and different mobile phase.

The curves shown in Fig. 2.15 are for a system devoid of stationary phase and the support will only contain "static" mobile phase. Under these conditions $K = 0$ and so the major factor affecting the value of h will be the convective mixing term, the longitudinal diffusion term only becoming significant at mobile phase velocities well below those used in the experiment. It follows that the curves should follow the form of the theoretical curve for convection mixing shown in Fig. 2.12. Indeed, the curves are of the same form and the value of h at any particular value of u falls as the particle diameter is reduced, which is also in accordance with (2.50).

The h/u curves shown in Fig. 2.16 are for columns packed with particles of the same size but of materials with differing porosity and using different solutes and mobile phases. As squalane is only retained slightly on silica gel when heptane is used as the stationary phase, $K = 0$ for both solute-solvent systems. Considering the effect of the different supports, the porosity of silica gel is much greater than brickdust and thus the residual resistance to mass transfer in the stationay ("static mobile") phase will be greater for silica gel than for brickdust whatever the solute-solvent systems employed. This theoretical prediction is evident from the curves. Similarly, the diffusivity of squalane in heptane is much smaller than salt in water; therefore it would be expected that, because of the residual resistance to mass transfer in the "static mobile" phase, the values of h for the squalane-heptane system would always be greater than the corresponding salt-water system irrespective of the support employed. This is also confirmed experimentally by the results in Fig. 2.16. Furthermore, the major factor affecting the value of h will still be convective mixing and thus the curves will maintain the general shape of the respective theoretical curve.

Huber and Hulsman [5] also examined the relationship of h to u for the solutes nitrobenzene and butylbenzene using columns of different support size. Their h/u curves are shown in Figs. 2.17 and 2.18 together with the column conditions. All columns show a minimum at very low mobile phase velocities and indicate the small contributions to h made by the longitudinal diffusion terms.

The partition coefficient of butylbenzene in the liquid-liquid system of 2,2,4-trimethylpentane, 1,2,3-tris-cyanoethoxypropane at 25°C is smaller than 0.1. Therefore (and because the particles do not contain eluent since there is optimum loading of stationary phase), the mass transfer term, in the expression

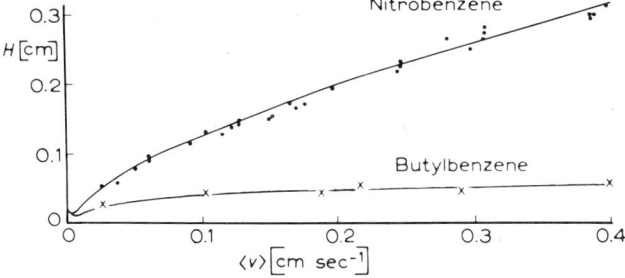

Fig. 2.17. Dependence of H on fluid velocity. Column: Dimension, 50 × 0.20 cm; particle size, 63-80 μ; particle loading, 1,2,3-triscyanoethoxypropane ($\epsilon\beta$ = 0.07); eluant 2,2,4-trimethylpropane ($\epsilon\alpha$ = 0.82). Sample: 0.6 μl (0.5% w/w butylbenzene (X) + 1.5% nitrobenzene (●) + eluent).

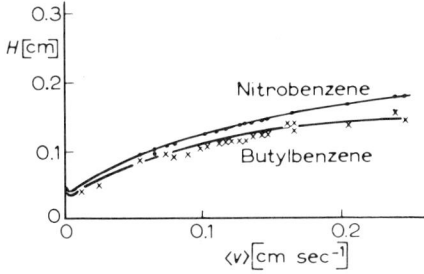

Fig. 2.18. Dependence of H on fluid velocity. Column: Dimension, 100 × 0.20 cm, particle size, <25 μ (37% w/w), 25-56 μ (28%), 56-80 μ (35%); particle loading 1,2,3-tris-cyanoethoxypropane ($\epsilon\beta$ = 0.11); eluant 2,2,4-trimethylpentane ($\epsilon\alpha$ = 0.60). Sample; 0.6 μl (0.5% w/w butylbenzyne (X) + 1.5% nitrobenzene (●) + eluent).

for h for butylbenzene, is negligible. The curves for butylbenzene in Figs. 2.17 and 2.18 are caused essentially by the mixing terms alone. At higher mobile phase velocities the convective mixing dominates; at lower ones, the diffusion process. On the column containing less uniform and smaller particles (Fig. 2.18), a considerably higher h/u curve is obtained for nonretarded compounds like butylbenzene, owing to stronger convective mixing. This observation is found to be significant, and it must be concluded that particles of greatly differing or very small size, unless special packing conditions are employed, tend to give a less regular packing.

Nitrobenzene has a partition coefficient of about 6.8, which gives rise to a considerable mass transfer term. Since the diffusion coefficients of butylbenzene ($D_m = 1.4 \times 10^{-5}$ cm^2/sec and nitrobenzene ($D_m = 1.8 \times 10^{-5}$ cm^2/sec) in 2,2,4-trimethylpentane differ only slightly, the mixing terms are also similar and the difference of the h curves of both substances gives an indication of the magnitude of the two mass transfer terms together. Figure 2.18 confirms that the mass transfer term is much smaller with smaller particles.

In Fig. 2.17 the h curve at higher velocities is determined mainly by the mass transfer term; in Fig. 2.18, it is determined by the convective mixing term. Figure 2.19 demonstrates how the overall curve can be built up from the four

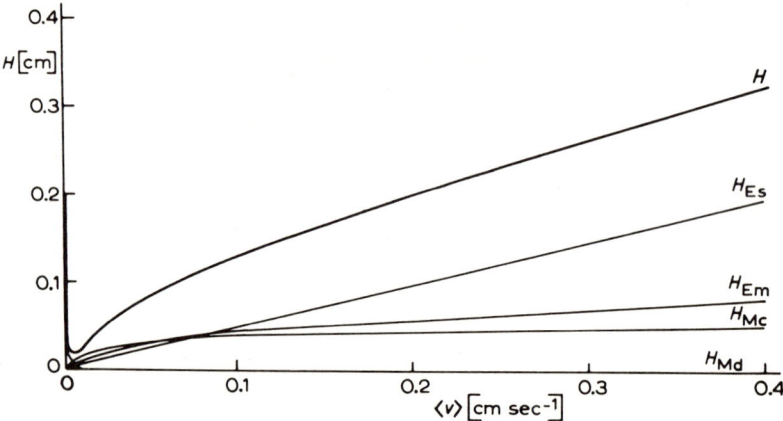

Fig. 2.19. Contributions to H; evaluated for the curve of nitrobenzene in Fig. 2.17.

partial curves that refer to the different dispersion process. The theoretical h curve obtained from the partial curves according to (2.50) fits the experimental results very well.

From (2.50) the resistance to mass transfer effect in the stationary phase would be expected to increase for an homologous series of compounds as their molecular weight increased. This would occur because K increases and D_s

decreases as the carbon number of the compound of the homologous series becomes larger. This effect was demonstrated by Scott et al. [16] from their h/u curves for the ^6C, ^{14}C, and ^{16}C fatty acids. These solutes were separated on a column employing n-octane as the stationary phase and a 65% acetone-35% water mixture as the mobile phase. The curves obtained are shown in Fig. 2.20.

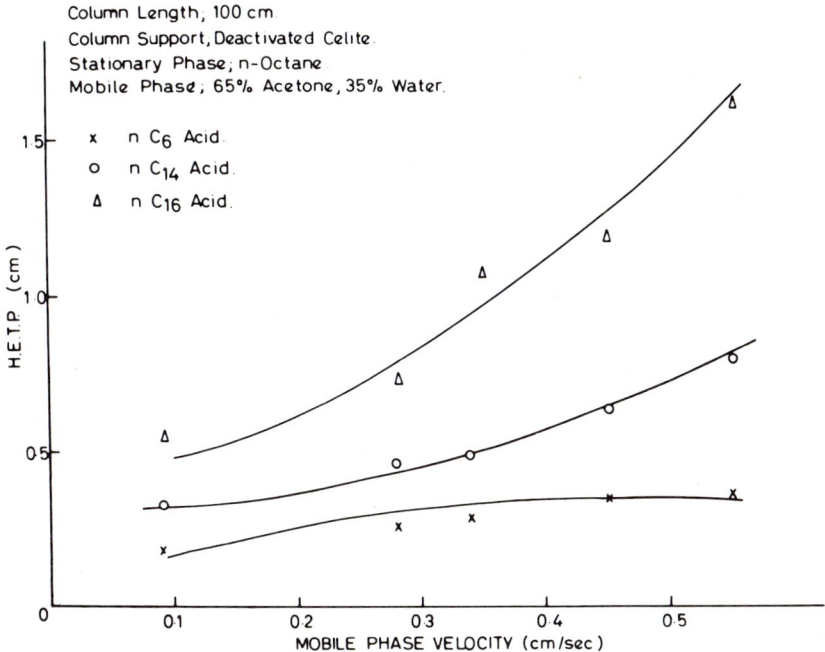

Fig. 2.20. Graph of HETP against mobile phase velocity for different solutes.

The curve for ^6C acid, which is virtually unretained, shows the typical shape for a system where the convective term is predominant. Conversely, the curve for the ^{16}C acid demonstrates that the major dispersion effect is due to the resistance to mass transfer in the stationary phase. The intermediate curve for the ^{14}C acid shows that at low mobile phase velocities the convection term is predominant, whereas at high mobile phase velocities the resistance to mass transfer in the stationary phase is the major dispersing effect.

The Empirical Approach to the Relationship Between h and u

An alternative equation for h was put forward by Snyder [17] relating h with a power function of the mobile phase velocity:

$$h = Du^{0.4}$$

where D is a constant. Synder clained that this function fitted the results obtained from most columns and Waters et al. [18] and Simpson and Wheaton [19] observed that the equation fitted a number of other liquid chromatography systems providing the index of the linear mobile phase velocity was allowed to take values ranging from 0.3 to 0.6. The equation is not expected to provide an exact fit to all experimental data, particularly at low ($<$ 0.1 cm/sec) and very high ($\gg$ 10 cm/sec) velocities. The equation will be occasionally imprecise, but seldom grossly wrong, as regards any qualitative or semiquantitative conclusions that may be derived from application to experimental systems. The equation affords a simple basis for comparing column efficiencies obtained by different workers, using any set of experimental conditions. For maximum accuracy in such comparisons, it is recommended that D be calculated from h values measured near u equal 1.0 cm/sec, using a solvent of low viscosity (e.g., 0.2-0.3 cP), and a solute in the 100-200 molecular weight range. Snyder also proposed the function $D = 18\, d_p^{0.8}$ on a bsis of experimental results and thus his equation h became

$$h = 18\, d_p^{0.8} u^{0.4}$$

This equation can be used to evaluate different columns providing they are operated within the experimental constraints that allow the relationship to be valid. It should be emphasized, however, that in the development of this equation the complex but rational equation of Huber is sacrificed for empirical mathematical simplicity and in doing so the basic relationship between the experimental results and the magnitude of the dispersing processes that take place in the column is lost.

3 RADIAL DISPERSION

The theory of radial dispersion has been elegantly treated by Horne, Knox, and McLaren [20]. When a stream of mobile phase carrying a solute impinges against a particle, the stream divides and flows around the particle. Part of the divided stream then joins other split streams from neigboring particles, implinges on another particle, divides again. This stream splitting process is depicted in Fig. 2.21a. If a particular molecule of solute is considered passing around a particle, it will suffer a laterial movement that, from Fig. 2.21b, can be seen to be given by

$$\text{Lateral movement/Particle} = \frac{\tfrac{1}{2} d_p}{\cos \theta}$$

It follows that the average lateral step will be

3 RADIAL DISPERSION   73

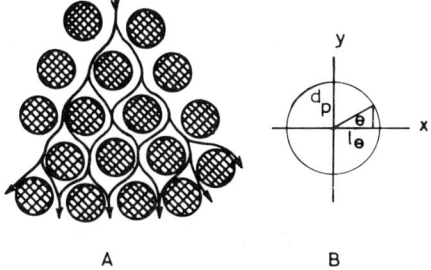

Fig. 2.21. The mechanism of radical dispersion. (a) The stream-splitting mechanism (b) Illustration of typical radial step.

$$\frac{d_p}{\pi}. \qquad (2.59)$$

Now using the "random walk" treatment described by Giddings [4], the radial variance is given by

$$\sigma_n^2 = \text{(Number of steps)} \times \text{(Step length)}^2 \qquad (2.60)$$

Assuming that one lateral step is taken for every distance $j\,d_p$ moved axially, then n, the number of steps, is given by

$$n = \frac{l}{jd_p} \qquad (2.61)$$

where l is the distance traveled axially by the solute band; thus substituting for n from (2.61) in (2.60) and using (2.59),

$$\sigma_n^2 = \frac{l}{jd_p}\left(\frac{d_p}{\pi}\right)^2 = \frac{ld_p}{j\pi^2}$$

In practice, the value of j will lie between 0.5 and 1.0 but for simplicity the value of j will be taken as unity. This implies that one lateral step will be taken by a given molecule for every distance equivalent to a particle diameter traveled axially. Thus

$$\sigma_n^2 = \frac{ld_p}{\pi^2}$$

or

$$\sigma_n = \frac{(ld_p)^{1/2}}{\pi} \tag{2.62}$$

Consider a solute sample injected at a point precisely at the center of a 4-mm column. Equation (2.62) permits the calculation of the distance traveled axially by the solute band before the radial standard deviation of the solute is numerically equal to the column radius. That is, the band has spread evenly across the column and the solute is in radial equilibrium with the column system. For the above conditions, $\sigma_R = 0.2$ cm. Thus substituting the value in (2.12),

$$0.2 = \frac{(ld_p)^{1/2}}{\pi}$$

or

$$l = \frac{(0.2\pi)^2}{d_p} \tag{2.63}$$

The values of l obtained from (2.63) plotted against particle diameter are shown in Fig. 2.22. It is seen that as the particle diameter decreases, the length of the

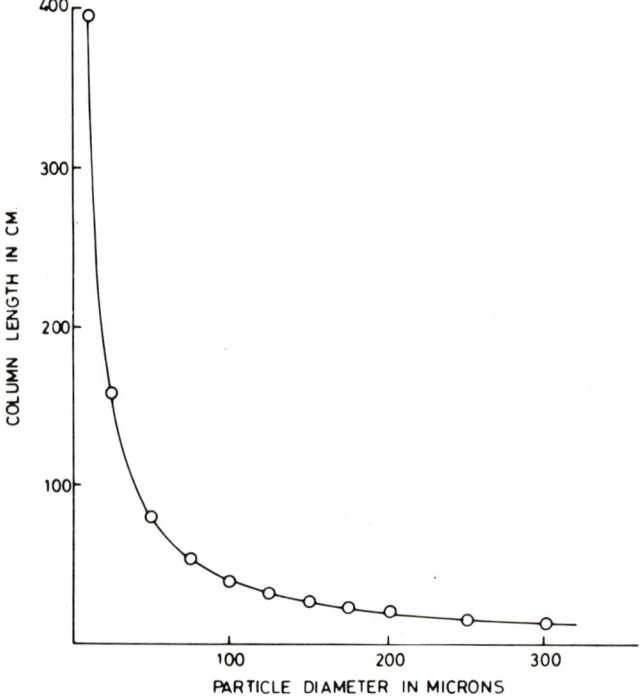

Fig. 2.22. Graph of distance traveled by solute along the column before radial equilibrium is reached against particle diameter.

3 RADIAL DISPERSION   75

column traversed by the solute, before radial equilibrium is attained, increases. Further, since it is highly desirable to use particles of minimum diameter to achieve maximum efficiency, the charge may never reach radial equilibrium before being eluted. If the column packing is completely homogeneous, then nonradial equilibrium would not impair the efficiency. In practice, however, it is often not possible to attain such ideal packing conditions and some channeling always occurs where there is a local variation in packing density. Thus, if radial equilibrium is desired it is necessary either to attain it before commencing development or to use very narrow columns packed with relatively large particles. Narrow columns would be difficult to pack and particles of large diameter would reduce column efficiency. Scott et al. [16] examined the effect of different injection conditions on efficiency by determining the spread of salt bands in a column conductometrically.

The column used consisted of an acrylic resin tube 100 cm long, 4 mm in diameter, packed with 100-120 BS mesh C22 firebrick. Five pairs of nichrome

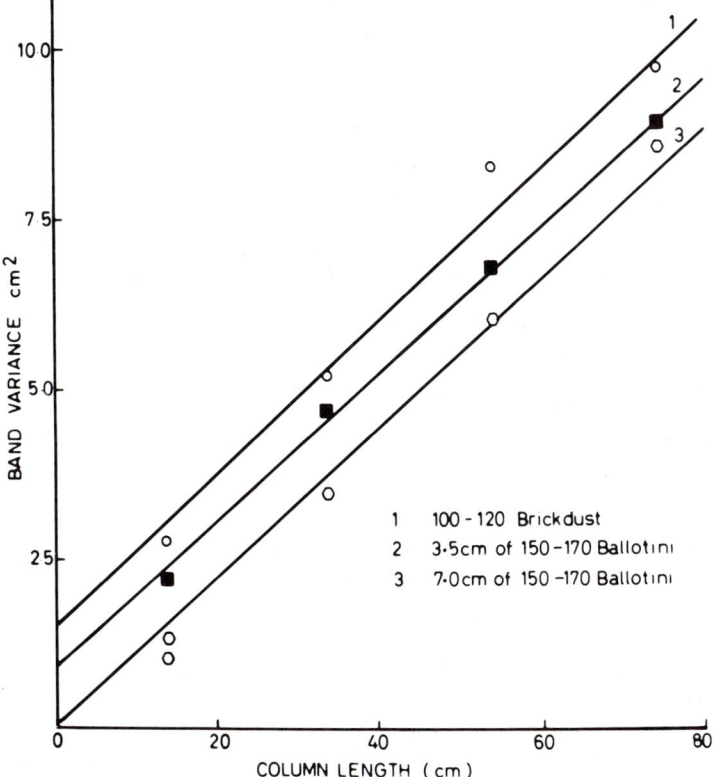

Fig. 2.23. Graph of variance against column length for solutes injected on different packing.

electrodes were sealed into the tube 20 cm apart. Distilled water was used as the mobile phase, and 1 l charges of 1% salt solution were injected on the column and the solute bands monitored as they passed each electrode.

The band varriances were then plotted against column length and the apparent variance of the injected sample, where $l = 0$, obtained by extrapolation. Graphs of variance against the column length are shown in Fig. 2.23 for the normal column and for brick dust columns carrying 3.5- and 7-cm layers of 150-170 BS mesh Ballotini on top of the column. It is seen from curve 1 that the initial variance of the peak was 1.5 cm^2 for brickdust alone. In curve 2 a layer of 3.5 cm of glass beads (Ballotini) has been place on top of the column and sample injected onto them. Initial variance of the sample under these conditions has been reduced to 1 cm^2 when the layer of glass beads was increased to 7.5 cm; the initial variance appeared to be reduced to zero.

The relatively coarse, uniform layers of glass beads ensured that radical equilibrium was reached rapidly and before meeting the actual packing. Thus the overall effect of channeling on the band variance during subsequent passage through the column was reduced to a minimum. An injection system designed to ensure that the sample is injected into the center of the column is described in the following section. Kirkland [21] also examined the effect of conditions of injection on plate height for columns packed with microparticulate silica gel. As a result of his experiments he recommended the use of a porous disk on the top of the packing, the sample being injected onto the center of the porous disks. Using this technique he maintained high column efficiencies under repeatable conditions.

4 DISPERSION IN CAILLARY CONNECTING TUBES

It is often necessary to connect columns to detectors, or possibly different column sections together, by capillary tubes and it is therefore desirable to estimate the contribution such connections will make to the solute band dispersion. The theory of dispersion in open tubular columns has been developed by Golay [3], and from his equation it can be shown that the only effective dispersion process operating in an open tube, containing *no* stationary phase, is the resistance to mass transfer in the mobile phase under conditions of Newtonian flow.

The contribution made by the resistance to mass transfer of solute in the mobile phase to the band dispersion in gas chromatography is usually negligible and has been examined by Grushka [24], but in liquid chromatography where the diffusivity of the solutes are four or five orders smaller, it can be significant.

Consider a solute band passing from a column into a capillary tube and thence to a detector. Let the variance of band leaving the column be σ_c^2, the contribution of the capillary tube to the band variance be σ_T^2, and the overall variance of the band entering the detector be σ^2. Thus

4 DISPERSION IN CAPILLARY CONNECTING TUBES   77

$$\sigma_c^2 + \sigma_T^2 = \sigma^2.$$

Assuming that a 5% increase in band width by dispersion in the connecting tube is the maximum acceptable, then

$$\sigma^2 = (1.05\ \sigma_c)^2 = 1.1\sigma_c^2.$$

Thus

$$\sigma_c^2 + \sigma_T^2 = 1.1\sigma_c^2$$

or

$$\sigma_T^2 = 0.1\sigma_c^2.$$

Now from the plate theory, (2.64), $\sigma_c = \sqrt{n_1}\,(v_m + Kv_s)$, where n_1 is the efficiency of the column. Thus

$$\sigma_T^2 = 0.1 n_1 (v_m + Kv_s)^2$$

and since

$$V_R = n_1(v_m + Kv_s), \qquad (2.15a)$$

(where V_R is the retention volume of the solute), therefore

$$\sigma_T^2 = \frac{0.1\ V_R^2}{n_1}. \qquad (2.65)$$

Now in a similar way,

$$\sigma_T^2 = n_2 (V_T)^2 \qquad (2.66)$$

where n_2 is the number of theoretical plates in the capillary tube and V_T is the volume of mobile phase per plate, there being no stationary phase present in the tube.

From the equation for h, the variance per unit length of a capillary column as derived by Golay [3] is given by

$$h = \frac{1}{24} \frac{r^2 u}{D_m}$$

where r is the radius of the capillary tube. Thus because $n_2 = \frac{1}{h}$, therefore $n_2 = \frac{24D_m l}{r^2 u}$, and because the flow rate $Q = \pi r^2 u$, then $n_2 = \frac{24\pi D_m l}{Q}$.

Hence

$$\sigma_T^2 = \frac{24\pi D_m l}{Q} (V_T)^2 \qquad (2.67)$$

Furthermore,

$$V_T = \pi r^2 h = \frac{\pi r^4 u}{24 D_m} = \frac{Qr^2}{24 D_m}. \qquad (2.68)$$

Substituting in (2.66) for V_T from (2.68),

$$\sigma_T^2 = \frac{24\pi D_m l}{Q} \left(\frac{Qr^2}{24 D_m}\right)^2 = \frac{r^4 l Q}{24 D_m} \qquad (2.69)$$

Substituting for σ_T^2 in (2.69) from (2.65),

$$\frac{\pi Q r^4 l}{24 D_m} = \frac{0.1 V_R^2}{n_1}$$

or

$$r = \left(\frac{2.4 D_m V_r^2}{\pi n_1 l Q}\right)^{1/4} \qquad (2.70)$$

Rearranging (2.70) to provide an expression for l,

$$l = \frac{2.4 V_R^2 D_m}{\pi Q r^4 n_1} \qquad (2.71)$$

Equations (2.70) and (2.71) show that to ensure that the increase in band width in the connecting tube is less than 5%, the values of r and l are conditioned among other factors by the solute retention volume (V_R). Now the smallest value for V_R will be the retention volume of an unretained peak (i.e., the void volume of the column V_0). Further, for packed columns, the void volume (i.e., the volume of mobile phase in the column) will be approximately one-third of the total column volume. Thus for a column length l_1 and radius r_1,

4 DISPERSION IN CAPILLARY CONNECTING TUBES 79

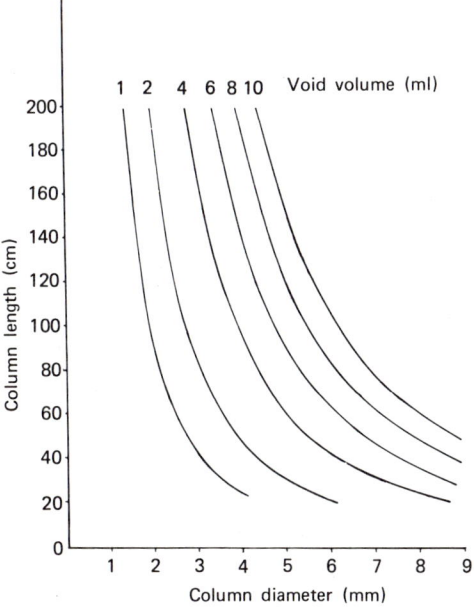

Fig. 2.24. Graphs relating column length and diameter for different void volumes.

$$V_R(\min) = V_0 = \pi r_1^2 l_1/3 \qquad (2.72)$$

In Fig. 2.24 the column length is plotted against column diameter for various values of V_0 calculated from (2.72). Using the curves in Fig. 2.24 the void volume of most practical columns can be obtained by interpolation from a knowledge of the column dimensions. Taking values for D_m and Q of 10^{-5} cm^2/sec and 0.2 ml/min. respectively, in (2.70), curves relating the diameter of a connecting tube 20 cm long and the void volume for columns of different efficiencies are shown in Fig. 2.25. In a similar way, using the same values of D_m and Q in (2.71), curves relating the lengths of a connecting tube, 0.020 in. in diameter, and void volume for columns of different efficiencies are shown in Fig. 2.26. Using the values for the void volume together with column efficiencies in conjunction with the curves shown in Fig. 2.25, the diameter of a connecting tube 20 cm long can be determined that will limit the maximum band spread that takes in the tube to 5%. In a similar way using the curves shown in Fig. 2.26, the maximum length of a connecting tube 0.020 in. in diameter that can be used with a given column can be determined that restricts the dispersion to the same limit for all eluted peaks. Figure 2.25 shows that for columns of low efficiency and high void volume, a 20-cm connecting tube can be about 0.040 in.

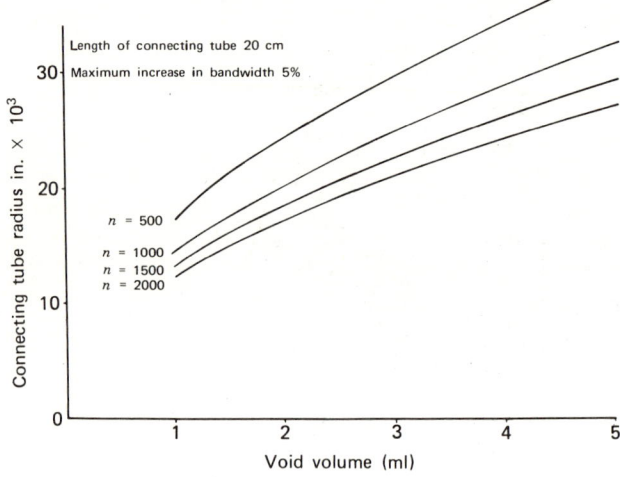

Fig. 2.25. Graph of connecting tube radius against column void volume.

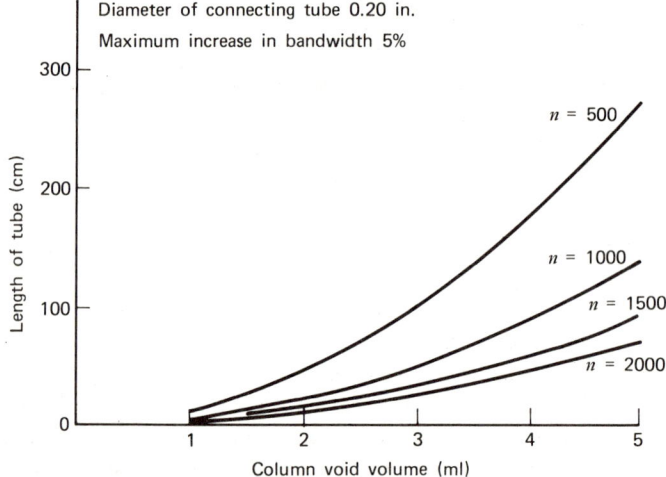

Fig. 2.26. Graph of connecting tube length against column void volume for different column efficiencies.

in diameter, whereas for a column having an efficiency of 2000 theoretical plates and a void volume of 1 ml, the diameter of the connecting tube must be reduced to 0.012 in. From Fig. 2.26 it is seen that if the diameter of the connecting tube is chosen to be 0.020 in. then a column having an efficiency of 500 theoretical plates and void volume of 5 ml can tolerate a connecting tube 270 cm long without adversely affecting the column performance. In contrast a column having an efficiency of 2000 theoretical plates and a void volume of 2 ml can only tolerate 10 cm of connecting tube before the band dispersion in the tube exceeds 5% of the original bandwidth of an unretained peak.

It follows that open tubes can be employed for column connections without affecting the column performance but the dimensions of the connecting tube must be chosen to suit the characteristics of the column, or range of columns, that will be employed in the chromatographic system concerned.

5 THE PEAK CAPACITY OF A CHROMATOGRAPHIC SYSTEM

Most workers in the field of liquid chromatography have experienced difficulties in the separation of multicomponent mixtures when using isopolar methods of development. If the early peaks of a multicomponent mixture are adequately separated then the later peaks are eluted at such low solute concentrations that they are hardly detectable. Conversely, if the charge is increased to improve the detection of the later peaks then the early peaks are no longer resolved and may be dispersed to such an extent that the individual components cannot be discretely eluted. Experience has shown that the extent to which these difficulties occur depends on the detector sensitivity, solubility of the samples in the mobile phase, and the column characteristics. The difficulties referred to are due to the limited peak capacity of the chromatographic system employed, and such a restriction reduces the efficacy of the technique. It follows that it is important to know, in detail, what factors affect the peak capacity of a given combination of column and detector. The theoretical treatment of peak capacity will be developed in the following manner:

1. The relationship between the capacity ratio k', column efficiency n, and the peak capacity r will be determined.
2. The limits imposed upon the capacity ratio k' by the detector sensitivity and solute solubility will be established.
3. From (1) and (2), the relationship between detector sensitivity, solute solubility, and peak capacity will be derived.
4. The limits imposed on the capacity ratio k' by the absorptive capacity of the stationary phase will be determined for both liquid-liquid and liquid-solid systems.
5. From (1) and (4) the effect of the absorptive capacity of the stationary phase on the peak capacity will be established.
6. The advantages of injecting the sample as a solution in the stationary phase will be evaluated.

From the plate theory the peak width at the base is given by

$$4\sqrt{n}(v_m + Kv_s)$$

where n is the column efficiency, v_m is the volume of mobile phase per plate, v_s is the volume of stationary phase per plate, and K is the distribution coefficient of the solute. Also from the plate theory the retention volume V_R of the solute is given by

$$V_R = n(v_m + Kv_s). \tag{2.15a}$$

Thus

$$S = \frac{n}{4\sqrt{n}} \frac{(v_m + Kv_s)}{(v_m + Kv_s)} = \frac{\sqrt{n}}{4}$$

S is commonly known as the peak capacity of the chromatographic system and is equivalent to the number of peaks of the same width that can be fitted into the chromatogram up to and including the peak concerned. Although this definition is suitable for comparing the peak capacity of columns having different plate numbers, it does not give a true value of the peak capacity. This is because the peaks in a chromatogram are not all of the same width and thus a considerably greater number of peaks can be fitted into a chromatogram than the figure for S suggests. In order to evaluate a true number for the peak capacity a different approach must be used. Consider the chromatogram diagramatically represented in Fig. 2.27. For all peaks in the chromatogram to

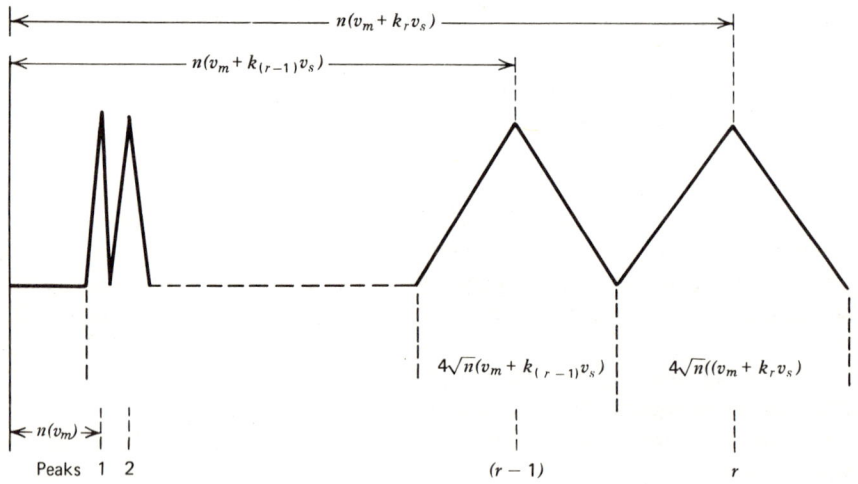

Fig. 2.27. Diagram of a chromatogram having r Completely Resolved Peaks.

be completely resolved, the width at the base of the rth peak for a given solute must be

5 THE PEAK CAPACITY OF A CHROMATOGRAPHIC SYSTEM

$$4\sqrt{n}\,(v_m + K_r v_s).$$

Consider point A where the last two bands r and $r-1$, merge. At this point,

$$n(v_m + K_{(r-1)}v_s) + 2\sqrt{n}(v_m + K_{(r-1)}v_s) = n(v_m + K_r v_s) - 2\sqrt{n}(v_m + K_r v_s)$$

or

$$(n+2\sqrt{n})(v_m + K_{(r-1)}v_s) = (n-2\sqrt{n})(v_m + K_r v_s).$$

Thus

$$(v_m + K_{(r-1)}v_s) = \frac{n-2\sqrt{n}}{n+2\sqrt{n}}\,(v_m + K_r v_s).$$

Therefore the width $w_{(r-1)}$ of peak $r-1$ will be

$$w_{(r-1)} = 4\sqrt{n}\left(\frac{n-2\sqrt{n}}{n+2\sqrt{n}}\right)(v_m + K_r v_s).$$

In a similar way it can be shown that the width $w_{(r-2)}$ of peak $r-2$ will be

$$w_{(r-2)} = 4\sqrt{n}(v_m + K_{(r-1)}v_s) = 4\sqrt{n}\left(\frac{n-2\sqrt{n}}{n+2\sqrt{n}}\right)^2 (v_m + K_r v_s).$$

Thus if the number of peaks that can be fitted into the chromatogram between the dead time and the time for the complete elution of the last peak is r, then

$$(n+2\sqrt{n})(v_m+K_r v_s) = \left[\left(\frac{n-2\sqrt{n}}{n+2\sqrt{n}}\right)^{(r-1)} + \left(\frac{n-2\sqrt{n}}{n+2\sqrt{n}}\right)^{(r-2)} + \cdots \right.$$

$$\left. + \left(\frac{n-2\sqrt{n}}{n+2\sqrt{n}}\right)^2 + \left(\frac{n-2\sqrt{n}}{n+2\sqrt{n}}\right) + 1 \right] 4\sqrt{n}(v_m+K_r v_s)+nv_m.$$

Noting that $kv_s/v_m = k'$ and rearranging,

$$\frac{k'}{1+k'} = \frac{4}{\sqrt{n}}\left[\left(\frac{n-2\sqrt{n}}{n+2\sqrt{n}}\right)^{r-1} + \left(\frac{n-2\sqrt{n}}{n+2\sqrt{n}}\right)^{r-2} + \cdots\right.$$

$$+ \left(\frac{n-2\sqrt{n}}{n+2\sqrt{n}}\right)^2 + \left(\frac{n-2\sqrt{n}}{n+2\sqrt{n}}\right) + 0.5 \Bigg.\Bigg]$$

Replacing the geometric series by the equation for its sum,

$$\frac{k'}{1+k'} = \frac{4}{\sqrt{n}} \left\{ \frac{1 - \left(\frac{n-2\sqrt{n}}{n+2\sqrt{n}}\right)^r}{1 - \left(\frac{n-2\sqrt{n}}{n+2\sqrt{n}}\right)} - 0.5 \right\} \qquad (2.73)$$

Equation (2.73) is a function similar to that described by Giddings [23], however, Giddings assumes that $(n-2\sqrt{n})/(n+2\sqrt{n})$ approaches unity and thus the peak capacities quoted by him for liquid chromatography are somewhat less than those derived from (2.73) where no such approximation is made.

Equation (2.73) shows that the peak capacity r depends on the column efficiency and the capacity ratio of the column for the last eluted peak. Using equation (2.73) the peak capacity r can be calculated for different values of k' and for columns of different efficiencies. Curves relating k' and r are shown in Fig. 2.28 and are similar in form to those given by Grushka [24].

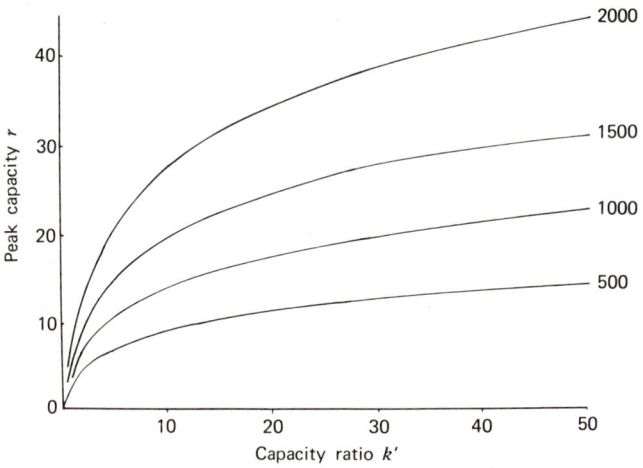

Fig. 2.28. Graph of peak capacity against k' for columns of different efficiencies.

It is seen that a column of 2000 theoretical plates eluting a series of solutes, the last of which has a k' value of 50, will provide a peak capacity of about 44 whereas a column of 200 plates eluting a series of solutes having a maximum

5 THE PEAK CAPACITY OF A CHROMATOGRAPHIC SYSTEM

k' value of 5 will only provide a peak capacity of 7.

The curves in Fig. 2.28 level off to an almost constant value and clearly indicate the futility of attempting to elute too many peaks under isopolar conditions. With regard to the 200-plate column, the elution of a solute in a volume of mobile phase equal to $8k'$ yields a peak capacity of 8, whereas elution by a further volume of $8k'$ would only yield a further three peaks. After passing a volume of mobile phase equal to $8k'$ it would therefore be advantageous to change to a more polar solvent and commence gradient elution.

For a column of given efficiency the peak capacity is solely dependent on the capacity ratio of the column for the most retarded solute. If, however, other characteristics of the chromatographic system place a limit on the maximum value for k', then they will also place a limit on the peak capacity. There are, indeed, two practical conditions that limit the value of k'.

1. The most retained solute nust be eluted at a concentration that is detectable.
2. The absorption isotherm of the most retained solute must be maintained linear.

Considering condition (1) the concentration of a solute at its peak maximum will decrease as its value of k' increases because of the dispersion processes that occur in the column. Now it has already been shown that if the sample is injected as a solution in the mobile phase, the maximum volume of this solution must be limited to restrict its effect on the bandwidth. Thus if the solute has a restricted solubility in the mobile phase, then the total mass of solute injected on the column is also limited. It follows, therefore, that for a detector of given sensitivity, the solubility of the solute in the mobile phase and thus the concentration of the solute in the solution injected will place a limit on the maximum value for k' at which the solute is detectable. Hence a limit will also be placed on the peak capacity.

Consider now the second condition, namely, to maintain peak symmetry and prevent peak broadening the absorption isotherm of the solute between the given phases must be maintained linear. This means that the stationary phase cannot be loaded excessively with solute. Now the capacity ratio k' of a column is numerically equal to the ratio of mass of solute in the stationary phase to that in the mobile phase. Thus because the maximum mass of solute in or on the stationary phase is limited by the linearity of the absorption isotherm and the minimum mass of solute in the mobile phase is limited by detector sensitivity, therefore the capacity ratio and hence the peak capacity will also have a maximum limit. The effect of both these factors on the capacity ratio will now be considered quantitatively; the effect of solute solubility and detector sensitivity will be dealt with first.

The Dependence of k' on the Sample Concentration and Detector Sensitivity

According to the plate theory and concentration X_n at the peak maximum from (2.31) is given by

$$X_n = \frac{X_0}{\sqrt{2\pi n}}$$

where X_0 is the initial concentration of the solute in the mobile phase placed on the first theoretical plate and n is the number of theoretical plates in the column.

Now the charge on the column is rarely placed on one theoretical plate, so to obtain a relationship between X_n and X_0, it is better to equate the total mass of sample placed on the column to the peak area of the eluted solute. Thus

$$V_S X_S = 2\sqrt{n}(v_m + K v_s) X_n f \tag{2.74}$$

where V_S is the volume of mobile phase that contains the sample when injected and X_S is the concentration of solute in volume V_S prior to injection. Here, f is a constant that normalizes the peak area given by the product of the peak height and twice the standard deviation to the total peak area. From statistical tables, f can be found to be 1.25.

Now the maximum sample volume that can be placed on the column to limit the increase in band width to 5% has already been determined; it is given by

$$V_S = 1.1\sqrt{n}(v_m + K v_s). \tag{2.75}$$

However, a sample will contain a mixture of solutes, and thus V_S will be conditioned by the minimum value of k', which for an unretained solute will be zero. Hence (2.75) becomes

$$V_S = 1.1\sqrt{n}(v_m). \tag{2.76}$$

Substituting for V_S in (2.74) from (2.76),

$$X_S \, 1.1\sqrt{n}(v_m) = 2.5\sqrt{n}(v_m + K v_s) X_n.$$

Thus

$$X_S = 2.27(1 + k') X_n. \tag{2.77}$$

Now if the detector sensitivity is D, where D is the concentration of solute that

5 THE PEAK CAPACITY OF A CHROMATOGRAPHIC SYSTEM

provides a signal from the detector equivalent to twice the noice level, then for practical analysis the smallest peak height that is useful will be defined as $5D$. Thus, at the limit of sensitivity,

$$X_n = 5D. \qquad (2.78)$$

Substituting for X_n in (2.77) form (2.78),

$$X_S = 2.27(1 + k')5D = 11.4D(1 + k').$$

Thus

$$k' = \frac{X_S}{11.4D} - 1$$

and

$$\frac{k'}{1+k'} = 1 - \frac{11.4D}{X_S} \qquad (2.79)$$

Thus the maximum value of $k'/(1+k')$ will be conditioned by the maximum value X_S. How the maximum value of X_S will be achieved if the sample injected contains the solute as a saturated solution in the volume of mobile phase given by (2.79). Thus

$$\frac{k'}{1+k'} = 1 - \frac{11.4D}{X_{(sat)}} \qquad (2.80)$$

Substituting in (2.73) for $\dfrac{k'}{(1+k')}$ from (2.80),

$$1 - \frac{11.4D}{X_{(sat)}} = \frac{4}{\sqrt{n}} \left\{ \frac{1 - \left(\dfrac{n-2\sqrt{n}}{n+2\sqrt{n}}\right)^r}{1 - \left(\dfrac{n-2\sqrt{n}}{n+2\sqrt{n}}\right)} - 0.05 \right\} \qquad (2.81)$$

Equation (2.81) allows the peak capacity of a column system to be calculated from a knowledge of the column efficiency, detector sensitivity, and the concentration of a saturated solution of the most retarded solute in the mobile phase that is placed on the column. At this stage it is assumed that there are no other factors limiting k' such as absorptive capacity. Using (2.81) the peak capacity can be calculated for a range of solute concentrations for columns having given efficiencies and for detectors of different sensitivities. The results from such calculations are shown in Fig. 2.29. Now assuming no other factor limits the value of k', it is seen that if a refractometer detector is employed and the concentration in the sample volume of mobile phase is in excess of 2.0% w/v a

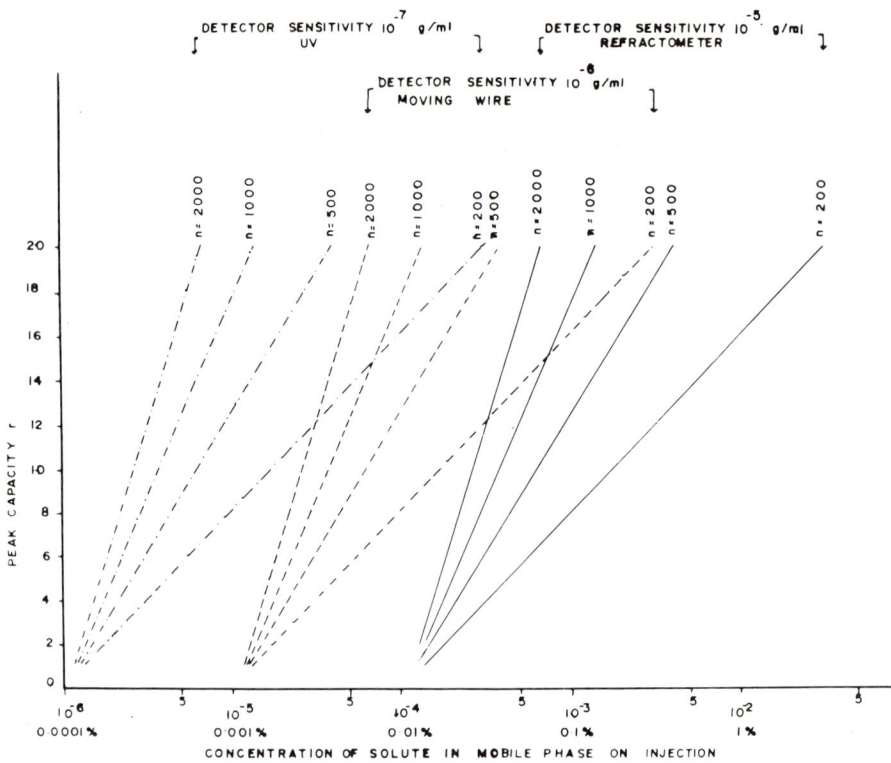

Fig. 2.29. Graph of peak capacity against sample concentration of the most retarded solute for different column efficiencies and detector sensitivities.

peak capacity of 20 can be realized even for columns having only 200 theoretical plates. However, if the solute concentration falls below 0.01% w/v, columns having efficiencies as high as 2000 plates are not usable. Conversely if the UV detector is employed then a peak capacity of 20 can be realized for solute concentrations as low as 0.0005% w/v for a column of little more than 2000 plates and with a concentration of 0.01% w/v only 400 theoretical plates are necessary. It should be stressed that the curves shown in Fig. 2.29 are for minimum values of X_s and apply equally to liquid-solid and liquid-liquid chromatography.

5 THE PEAK CAPACITY OF A CHROMATOGRAPHIC SYSTEM

The Dependence of k' on the Maximum Absorptive Capacity of the Stationary Phase

The adsorptive capacity of the stationary phase is defined as the maximum quantity of solute that can be contained by the stationary phase while maintaining a linear adsorption isotherm for the solute with respect to the mobile phase. Thus because the mass of solute held by the stationary phase is limited to its absorptive capacity and the minimum mass of solute in the mobile phase is limited by the detector sensitivity, the absorptive capacity will limit the maximum value of k'. The capacity ratio k' can also be defined as the ratio of the mass of solute in the stationary phase (m_s) to the mass of solute in the mobile phase (m_m) in contact with it. Thus

$$k' = \frac{Kv_s}{v_m} = \frac{Kv_s X_m}{v_m X_m} = \frac{m_s}{m_m} \quad (2.82)$$

However for a maximum value for k', m_s will be the absorptive capacity of the stationary phase and X_m will be X_s as given in (2.77). Now k'_{max} will differ for liquid-liquid chromatography and liquid-solid chromatography.

Liquid-Solid Systems

From (2.82)

$$k'_{max} = \frac{m_s}{V_m X_s}$$

where V_m is the interstitial volume of mobile phase associated with the adsorbent when packed in a column. Snyder [25] shows a value for m_s of about 3×10^{-3} g/g for a water deactivated silica gel, and this figure, although approximate, will be taken as representative of the adsorptive capacity of silica gel when used as a stationary phase in liquid chromatography. Thus for silica gel,

$$k'_{max} = \frac{3 \times 10^{-3}}{V_m X_s}$$

Now for 65-μ particle diameter silica gel packed in a chromatographic column,

$$V_m \simeq 1.15 \text{ ml/g}.$$

Thus

$$k'_{max} = \frac{2.61 \times 10^{-3}}{X_S} \quad (2.83)$$

From (2.79) k'_{max} is also conditioned by the following expression due to dector sensitivity:

$$k'_{max} = \frac{X_S}{11.4D-1}. \qquad (2.84)$$

Substituting for X_S from (2.83) in (2.84) and rearranging,

$$(k'_{max})^2 + k' \frac{-2.3 \times 10^{-4}}{D} = 0$$

or

$$k'_{max} = -0.5 + \sqrt{\frac{1 + 9.2 \times 10^{-4}/D}{2}} \qquad (2.85)$$

The maximum values of k' for the refractometer, moving wire, and UV detectors having values for D of 10^{-5}, 10^{-6}, and 10^{-7} g/ml, respectively, are shown in Table 2.1 together with the respective peak capacities for columns of different

Table 2.1 Limiting Values of k' and Peak Caoacity for Liquid-Solid Chromatography using Silica Gel as the Stationary Phase

Detector Sensitivity (g/ml)	k'_{max}	Peak Capacity			
		n=200	n=500	n=1000	n=2000
10^{-5}	4.8	7	10	14	20
10^{-6}	14.7	10	16	22	31
10^{-7}	47.5	14	22	31	44

efficiencies. Referring to Fig. 2.29 it is seen that solute concentration is the only limiting factor and the solute concentration in the sample is 1% w/v; hence with the refractometer detector and a column of 200 theoretical plates, a peak capacity of 16 could be realized. However, as seen in Table 2.1, if the absorptive capacity of the stationary phase is only 0.003 g/g, then the maximum peak capacity that can be obtained is only 7. Thus, using silica gel as the stationary phase with columns of low efficiency and detectors of low sensitivity, the absorptive capacity plays a dominant role in limiting the peak capacity of the chromatographic system. Conversely, with high efficiency columns and the UV detector, the solute solubility is the major factor controlling peak capacity. The absorptive capacity of alumina is about an order less than that of silica gel and thus the peak capacity from alumina would be expected to be commensurably less than silica gel.

It is of interest to consider the performance of the special liquid chromatography supports, for example, coated galss beads and the so-called brushes. Values for the adsorptive capacity of such supports do not appear to be available, but the adsorptive capacity will not only be a function of the nature of the adsorbant but also the total quantity of it present. If the effective adsorbent is coated over a support of limited surface area then the overall absorptive capacity may well be quite low. Thus, although by design these supports should provide columns of high efficiency and thus increase the peak capacity of the system, the advantages may be lost if accompanied by a reduction in the absorptive capacity of the material. The relative efficiency of these stationary phases, from the point of view of peak capacity, remains to be demonstrated.

An example of a chromatogram obtained from a column packed with resincoated glass beads separating a series of nucleic acid bases is shown in Fig. 2.30. The column has an efficiency of 280 theoretical plates and the last peak was eluted at a k' factor of 12.3. A UV detector was used with a sensitivity of about

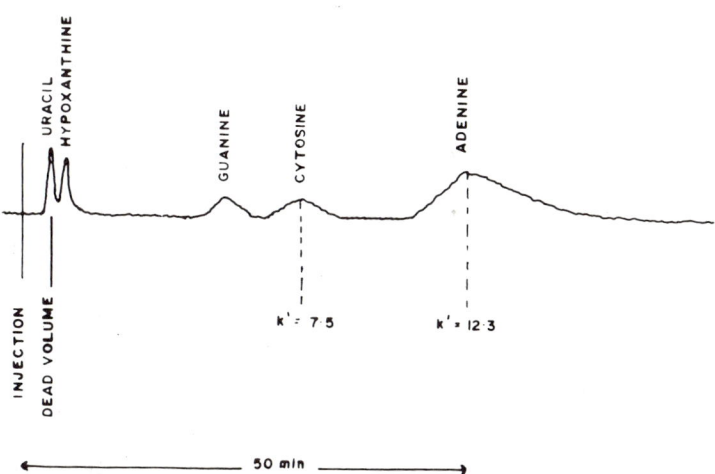

Fig. 2.30. Chromatogram of nucleic acid bases.

10^{-7} g/ml. The last peak height corresponds approximately to $5D$ (ten times the noise level), and is obviously asymmetric, indicating that the absorptive capacity of the support has been exceeded. The concentration of the most retained solute in the mobile phase on injection was 0.004% and a total volume of 5 μl was used. From Fig. 2.29 it is seen that a concentration of 0.004% of sample on a 280-theoretical-plate column will limit the peak capacity for a UV detector to about 18, and thus sample solubility would not explain the low peak capacity exhibited by the chromatogram. Referring to Fig. 2.28, a k' factor of 12.3 for a column of 280 theoretical plates would limit the peak capacity to about 12, which is still in excess of the peak capacity shown. It follows that the absorptive capacity has been exceeded, and this is substantiated by the gross asymmetry of the last peak. The penultimate peak has a k' factor of about 7.5 and appears symmetrical, and thus it can be taken that the maximum value of k' permitted by the absorptive capacity of the absorbent is greater than 7.5 and less than 12. Using (2.82),

$$k'_{max} = \frac{m_s}{V_m X_s}$$

Now X_S for the penultimate peak was also 40 μg/ml (the difference in peak height between the last two peaks being accounted for by the different extinction coefficient of the two solutes), and V_m was experimentally found to be 0.31 ml/g. Thus

$$7.5 = \frac{m_s}{0.31 \times 4 \times 10^{-5}},$$

whence

$$m_s = 0.1 \text{ mg/g}.$$

It is seen that the absorptive capacity of the resin-coated beads is 30 times less than that of the silica gel; this explains the poor peak capacity of the adsorbant. In the design of stationary phases for liquid chromatography it is of little advantage to reduce the film thickness of the partitioning substance to improve column efficiency if the absorptive capacity of the support is also reduced. If the film of resin could be placed on an alternative support, for example Celite, then a reasonably high column efficiency might be obtained and at the same time a satisfactory absorptive capacity might be maintained. It should be noted that the above procedure provides a simple method for determining the absorptive capacities of different liquid chromatography absorbents.

Liquid-Liquid Systems

From (2.82),

$$k' = \frac{Kv_s X_m}{v_m X_m} = \frac{v_s X_{st}}{v_m X_m} = \frac{X_{st}}{aX_m} \qquad (2.86)$$

where a is the phase ratio of the column, X_{st} is the concentration of solute in the stationary phase, and X_m is the concentration of solute in the mobile phase. The maximum concentration of solute that is permissible in the stationary phase is conditioned by the form of the absorption isotherm of the solute between the two phases. The limiting concentration in the stationary phase will be that where the absorption isotherm tends to become nonlinear. The concentration at which this occurs will vary with both the solute and solvent, but generally concentrations of 2.0% can usually be tolerated before the isotherm become nonlinear.

Thus $X_{st} = 0.02$ g/ml and for an optimally loaded support carrying about 40% w/v of stationary phase, $a = 3$. Thus

$$k' = \frac{0.0067}{X_S} \qquad (2.87)$$

where at injection X_m is replaced by X_S. Now the detector sensitivity D also conditions X_S according to (2.79). Thus

$$k' = \frac{X_S}{11.4D} - 1$$

Substituting for X_S in (2.87) from (2.79) and rearranging, the equation for k'_{max} is given by

$$k'^2_{max} + \frac{k'_{max} - 5.9 \times 10^{-4}}{D} = 0$$

or

$$k' = -0.5 + \frac{\sqrt{1 + 2.36 \times 10^{-3}/D}}{2} \qquad (2.88)$$

Values of k'_{max} for liquid systems and for detectors having sensitivities of 10^{-3}, 10^{-6}, and 10^{-7} g/ml, respectively, are shown in Table 2.2.

94   THE THEORY OF CHROMATOGRAPHY

Table 2.2 Limiting Values of k' for Liquid-Liquid Chromatography

Detector Sensitivity (g/ml)	k'
10^{-5}	7.2
10^{-6}	23.8
10^{-7}	72

Comparing the k'_{max} values given in Tables 2.1 and 2.2 for liquid-solid and liquid-liquid systems it is seen that, over the peak capacity ranges considered, the absorptive capacity of a liquid-liquid system does not restrict the peak capacity to the extent it does in liquid-solid chromatography. However, the advantages of the higher absorptive capacity of the liquid-liquid system over the liquid-solid sistem can only be relized under certain circumstances. The limiting values of k' and r for the two systems for a column of 1000 theoretical plates are shown in Table 2.3. It is seen that although the liquid-liquid system has a 50%

Table 2.3 Limiting Values of k' and r (N = 1000)

Detector Sensitivity (g/ml)	Liquid-Solid Silica Gel		Liquid-Liquid 40% Phase on Celite	
	k'_{max}	Peak Capacity	k'_{max}	Peak Capacity
10^{-5}	4,8	14	7.2	17
10^{-6}	15	22	24	26
10^{-7}	48	33	72	31

greater capacity ratio than the liquid-solid system the increase in peak capacity is only 10-15%. This is due to the shape of curves shown in Fig. 2.29, where it is seen that at a high capacity ratio the increase in peak capacity with k' is very small because of the large dispersion of the late eluted peaks. If the column efficiency were less, the difference between the peak capacity of the two systems would be hardly significant. Conversely, however, if the efficiencies are increased then the advantages of the liquid-liquid system over the liquid-solid system are more pronounced. It follows that high peak capacities for liquid-liquid systems relative to liquid-solid systems can only be realized when employing columns of very high efficiencies.

Injection of Sample as a Solution in the Stationary Phase

In liquid-liquid systems there is the possibility of injecting the sample as a solution in the stationary phase as opposed to the mobile phase, and the effect of using this alternative method of injection on peak capacity also needs to be studied.

The maximum volume of sample dissolved in the stationary phase (measured in plate volumes of the respective phase) would, at first sight, be thought to be smaller than that for the sample when dissolved in the mobile phase. It would seem that the extraction process afforded by the stationary phase in the column that would compress the solute band into a smaller volume (when measured in plate volumes of the respective phase) will be the same whether injected as a solution in the stationary phase or the mobile phase. Thus for a sample injected in the stationary phase,

$$V_s = 1.1\sqrt{n}(v_s). \qquad (2.89)$$

Note that v_s is now the volume of stationary phase per plate. Thus substituting for V_S in (2.74) from (2.89),

$$X_S 1.1\sqrt{n}(v_s) = 2.5\sqrt{n}(v_m + Kv_s)X_n$$

or

$$X_S = 2.27(1 + k')aX_n$$

where $a = v_m/v_s$ is the phase ratio of the column. Again taking $X_n = 5D$,

$$X_S = 11.4\ Da(1 + k')$$

or

$$k' = \frac{X_S}{11.4Da} - 1$$

and

$$\frac{k'}{1+k'} = 1 - \frac{11.4Da}{X_S} \qquad (2.90)$$

The maximum concentration of solute that is permissible in the stationary phase is conditioned by the form of the absorption isotherm of the solute between the two phases. As a first approximation this can be taken to be at a solute concentration of about 2% w/v. Thus X_S in (2.90) is replaced by a value of 0.020 g/ml. Thus (2.90) becomes

$$\frac{k'}{1+k'} = 1 - 570\, Da. \qquad (2.91)$$

Substituting in (2.73) for $\dfrac{k'}{(1+k')}$ from (2.91),

$$1 - 570\, Da = \frac{4}{\sqrt{n}} \left\{ \frac{\left(\dfrac{n-2\sqrt{n}}{n+2\sqrt{n}}\right)^r}{\left(\dfrac{n-2\sqrt{n}}{n+2\sqrt{n}}\right)} - 0.5 \right\} \qquad (2.92)$$

From (2.92) it is seen that if the solute is injected as a 2% solution in the

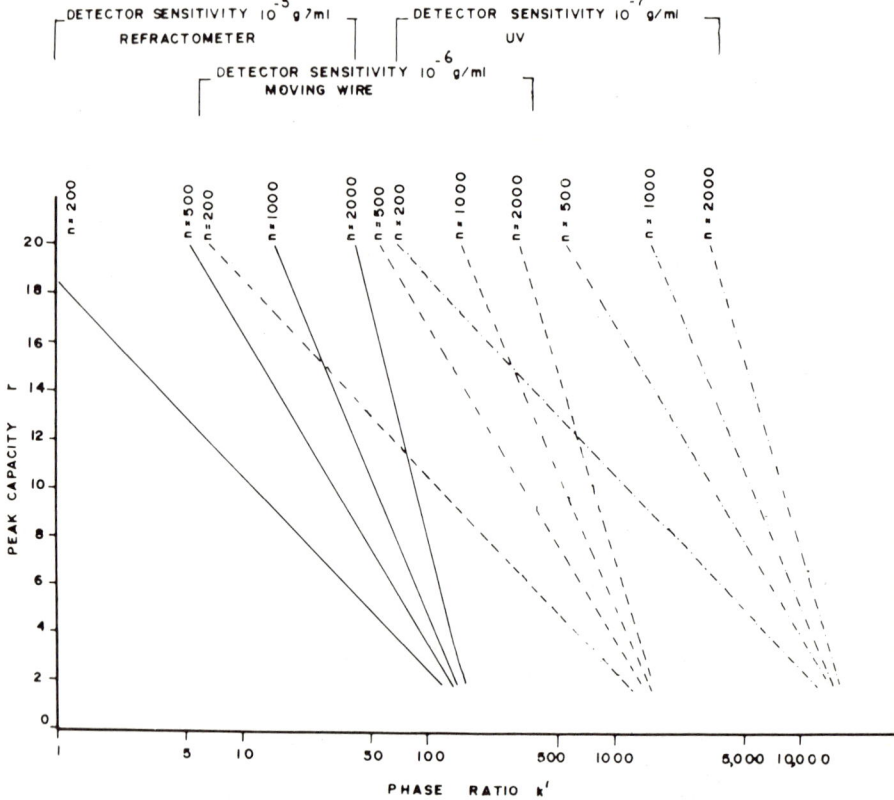

Fig. 2.31. Graph of peak capacity against phase ratio for different column efficiencies and detector sensitivities.

stationary phase on to a give column, the peak capacity is dependent on a, the phase ratio of the column, and the detector sensitivity. This is because the peak capacity depends on a given value of $k'/(1+k')$, which demands a certain mass of solute to be placed on the column to provide a detectable peak. Since the concentration of solute in the stationary phase is set at a value of 2%, the only way to meet the demands of a larger mass of solute is to increase the volume of stationary phase per plate (v_s), that is, decrease a and thus increase the absolute value of V_S while maintaining the same number of plate volumes of charge, namely, $1.1\sqrt{n}$. Values of the plate capacity for different values of a for columns of different efficiencies employed with different detectors are shown in Fig. 2.31. It should be pointed out that the curves represent the maximum values for a that can be tolerated to provide the respective values of r. It is seen that, providing the phase ratio of the column is adequate for the respective detector employed, the maximum permissible peak capacity can be realized. Furthermore, because solute solubility would be expected to increase with the k' factor, sample solubility should not present a limiting effect. It should be pointed out, however, that injection of the sample dissolved in the stationary phase will limit the life of the column as eventually it will produce excessive stationary phase "bleed."

List of Symbols

a	The phase ratio of the columns, i.e., v_m/v_s
c_a	Concentration of solute in phase (g/cm^3)
c_s	Concentration of solute on adsorbent surface (g/cm^2)
d_f	Stationary phase film thickness (cm)
d_p	Particle diameter (cm)
D_A	Dynamic Adsorption coefficient (cm/sec)
D_e	Dispersion coefficient (cm^2/sec)
D_m	Diffusivity of the solute in the mobile phase (cm^2/sec)
D_s	Diffusivity of the solute in the stationary phase (cm^2/sec)
D_{AS}	A "weighted" combination of D_A and D_S (cm^2/sec)
h	Height equivalent to a theoretical plate, or variance per unit length (cm)
j	Dimensionless constant
K	Distribution (partition) coefficient
k^1	Capacity ratio (K/a)
K^1	Ratio of the concentration of the solute in the stationary bed to that in the moving fluid
K_e	Kinetic energy of a solute molecule in the moving phase (g-cm^2/sec^2)
l	Length (cm)

n		Number of theoretical plates
p		A selected theoretical plate in a column
P_e		Potential energy of an active site on an adsorbent (g-cm^2/sec^2)
Q		Flow rate of mobile phase (cm^3/sec)
r		Radius of column (cm)
R		Overall transfer coefficient (cm/sec)
R_m		Transfer coefficient for the moving phase (cm/sec)
R_s		Transfer coefficient for the stationary phase (cm/sec)
T_m		Interparticle tortuosity factor
T_s		Intraparticle tortuosity factor
t_R		Retention time (sec)
u		Linear velocity of the mobile phase (cm/sec)
v		Volume of mobile phase (plate volumes)
v_m		Volume of mobile phase in a theoretical plate (cm^3)
v_s		Volume of stationary phase in a theoretical plate (cm^3)
V		Volume of mobile phase (cm^3)
V_m		Total volume of mobile phase in column (cm^3)
V_s		Total volume of stationary phase in column (cm^3)
V_S		Volume of charge
V_1		Column dead volume (cm^3)
V_R		Retention volume (cm^3)
V'_R		Adjusted retention volume (cm^3)
V_{rA}		Retention volume of solute A (cm^3)
V_{BA}		Ratio of retention volume of solute B to that of solute A
V'_{BA}		Ratio of adjusted retention volume of solute B to that of A (retention ratio)
V_{aA}		Ratio of dead volume to retention volume of solute A
X_0		Initial concentration of solute in the mobile phase when placed on the column (g/cm^3)
X_m		Concentration of solute in the mobile phase (g/cm^3)
X_s		Concentration of solute in the stationary phase (g/cm^3)
Z		Velocity of solute band (cm/sec)
γ		Dimensionless constant
ϵ_a		Fraction of column volume occupied by phase a
ϵ_β		Fraction of column volume occupied by phase β
ϵ_m		Fraction of column volume occupied by moving phase
ϵ_s		Fraction of column volume occupied by static phase
ϵ		Fraction of column volume occupied by the two phases
$1-\epsilon$		Fraction of column volume occupied by the support material
λ_1, λ_2		Dimensionless constants
ν		Kinematic viscosity (cm^2/sec)

σ Standard deviation of solute band in appropriate dimensions
ϕ Ratio of total surface area of adsorbent to column volume (cm^{-1})
ψ Dimensionless constant

References

1. A. J. P. Martin and R. L. M. Synge, *Biochem. J.*, 35, 1358 (1941).
2. J. J. Van Deemter, F. J. Zuiderweg, and A. Klinkenberg, *Chem. Eng. Sci.*, 5, 271 (1956).
3. M. Golay, *Gas Chromatography 1958*, Butterworth, London, 1958, p. 36.
4. J. C. Giddings, *Dynamics of Chromatography*, Part 1, Marcel Dekker, New York, 1965.
5. J. F. K. Huber and J. A. R. J. Hulsman, *Anal. Chim. Acta*, 38, 305 (1967).
6. J. C. Giddings, *J. Chem. Ed.*, 35, 588 (1958).
7. C. Horvath, B. Preiss, and S. R. Lipsky, *Anal. Chem.*, 39, 1422 (1967).
8. G. C. Kennedy and J. H. Knox, *J. Chromatog. Sci.*, 10, 549 (1972).
8a. S. Karger et al., *An Introduction to Separation Science*, Wiley, New York, 1973.
9. J. H. Purnell and J. Bohemen, *J. Chem. Soc.*, 2030 (1961).
10. D. H. Desty and A. Goldup, *Gas Chromatography 1960*, R. P. W. Scott, Ed., Butterworth, London, 1960, p. 162.
11. R. P. W. Scott, *Nature*, 183, 1753 (1959).
12. J. C. Giddings, Ed., *Dynamics of Chromatography*, Marcel Dekker, New York, 1965, p. 265.
13. R. P. W. Scott and P. Kucera, *J. Chromatog. Sci.*, 12, 473 (1974).
14. J. W. Hiby, *Proceedings of the Symposium on Interaction Between Fluid and Particles*, Inst. of Chem. Eng., London, 1962, p. 312.
15. J. F. K. Huber and Quaadgrass, *J. Chromatog.*, in press.
16. R. P. W. Scott, D. W. J. Blackburn, and T. Wilkins, *J. Gas Chromatog.*, 183 (1967).
17. L. R. Snyder, *J. Chromatog. Sci.*, 7, 352 (1969).
18. J. L. Waters, J. N. Little, and D. F. Horgan, *J. Chromatog. Sci.*, 7, 293 (1969).
19. D. W. Simpson and R. M. Wheaton, *Chem. Eng. Progr.*, 50, 45 (1954).
20. D. S. Horne, J. H. Knox, and McLaren, *Separation Techniques in Chemistry and Biochemistry*, R. A. Keller, Ed., Marcel Dekker New York, 1900, pp. 97-120.
21. J. J. Kirkland, *Gas Chromatography 1972*, S. G. Perry, Ed., Applied Sciences Ltd., London, 1972, p. 43.
22. V. Maynard and E. Grushka, *Anal. Chem.*, 44, 1427 (1972).
23. J. C. Giddings, *Anal. Chem.*, 39, No. 8, 1027 (1967).

24. E. Grushka, *Anal. Chem.*, **42**, 1142 (1970).
25. L. R. Snyder, *Principles of Adsorption Chromatography*, Marcel Dekker, New York, 1900, p. 89.
26. J. F. K. Huber, *Advances in Chromatography*, A. Zlatkis, Ed., Preston Technical Abstracts, City, 1900, p. 283.

Chapter III

LIQUID CHROMATOGRAPHY APPARATUS

1 The Mobile Phase Supply System   103

Solvent Reservoirs   104
Solvent Degassing System   105
Gradient Elution Facilities   106
Gradient Elution by Solvent Mixing   106
Incremental Gradient Elution   111
Gradient Elution by Displacement   116
Simulated Gradient Elution by Temperature Programming   116
Other Methods of Gradient Elution   118

2 Solvent Pumps   120

Mechanical Pumps   121
Servicing Mechanical Pumps   122
Pneumatic Pumps   122
Choice of Pump for Liquid Chromatography   123
Removal of Pressure Pulses from Reciprocating Pump Systems   124

3 Flow Programing Systems   124

4 Mobile Phase Equilibrium Systems   126

5 Sample Injection Systems   126

Injection with Solvent Flow   127
Stop-Flow Injection   130
Sample Valve Injection   131

6 Liquid Chromatography Columns   133

7 Column Ovens   136

8 Column Detector Connections   137

9 Liquid Chromatography Detectors: Classification and Specifications   137

10 Calibration of Detectors-Measurement of Sensitivity and Linearity   138

The Incremental Method   138
The Logarithmic Dilution Method   141

11 Bulk Property Detectors   143

The Refractive Index Detector   144
The Dielectric Constant Detector   148
The Electrical Conductivity Detector   152
Other Bulk Property Detectors   153

12 Solute Property Detectors 155

The UV Absorption Detector 155
Solute Transport Detectors 158
  The Moving Wire Type 159
  The Moving Chain Type 163
The Mass Detector 165
The Heat of Adsorption Detector 167
The Polarographic Detector 173
The Electrochemical Detector 176
The Fluorometric Detector 177
The Spray Impact Detector 179

13 Choice of Detector 181

14 Ancillary Equipment 182

Microsyringes 185
Flow-Rate Measuring Apparatus 185
Compression Fittings and Unions 185
Appendix 186

There have been many changes in the apparatus used for liquid chromatography in the past few years; the traditional "gravity-feed" mobile phase supply has been replaced by various forms of solvent pumps, some operating at pressures up to 16,000 psi and including flow and gradient programming facilities. In place of the 2-ft-long, 2-cm-diameter glass column, those used today may be only 3 mm in diameter and may be up to 10 or even 15 ft long. Instead of being packed with coarse particles of adsorbent the modern liquid chromatography column may contain supports whose average particle diameter may be only 2-5 μ. Some separations are now carried out at temperatures above ambient or under temperature programming conditions and thus columns are situated in well-thermostated ovens that in the future may be fitted with elaborate temperature programming facilities. Samples are placed on the column by hypodermic syringe injection devices or with on-line sample valves. Improved methods of injection have resulted in improved resolving power and more accurate quantitative analysis. Probably the greatest change, however, has taken place in the field of detectors. The original method of monitoring column eluents was to collect a large number of individual fractions and analyze each fraction separately. Modern detectors monitor the column eluent continuously during development of the chromatogram using an in-line system and produce a continuous record on chart paper of the solute concentrations leaving the column. Much of the advances made in liquid chromatography has been directly or indirectly due to the development of high-sensitivity linear detectors. The apparatus available for liquid chromatography will now be described in logical sequence, commencing with the mobile phase supply system.

1 MOBILE PHASE SUPPLY SYSTEM

The mobile phase supply system can be considered as comprising those parts of the liquid chromatograph that are prior to the injection system. A fully com-

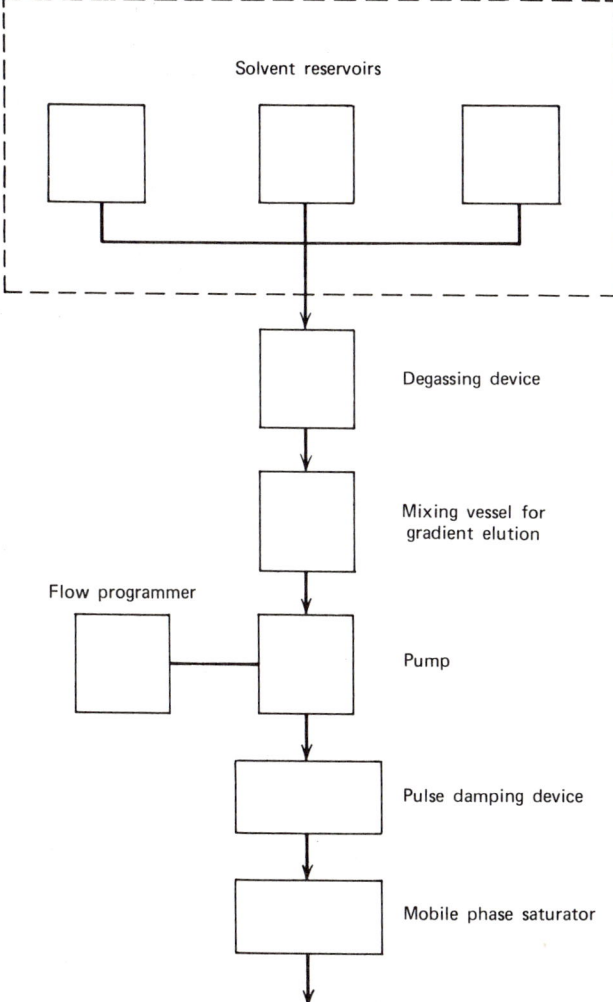

Fig. 3.1. The mobile phase supply system.

prehensive system is depicted in Fig. 3.1 and would consist of the following units:

1. A series of solvent reservoirs to provide mobile phases of different polarity and ability for gradient elution.
2. A degassing device to remove dissolved air from the mobile phase.
3. An appropriate mixing system that will provide a gradient elution facility.
4. A suitable pump or pumps to provide mobile phase flow at the required column inlet pressure.
5. Where mechanical or diaphragm pumps are employed a pulse smoothing system is required to maintain a steady flow of mobile phase.
6. Suitable device to provide flow programming facilities.
7. A column presaturator that ensures that the mobile phase is saturated with stationary phase before entering the column. This device is only necessary for liquid-liquid chromatography.

In most applications of liquid chromatography such a comprehensive system is seldom necessary. However, all can be considered as part of the mobile phase supply system and there are certain requirements for each part that must be met. If the liquid chromatography is to reasonably versatile, all parts of the system must be inert to solvents of all polarities and also to acid and alkaline media. This means that the material of fabrication must be either stainless steel, polytetrafluoroethylene (PTFE), or glass. There are a number of plastic materials that can be used with a restricted range of solvents, and if this range of solvents is adequate for the applications in mind such materials can be used. However, it is strongly recommended that glass, PTFE, or stainless steel be the only materials used, if at all possible. As a result of the developments that have taken place in column technology the system may be required to maintain high column inlet pressures of 5000-10,000 psi to provide adequate flow rates. However, many separations can be realized by liquid chromatography employing pressures of less than 500 psi. The choice of suitable equipment for mobile phase systems is relatively large if the maximum operating pressure is restricted to 500 psi. Under some circumstances it may well be worthwhile in many laboratories to restrict the maximum operating pressure to 500 psi and to tolerate the longer analysis times necessary to achieve the required resolution so as to take advantage of a wider range of equipment.

Solvent Reservoirs

In a versatile liquid chromatographic system, a number of solvent reservoirs should be available to provide gradient elution facilities. With some methods of gradient elution (which will be discussed later) supplies of solvents can then be drawn sequentially from the reservoirs, by means of a multiway valving system, which must also be made of inert materials. Such a multiport valve is often included in the solvent reservoir system to allow rapid selection of specific solvents for differnt analyses or for column cleaning purposes. In order to draw the solvent into the pump and thence to the chromatographic column, the reservoirs must either be open to air or have a small pressure of nitrogen applied to them. It should be noted, however, that trace quantities of water in the solvent can significantly change its polarity or can effect the retention properites of any solid absorbent in the column. It is, therefore, important to exclude moisture from the solvent resevoir, and thus a drying tube containing molecular sieve or other suitable dessicant should be placed in series with the air vent wherever possible. The inclusion of a dehydrating agent in contact with the solvent itself can also be used. To exclude the possibility of particulate matter entering the pump a filter should be placed in the end of each reservoir, this filter can take the form of a glass or stainless steel fritted disc. If glass is used for the solvent reservoir, it should be brown in color to eliminate solvent decomposition by UV light (e.g., chlorinated hydrocarbons).

Solvent Degassing System

Owing to the viscosity of liquid mobile phases, the liquid chromatographic column requires a significant pressure drop across it to maintain suitable flow rates. As the mobile phase progresses along the column, its pressure is continually reduced and any dissolved gases present in the mobile phase tend to form bubbles in the column. This can be particularly serious where alcohol-water mixtures are being used. Such bubbles cause serious band spreading and can completely destroy the column resolution and, furthermore, can produce serious detector instability where flow through cells are employed. The evolution of bubbles in the column packing, due to dissovled gas, becomes a progressively greater problem as the inlet pressures employed are increased. In order to eliminate this effect, it is sometimes necessary to use some form of degassing device. One of the most commonly used is shown in Fig. 3.2. The solvent from the reservoir passes into a small flask fitted with a reflux condenser and situated on a hot-plate. The solvent is arranged to boil continuously and the solvent is fed into the flask from the reservoir and sucked from the flask by the solvent pump. A cooling jacket is placed between the solvent pump and the degassing device to bring the temperature of the solvent back to ambient before entering the pump.

There are several varieties of this type of degassing system, all of which are based on the principle of heating the solvent to remove the dissolved gas. However, in some instances the removal of the gas is aided by heating the solvent under reduced pressure. If columns of 3 mm or more in diameter are used, column inlet pressures may be only about 500 psi or less. At these pressures and below, it is often possible to operate the chromatograph without the use of a degassing device.

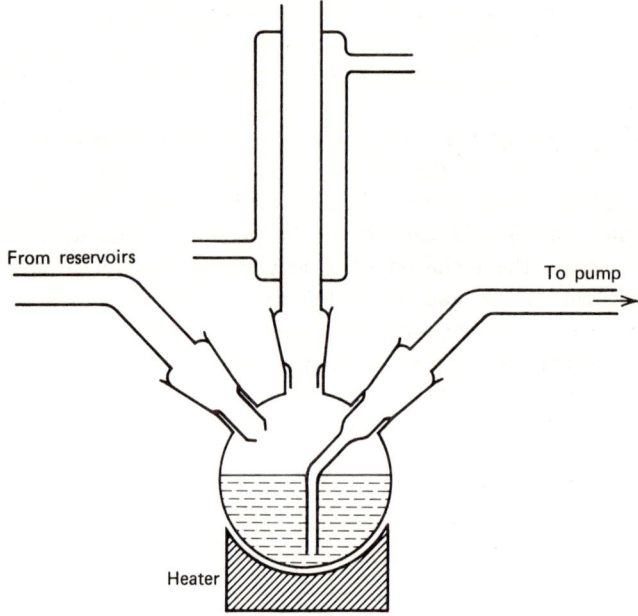

Fig. 3.2. The solvent degassing system.

Gradient Elution Facilities

Gradient elution is the counterpart in liquid chromatography of temperature programming in gas chromatography. Discussion of the uses of gradient elution in liquid chromatographic separations is dealt with in the section on chromatographic procedure, but the apparatus necessary for performing gradient elution is described here. There are four general methods used to achieve gradient elution:

1. Gradient elution by solvent mixing.
2. Incremental gradient elution.
3. Gradient elution by displacement.
4. Simmulated gradient elution by temperature programming.

Gradient Elution by Solvent Mixing

This method involves the continuous addition of a polar solvent to another, nonpolar solvent or vice versa before its entry into the chromatographic column. During the development of the chromatogram the proportion of the second solvent is progressively increased. This process can be achieved by the use of a rather elaborate twin pump device that is fitted with an automatic programming system. The total flow from the pump is maintained constant and the delivery

1 THE MOBILE PHASE SUPPLY SYSTEM

from one pump is continuously increased at the expense of that from the other. The form of the program required is cut out on a template and fitted to a revolving drum. During the program that results in gradient elution the edge of the template is sensed by a device that controls the relative delivery rates of each pump. The system is effective but a little expensive; it suffers from the disadvantage that a new template has to be cut for each different program used. The more usual method of achieveing gradient elution by mixing is to draw the the mobile phase by means of the pump from a mixing vessel into which the second solvent continuously flows. Such a system is shown in Fig. 3.3. The

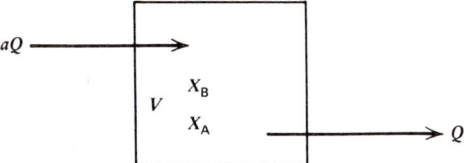

Fig. 3.3. Dilution vessel.

theory of the mixing process is as follows:

Consider the mixing conditions shown in Fig. 3.3. Let a solvent A flow into the mixing vessel, which contains volumes V_A and V_B of solvents A and B respectively, at time t. Let the concentration of solute B in the vessel be X_B, the flow of the mixture to the pump be Q, and the flow of solvent A into the vessel be aQ. It will be assumed that the volume change on mixing for the two solvents is negligible and that a can be greater or less than unity. All concentrations will be measured in units of volume per volume, that is, milliliter per milliliter. In time dt the change in volume dv in the mixing vessel is given by

$$dv = (aQ - Q)dt.$$

Integrating,

$$v = (aQ - Q)t + k$$

where k is a constant. Now when $t = 0$, $v = V$, where V is the initial volume of solvent contained in the vessel. Therefore

$$k = V.$$

Thus

$$v = Q(a-1)t + V.$$

The flow rate Q through the column is conditioned by the properties of the column and can be considered a constant. Thus if the maximum retention time that is acceptable is T, the volume of solvent removed from the vessel in time T will be QT. If the apparatus is to be versatile, the mixing vessel must be capable of providing a volume QT of solvent B for isopolar operation and a little excess to prevent air entering the pump. Assuming P ml is adequate excess, then the initial volume of solvent to be placed in the vessel will be

$$V = QT + P. \qquad (3.2)$$

Equation (3.2) allows the calculation of the maximum volume of solvent that most be placed in the vessel that will provide the necessary flexibility of elution conditions. It now remains to determine the relationship between X_B, the concentration of solvent B in the vessel, and time when solvent A is pumped continuously into the mixing vessel. Now by definition,

$$X_B = \frac{V_B}{V_A + V_B}.$$

Differentiating,

$$\frac{dX_B}{dt} = \left[(V_A + V_B)\frac{dV_B}{dt} - V_B\left(\frac{dV_A}{dt} + \frac{dV_B}{dt}\right)\right]/(V_A + V_B)^2$$

$$= \frac{1}{V_A + V_B}\frac{dV_B}{dt} - \frac{V_B}{(V_A + V_B)^2}\left(\frac{dV_A}{dt} + \frac{dV_B}{dt}\right). \qquad (3.3)$$

Now $\frac{V_B}{(V_A + V_B)} = X_B$ and from (3.1), $V_A + V_B = Q(a - 1)t + V$. Substituting for $\frac{V_B}{(V_A + V_B)}$ and $V_A + V_B$ in (3.3),

$$\frac{dX_B}{dt} = \frac{1}{Q(a - 1)t + V}\left[\frac{dV_B}{dt} - X_B\left(\frac{dV_A}{dt} + \frac{dV_B}{dt}\right)\right] \qquad (3.4)$$

Furthermore,

$$\frac{dV_B}{dt} = -X_B Q \quad \text{and} \quad \frac{dV_A}{dt} = aQ - (1 - X_B)2.$$

Thus substituting for $\frac{dV_B}{dt} + \frac{dV_A}{dt}$ in (3.4),

$$\frac{dX_B}{dt} = \frac{1}{Q(a-1)t + V} \left\{ -X_B Q - X_B a Q - (1-X_B)Q - X_B Q \right\}$$

$$= \frac{1}{Q(a-1)t + V} [-X_B Q - X_B(aQ - Q)]$$

$$= \frac{-aQX_B}{Q(a-1)t + V}$$

Thus

$$\frac{dX_B}{X_B} = \frac{-aQ\,dt}{Q(a-1)t + V}.$$

Integrating,

$$\operatorname{Log} X_B = \frac{-a}{a-1}[\operatorname{Log} Q(a-1)t + V] + K$$

Where K is a constant or

$$X_B = K'[Q(a-1)t + V]^{-a/(a-1)}$$

Now when $t = 0$, $X_B = 1$. Thus $K' = 1/V^{-a/(a-1)}$. Thus

$$X_B = \left(\frac{Q(a-1)}{V} + 1\right)^{-a/(a-1)} \qquad (3.5)$$

and

$$X_A = 1 - X_B = 1 - \left(\frac{Q(a-1)t}{V} + 1\right)^{-a/(a-1)} \qquad (3.6)$$

From (3.5) it can be seen that when the flow rate into the vessel is equal to the flow rate to the column (i.e., $a = 1$) then the function becomes indeterminate. Now under these circumstances the system acts as a log dilution system and it has been shown that under such circumstances,

$$X_B = e^{-Qt/V}$$

If (3.4) is valid then, when a tends to 1, $[Q(a-1)t/V + 1]^{-a/(a-1)}$ tends to

$e^{-QT/V}$. For convenience let $Qt/V = b$ and $a - 1 = x$. Then it is required to show that as X tends to 0, $(bx + 1)^{-a/x}$ tends to e^{-b}. Expanding by the binomial theorem,

$$(bx + 1)^{-a/x} = 1 - \frac{abx}{x} + \frac{a}{x}\frac{a+x}{x}\frac{b^2x^2}{2!} - \frac{a}{x}\frac{a+x}{x}\frac{a+2x}{x}\frac{b^3x^3}{3!} + \cdots$$

$$(bx + 1)^{-a/x} = 1 - ab + a(a + x)\frac{b^2}{2!} - a(a + x)(a + 2x)\frac{b^3}{3!} + \cdots$$

Now as a tends to unity and x tends to 0,

$$(bx + 1)^{-a/x} = 1 - b + \frac{b^2}{2!} - \frac{b^3}{3!} + \cdots$$

$$= e^{-b}.$$

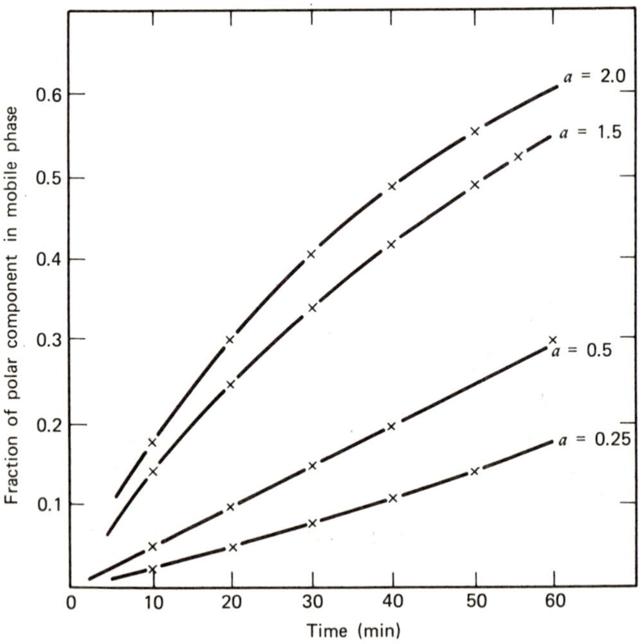

Fig. 3.4. Graph of fraction of polar component in the mobile phase against time for different values of a.

Equation (3.5) permits the concentration of the polar solvent in the mobile phase leaving the diution vessel (X_A) to be calculated using the flow rates to and from the dilution vessel and the initial volume of solvent B contained by it. Using (3.5), concentration/time curves have been calculated for solvent A over a 60-min time period for a column flow rate of 0.2 ml/min (Q), an initial volume of solvent B of 20 ml (v), and using values of 0.25, 0.5, 1.5, and 2 for a. The curves obtained are shown in Fig. 3.4; they include points obtained experimentally for the conditions given.

Incremental Gradient Elution

The incremental method of gradient elution is achieved by changing the mobile phase supply through a sequence of different solvents or solvent mixtures in a stepwise manner. This is accomplished by the use of a number of mobile phase supplies that can be connected to the pump selectively by means of a multiport valve. An appropriate valve for this purpose is shown in Fig. 3.5; it is manufactured by Chromatronix Incorporated. The valve has a total of 20 ports that can be manually selected by means of a switch or automatically actuated by a suitable programming device. All parts of the valve that are in contact with the solvents are made from PTFE or Kel-F and thus the valve is completely inert to all types of mobile phases. This valve has a maximum operating pressure of 500 psi. If the valve is connected directly to the solvent pump, the profile of the resulting polarity gradient will be a step function, each step representing the polarity of each solvent. However, if a significant dilution volume is situated between the valve and the pump, then the step function will be modified by the logarithmic decay effect of this dilution volume.

In order to determine the form of the gradient produced by this method it is necessary to treat the system as a series of discontinuous processes. Let the gradient elution take place in three stages between times t_0 and t_1, t_1 and t_2, and t_2 and t_3. Let the flow rate of solvent be Q and the volume of the mixing vessel, V. Let the concentration of the polar component of the solvent mixture contained in the mixing vessel at time t_0 be X_0 and let this concentration be also that of solvent mixture flowing into the vessel between t_0 and t_1. Let the concentration of the polar component of the solvent mixture flowing into the vessel between time t_1 and t_2, and t_2 and t_3 be X_2 and X_3, respectively. Now during times t_0 to t_1 the concentration of the polar component of the solvent will remain at X_1; however, at t_1 a different concentration X_2 enters the mixing vessel. Thus between t_1 and t_2 the effect of the solvents having concentration X_1 and X_2 on the net concentration of the polar solvent in the vessel can be treated directly and it follows that:

Between t_1 and t_2 the concentration of the polar solvent in the vessel due to X_1 will decay according to the equation

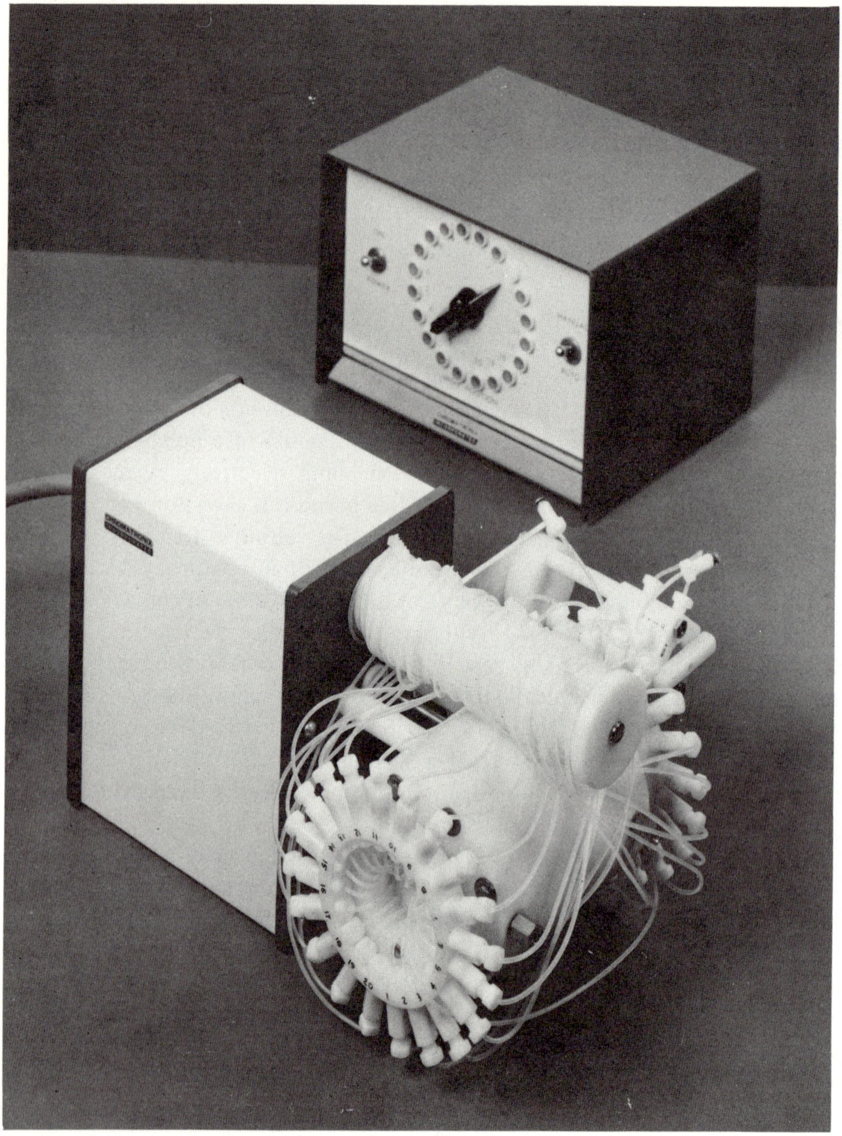

Fig. 3.5. The Chromatronix 20-port valve.

$$X_t = X_1 e^{-Q(t-t_1)/V}$$

and the concentration of the polar solvent in the vessel due to X_2 will increase

$$X = X_2(1 - e^{-Q(t-t_1)/V}).$$

Because solvent of concentration X_2 will only enter until time t_2, the concentration of polar component in the vessel due to this solvent alone will reach a maximum at t_2, and is given by,

$$X = X_2(1 - e^{-Q(t_2-t_1)/V}).$$

Thus between t_1 and t_2 the concentration of polar component X will be given by

$$X = X_1 e^{-Q(t-t_1)/V} + X_2(1 - e^{-Q(t-t_1)/V}).$$

Between times t_2 and t_3 the residue of polar component from the solvent of concentration X_1 will continue to decay. During this period the concentration of polar component due to X_2 will, having reached its maximum at t_2, also commence to decay while the introduction of the new solvent having concentration X_3 will increase the polarity of the solvent mixture. Thus between t_2 and t_3,

$$X = X_1 e^{-Q(t-t_1)/V} + X_2(1 - e^{-Q(t_2-t_1)/V}) e^{-Q(t-t_2)/V}$$
$$+ X_3(1 - e^{-Q(t-t_2)/V}).$$

Combining the above equations to describe the change in concentration of polar component in the vessel with time during the total period between t_0 and t_3 it follows that:

From t_0 to t_1,

$$X = X_1$$

From t_1 to t_2,

$$X = X_1 e^{-Q(t-t_1)/V} + X_2(1 - e^{-Q(t-t_1)/V})$$

From t_2 to t_3,

$$X = X_1 e^{-Q(t-t_1)/V} + X_2(1 - e^{-Q(t_2-t_1)/V}) e^{-Q(t-t_2)/V}$$
$$+ X_3(1 - e^{-Q(t-t_2)/V}).$$

The correct values for t_0, t_1, t_2, ··· and X_1, X_2, X_3, ··· and volume v to provide a specific polarity program are best determined with the aid of a computer, and a simple computer program for this purpose, written in basic language, together with examples of its use is given in the Appendix. It can be seen in the examples that the incremental mixing system with the appropriate logarithmic dilution volume can provide a wide range of solute profiles, and with the aid of a computer program the necessary operating parameters are easily determined. It should be pointed out that this program will be effective for buffer solutions and two-component mixtures only. It is not appropriate if multisolvent systems are employed.

A diagram of a suitable form of the apparatus for gradient elution by the incremental method of mixing is shown in Fig. 3.6. The multiport value can have

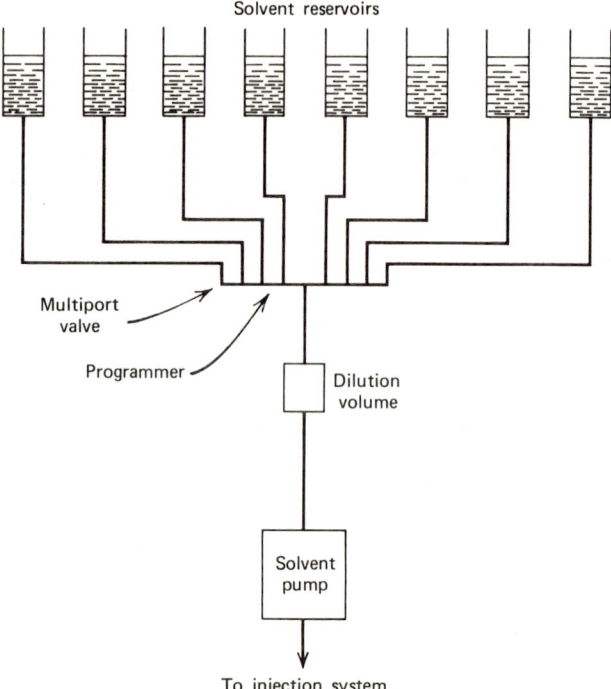

Fig. 3.6. Apparatus for gradient elution using the incremental method of mixing.

any number of ports, but a minimum of 12 should probably provide satisfactory performance. Because the multiport valve is operating at atmospheric pressure, its specifications are not stringent and such a valve need not be expensive. The magnetically stirred dilution vessel can take the form of a simple adjustable piston similar to a serum syringe or a stainless steel bellows that can be compressed by a calibrated screw head. The great advantage of the system lies in its simplicity, predictable performance, and the need for only one pump for the mobile phase supply to the chromatograph. When a solvent program has been completed, the volume of the mixing vessel can be reduced to its minimum value and the multiport valve returned to the position for the first solvent. The chromatographic column can thus be rapidly brought back to its starting condition in readiness for the next analysis. The multiport valves could be operated by a simple cam shaft/timer having the same number of cam switches as ports to the valve. The position of the cams can be set from the values of A to G obtained from the computer program.

A commercially available programer suitable for multiple component solvents is manufactured by Analabs Inc., and a photograph of the unit is shown in Fig. 3.7.

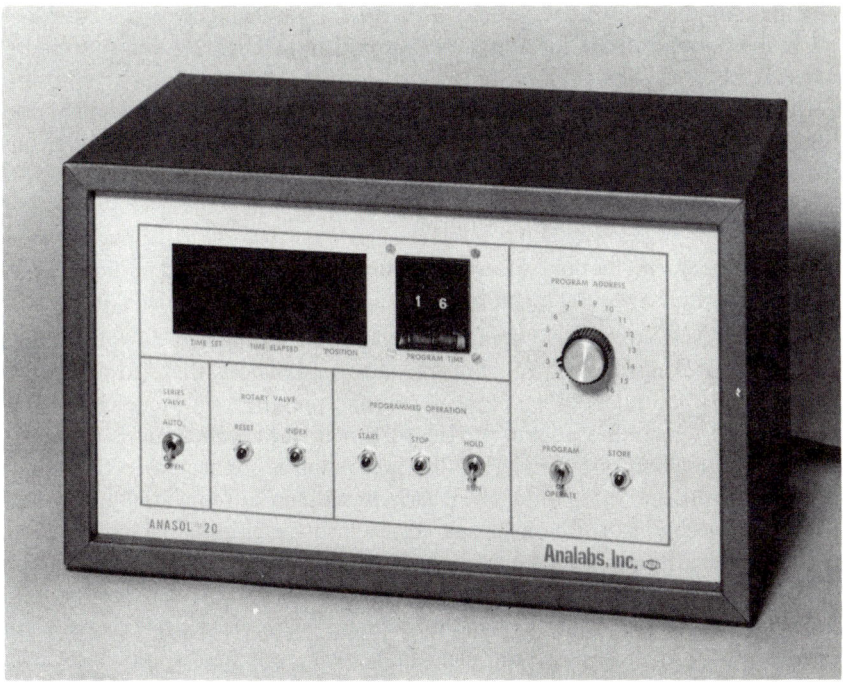

Fig. 3.7. The Analabs programmer for incremental gradient elution.

The apparatus is an all-electric timing instrument capable of controlling as many as 16 valves in sequence for precisely programmed time periods. The device in fact is designed to provide, first, a development program, and if required, a second reverse program that can be chosen to recondition the column in readiness for the next analysis. The programer uses a precise, two-digit, digital electronic clock unit, a sequential programming unit, and an electronic memory unit that stores 16 numbers of two digits each.

In operation, the sequential programming unit opens a valve and at the same time "reads" a two-digit number from the memory location associated with that valve. Simultaneously the electronic clock is started. The number from memory, representative of the operating period for that valve, is continuously compared with the contents of the clock. When the numbers are equal, the opened valve is closed; the clock is reset; the sequential programming unit steps, opens the next valve, and the cycle repeats. After completion of the entire 16-period operating cycle the instrument indexes a further four positions for reconditioning purposes. Each reconditioning solute has the same constant time period. Two-digit numbers, representing the valve operating periods, can be easily placed in memory whenever the instrument is not in the running mode.

In the initial set up the valves can, if necessary, be manually operated. Where particular solvents from specific valves are not required, the number stored in memory can be made 0.

This type of equipment allows complex programs to be employed and to take the form of any function desired. It also allows the rapid selection of any particular solvent for isocratic operation.

Gradient Elution by Displacement

The displacement method of gradient elution was introduced to liquid chromatography by Snyder [1] and employs a series of packed precolumns, each filled with solvents of different polarities and connected in series. The precolumn packing consists of celite or glass beads and is there to maintain a relatively sharp front between each solvent and prevent each precolumn acting as a dilution vessel. The polarity of the solvent in each precolumn is chosen to provide the required gradient. The first precolumn is connected to the mobile phase reservoir and the contents of each precolumn is sequentially displaced from one precolumn to the next and finally into the chromatographic column itself. The polarity profile is stepped, the height of each step being dependent on the polarity of the solvent in the respective precolumn and the length of the step is proportional to the volume of the precolumn. Because some mixing occurs at each solvent front as it passes through the system, each step is not sharp. By a suitable choice of the design of precolumns and appropriate solvents, any form of gradient elution can be obtained. However, one disadvantage of this system lies in the fact that after each separation the precolumns have to be individually emptied and refilled with the correct solvent.

Simulated Gradient Elution by Temperature Programming

This method of gradient elution employs the moderated silica gel system described by Maggs and Young [2]. Maggs and Young used a nonpolar solvent (heptane) containing a small quantity of alcohol (the moderator) at levels between 0.5 and 2% w/v. The moderator was allowed to come into equilibrium with the silica gel, the stationary phase. Depending on the concentration of the moderator, the activity of the silica gel was reduced owing to the active sites being covered with alcohol molecules. With this system Maggs and Young [2] and Scott and Lawrence [3] showed that by varying the level of the moderator between limits of 0.05 and 2.0% w/v, substances ranging in polarity from paraffins to free fatty acids could be chromatographed. Scott and Lawrence [4] also showed that on raising the temperature of a given moderated silica gel system, the moderator desorbed from the silica gel to form a new equilibrium system and in doing so caused an increase in the concentration of the moderator in the mobile phase, thus increasing its polarity. From these experiments it became apparent that the temperature programing of a modified silica gel system

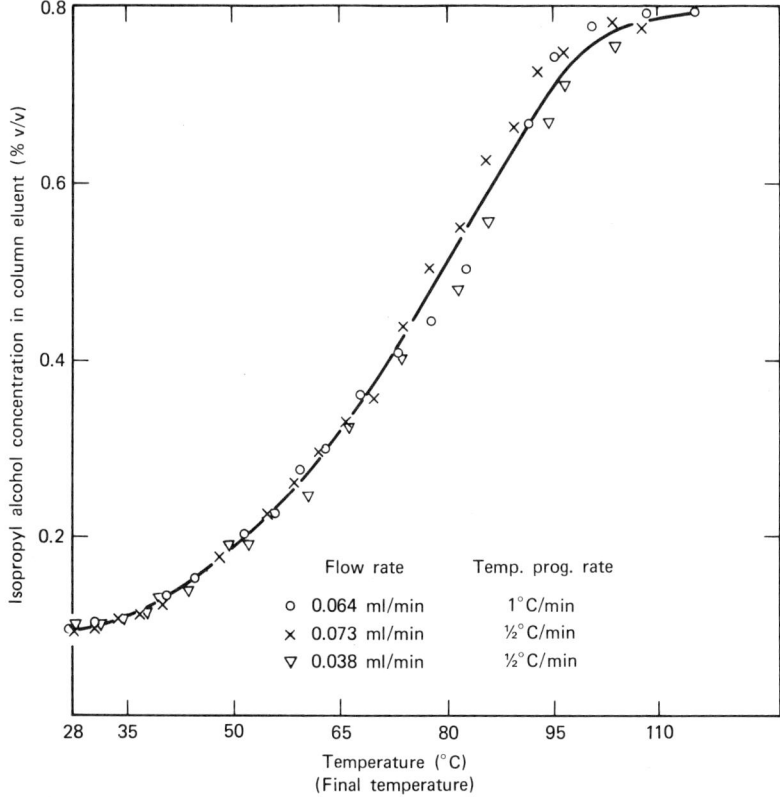

Fig. 3.8. Effect of tempeature and flow rate on the isopropyl alcohol concentration in the column eluent.

could simulate the effect of gradient elution but with one important difference: the polarity of the mobile phase would be increased consistently along the length of the column. The polarity of the mobile phase entering the column was maintained at the equilibrium moderator concentration in the column by supplying it from a large-capacity precolumn, also packed with silica gel, and programed in the same oven. The disadvantage of this system is the time required to reequilibrate the precolumn between each separation. A curve relating moderator concentration and temperature for different column flow rates and program rates is shown in Fig. 3.8. It is seen that the polarity of the mobile phase is independent of flow rate and program rate but solely dependent on the column temperature.

Other Methods of Gradient Elution

A method introduced by Byrne, Schmit, and Johnson [5] based on the dilution vessel system, but in such a form that it can be used with pneumatic pumps of large reservoir volume, is shown in Fig. 3.9. A single pump (P) provides the

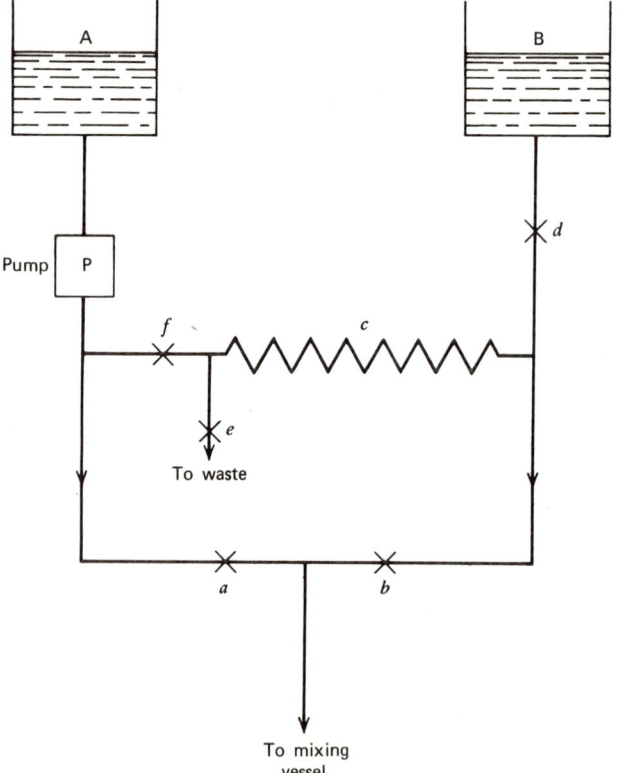

Fig. 3.9. Gradient elucion system for use with pulseless pumps.

pressure to drive A and B liquids to the on/off solenoid valves a and b. In order to use a single pump the B liquid is first introduced to the holding coil C through valves d and e with f closed. The holding coil is made of ¼-in.-o.d., stainless steel tubing, 25 ft long. The relatively narrow bore of the tubing minimizes the mixing between the A and B liquids due to diffusion and viscosity differences. The volume of the coil can be adjusted to any value by increasing the length of tubing in the coil. A 140-ml volume was found to be optimum for high-speed liquid chromatography and of this, about 95% can be used. Under most chromatographic conditions many gradients can be run before the coil requires refilling. Since diffusion in liquids is slow, the flow can be stopped, the coil filled, and then the gradient can continue from the point where it stopped.

When the system is started, liquid from the A reservoir is pumped both to proportioning valve a and to the holding coil C. (The flow path is indicated on the diagram by the small arrows). From the coil, the B liquid is then delivered to the proportioning valve b. These proportioning valves are alternately cycled to allow a prescribed amount of each liquid to flow into the mixing chamber. Flow is continuous since either valve a or valve b is always open. By controlling the on-time of each valve during the cycle, any concentrations of A and B liquids can be produced. For example, if the valve period is set at 30 sec and 75% A, 25% B solvent mixture is desired, then valve a is on for 22.5 sec and valve b is on for 7.5 sec.

The theory of this system is basically the same as that given for gradient elution by the method of mixing, and very similar gradient profiles can be obtianed. The main advantage of this system is that it can employ syringe-type pumps and does not require pulsating pumps having small stroke volumes.

Another gradient elution system for use with pulseless pumps of large capacity is that developed by Nestor and Faust and now marketed by Perkin Elmer. A diagram of their apparatus is shown in Fig. 3.10.

The pumps employed are a positive displacement design that provides for smooth, constant, repeatable flow rate regardless of back pressure or changes in back pressure.

The constant-speed digital motors driving the pumps maintain a constant flow rate within ± 0.1%. Flow rates from 0.01 to 6 ml/min can be provided.

The body or cylinder of the pump is constructed of precision-bore glass tubing (or inert metal for presures above 100 psi) and the piston is PTFE with sealing rings that are inert to the wide range of organic solvents employed in liquid chromatography. When the pump is exhausted, forward travel of the piston stops automatically. The pumps can be filled from the reservoirs by reversing the motors, and the travel of the piston on the fill cycle stops automatically when the pumps are charged. The pumps can be connected so that pump 1 discharges into pump 2, where the two liquids are kept will mixed by means of a magnetic stirrer. This ability to use one pump as a mixing chamber makes the generation

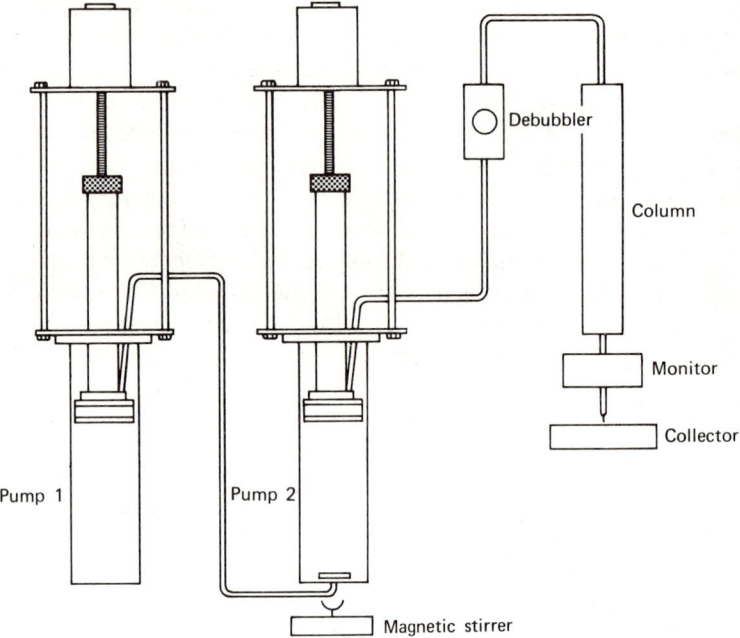

Fig. 3.10. Gradient elution system for use with twin pulseless pumps.

of linear and nonlinear gradients quite simple.

Again, the theory of this pump is similar to that of the method of mixing, and solvent gradients of the same form as those previously described can be obtained.

2 SOLVENT PUMPS

There are essentially two types of pumps used in liquid chromatography, the mechanical pump and the pneumatic pump. The mechanical reciprocating pump can be piston or diaphragm operated, or the single-stroke syringe type where the mobile phase is contained in a large stainless steel cylinder, the piston being slowly driven forward by a screw thread. Generally the piston-type pumps are capable of providing the highest pressures.

The maximum volume delivery of the pump that is required will vary with the dimensions of the column employed. For general analytical work (column diameter 1-3 mm) a range of 0-10 ml/min is usually adequate. For large-bore columns the required column delivery of the pump will need to be proportionally increased. Any pump designed for use with liquid chromatographs must be capable of maintaining a constant flow rate to at least ±2%.

2 SOLVENT PUMPS

Unless a very limited range of solvents is to be employed, the pump and valve glands should not be made of polymeric materials except PTFE or Kel-F. An excellent review of pumps available for liquid chromatographs together with their operating characteristics has been given by Berry and Karger [5b].

The following are some examples of solvent pumps that have been successfully employed in liquid chromatography.

Mechanical Pumps

Whitey Model LP 10

This pump is a diaphragm type that can operate at pressures up to 5000 psi. provision is made so that a jacket can also be fitted to the pump which can be used for heating or cooling the fluid being pumped.

Milton Roy Milroyal Pump

This is a plunger operated pump that is rated for continuous operations at 1000 psi but under favorable circumstances has been used up to 3000 psi. The plunger and valve system is fitted with PTFE glands, which makes it impervious to all solvents and allows it to be adjusted to prevent leaks.

Milton Roy Minipump

This is also a plunger operated pump but is fitted with O-ring seals. It is rated at 5000 psi and this is its absolute maximum operating pressure. This pump is much smaller in bulk than the others listed.

Ruska Pumps

These pumps are syringe type in operation and one model can provide pressures up to 25,000 psi. The cylinder volume has about 250 ml and thus with a 0.25-mm/min column flow, it will provide for 1000 min of chromatographic development. The gland systems in this pump are made of PTFE and models can be obtained that will handle any solvent.

Waters Model 6000 Pump

The Waters 6000 pump is a dual piston type pump where the individual pistons are driven by carefully designed cams to provide a virtually pulse-free flow. The pump can operate a pressures up to 6000 psi and is fitted with an automatic cut out that can be adjusted to operate at any pressure up to 6000 psi. The pump has a flow rate range of 0.1 to 9.9 ml/min and any particular flow can be selected by a thumb switch controlling a stepping motor that drives the twin piston pumps. As a result of the use of a stepping motor drive system, the pump can easily be adapted for flow programing. At the present time the Waters 6000 pump is probably one of the best for use in liquid chromatography, although it is one of the most expensive.

Servicing Mechanical Pumps

One of the most frequent causes of breakdown in a liquid chromatograph is the mobile phase supply system and the probelms are usually associated with the pump, particularly if a mechanical pump is employed. Use of solvents outside the specification for the pump can result in swelling and disintegrating of the gland packing material and in leakage or blockage of the nonreturn values. Periodically the pump should be stripped down and the nonreturn valves cleaned since any particulate material accumulating in these values will result in low pump pressure, increased pulsing, or complete breakdown of the supply. The gland packing material should also be examined periodically and replaced if there are any signs of wear. In some instances when high pressures are used on mechanically weak supports, for example, large-partical-diameter silica gel, the packing fractures and results in an impermeable bed of adsorbent being formed and very high back pressure on the pump. In most instances when this occures the gland backing material will be deformed or damaged and in some instances the nonreturn valves (check valves) may be broken. If column packing does breakdown in this way it is essential that the pump be stripped down and examined for damage before proceding to replace the column with a new one. It is wise to maintain a supply of spare pump glands and nonreturn valves in case such a situation arises. All union connections to the pump must be maintained tight or again high pressures will not be realized. Furthermore, a leak, however indiscernible, can provide a passage for dissolved air and produce air bubbles in the column and consequent loss of resolution and interfere with the detecor performance. Some pumps, for example, the Waters 6000 pump, are fitted with 2-3-μ filters. These can easily become blocked unless the solvents employed are carefully filtered free from particulate matter.

Pneumatic Pumps

Pneumatic pumps are operated by gas pressure acting on a suitable collapsable container or a cylinder that holds the mobile phase. In the absence of pneumatic amplifiers, the mobile phase passes from the container (isolated from the gas) to the columns at the same pressure as the applied gas. In its simplest form, it can be a polythylene wash bottle situated in an airtight container. The wash bottle contains the mobile phase and the pneumatic pressure is applied to the inside of the container and thus the outside of the plastic bottle. Such a system, however, will only operate up to a few hundred pounds per square inch of pressure and has obvious limitations with respect to the solvents that can be used. The plastic bottle can be replaced by PTFE or stainless steel bellows or the pneumatic side of a liquid-gas piston. Again it must be emphasized that the material of construction must not be affected by the solvent employed as the mobile phase.

Using pneumatic amplification, pressures up to 75,000 psi can be obtained and the Haskel Engineering Company can supply such pumps having outlet pressures from a few hundred to 100,000 psi. As a result of pneumatic amplification, gas pressures of only 250 psi are necessary in provide these high output pressures. Today such high pressures are not used in liquid chromatographs except possibly for some methods of slurry packing; nevertheless, such pressures may be employed in the future if the submicroparticulate packings are developed. However, many successful separations by liquid chromatography can be achieved using inlet pressures of less than 500 psi so the simple pneumatic pump can be a very useful mobile phase supply system. Another advantage of the pneumatic pumps is that it can be easily employed for flow programing; however, the main disadvantage of such pumps is the inability to provide a flexible gradient elution system.

Choice of Pump for Liquid Chromatography

New and improve pumps for liquid chromatography are continually being introduced by manufacturers, and for this reason it would be misleading to recommend specific pumps; such recommendations are likely to be obsolete by the time this book is in print. However, specifications can be given for pumps to be used under given chromatographic conditions to help the chromatographer to choose a suitable pump for his particular needs. Pumps to be used with apparatus for analytical or semipreparative operation should, as stated before, provide flow rates from 0.1 to 10 ml/min, and any particular flow rate should be maintained constant within ±2% variation. For isocratic operation pneumatic or single-stroke piston-type pumps are recommended. Such pumps will be pulseless in their function and thus produce minimum noise on a detector employing flow through cells. For preparative work flow rates of up to 400 or 500 ml/min may be necessary, and for such high flow rates reciprocating piston pumps will be required. For gradient elution reciprocating piston-type pumps are also recommended, but they must be designed to provide minimum pulse effects with built-in pulse damping if possible and they must have minimum stroke and dead volumes. For gradient elution, the combined stroke and dead volume should not exceed 0.25 ml; otherwise the pump itself will modify the solvent concentration profile in much the same way as a mixing vessel, which will be discussed later.

If only two solvents are being employed in the gradient procedure, two single-stroke piston-type pumps can be used, but this necessitates the purchase of two pumps and for multisolvent gradient elution a pump for each solvent will be required. In general the low-volume reciprocating pump is recommended for gradient elution.

Some pumps are fitted with pressure transducers that provide direct read out of pressure on a suitable meter, and other provide means for recycling procedures. Such devices can be extremely useful but are not always essential.

The Removal of Pressure Pulses from Reciprocating Pump Systems

A steady flow of mobile phase through the column free of pressure pulses is essential to obtain maximum column efficiency. Jerky movements of mobile phase through the column, which will be accentuated for columns of high permeability, may cause very serious band spreading. Furthermore, any pulsation of mobile phase experienced by the detector can produce instability and result in a very noisy chromatogram. It is therefore essential where reciprocating mechanical pumps are employed to eliminate these perturbations by a suitable pulse-damping device. One method is to employ a coil of capillary tube between the pump and the column inlet that acts in much the same way as an inductance smooths the alternating current ripple on a direct current electrical supply. The dimensions of the coil of tubing depends largely on the magnitude of the pulses produce by the pump. However, a 10-ft length of 0.050-in.-diameter capillary tube formed into a coil about 4 c in diameter will provide adequate damping for most pumps. The coils expand with each pulse from the pumps and take up the pressure surge, and it is therefore essential that the coil of tubing hang freely from its ends and not in any way be clamped in a rigid position. Such a system unavoidably takes up some space in the instrument, and adequate accommodation should be allowed for it. Other systems that have been suggested and are commercially available employ "damping pots" in the form of stainless steel bellows, Bourdon springs, and various other devices, but for a wide range of operating pressures and for most pumps the simple capillary tube coil probably gives the most effective damping with the minimum dead volume.

3 FLOW PROGRAMING SYSTEMS

Flow programing is a method of chromatographic development similar to gradient elution except that the column flow is progressively increased during elution instead of mobile phase polarity. However, flow programing is less effective than polarity programing in that it reduces the retention of a solute linearly with respect to flow rate, whereas polarity programing reduces the retention of a solute exponentially with respect to the mobile phase polarity. Flow programing can be very effective for reducing analysis times where gradient elution is impractical or difficult, and it can be also used subsequent to gradient elution where retarded peaks would take an unacceptably long time to elute. The increase of mobile phase flow rates, however, results, in an increased band dispersion and thus loss of column efficiency, although this can be often tolerated with strongly retained peaks. Flow programing with mechanical pumps is difficult since it requires programing either the stroke of the pump or the speed of the motor. Programing with pneumatic pumps, however, is very easy since it can be achieved by supplying the operating gas from a pressure programer of the form normally used in gas chromatography. The diagram of a pressure programer that can be used with a pneumatic pump is shown if Fig. 3.11. The operation of the programer

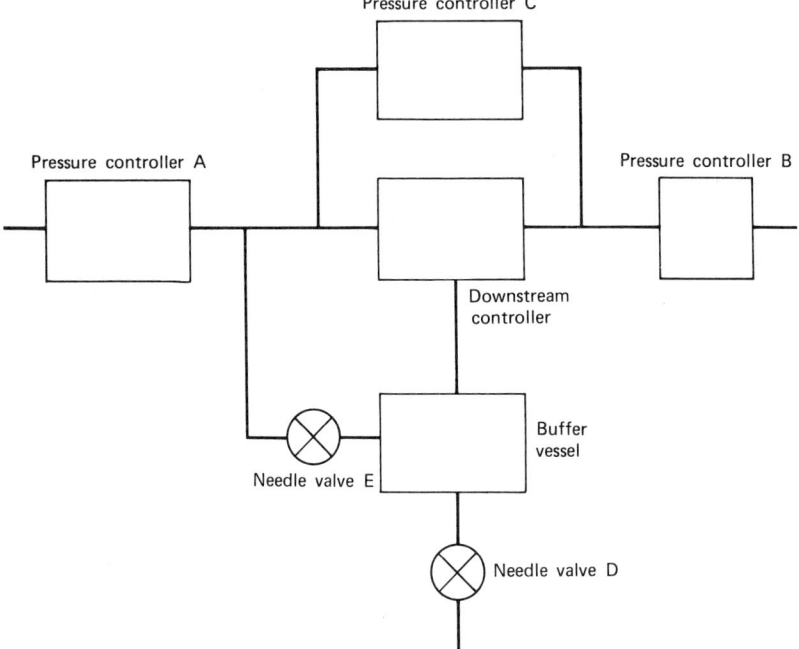

Fig. 3.11. Diagram of pressure programmer.

depends on the use of a down-stream pressure controller, the output of which is controlled by the gas pressure inside the buffer vessel. The final outlet pressure and thus the maximum column flow is set by pressure controller B and the initial pressure and column flow set by pressure controller C. The pressure controller A is adjusted to a value that is approximately 20% in excess of that set by B. Needle valve D is closed and E opened to allow gas to flow into the buffer vessel and thus build up the necessary pressure to actuate the down-stream controller. As the pressure increases in the buffer vessel, so the output pressure to the pneumatic pump also increases until it reaches the maximum set by controller B. At the end of the program, needle valve D is opened and E closed and the gas in the buffer vessel released. The system is then ready to start the next program. It is obvious that needle valve E controls the program rate and for this reason has to be a good quality valve that can be set reproducibly.

4 MOBILE PHASE EQUILIBRIUM SYSTEMS

Some liquid chromatography phase systems require the mobile phase to be brought into equilibrium with the stationary phase before entering the column. For example, when using liquid-liquid systems, unless the mobile phase is saturated with stationary phase before entering the column, the stationary phase will be progressively stripped from the support resulting in reduced retention and loss of resolution. Even so-called immiscible liquids still have a small mutual solubility and thus unless the mobile phase is saturated with stationary phase it will eventually completely strip the column. The best method of ensuring mobile-stationary phase equilibrium is to pass the solvent through a precolumn packed in an identical manner to the chromatographic column. The precolumn should have an adequate capacity and should be renewed frequently to ensure that there is always sufficient stationary phase to maintain saturation. However, solubility will vary with temperature, and therefore the precolumn should be maintained at the column temperature and preferably be situated in the same thermostat.

With liquid-solid chromatography the same problem of mobile phase equilibrium under certain circumstances also arises. For example, when using a silica gel column the silica gel may be modified with water or in the moderated phase systems used by Maggs and Young [2], a delicate equilibrium must be maintained between the silica gel and the very small concentration of alcohol in the mobile phase. Under these conditions the mobile phase should be prepared to contain the required water or alcohol concentration, the composition of which must be made as close to the required equilibrium concentration as possible. However, precise equilibrium must still be obtained by passing the mobile phase through a precolumn containing the same phase system as the chromatography column prior to entering the chromatographic column itself. Experience has shown that the need to frequently replace the precolumn is as great as that for liquid-liquid systems.

5 SAMPLE INJECTION SYSTEMS

The basic requirements of a satisfactory injection system are that it permits a discrete sample to be placed on the column such that the dispersion of the band is kept to a minimum and that the procedure is simple and repeatable from a practical point of view.

There are three general methods for sample injection in liquid chromatography:

1. Injection with solvent flow.
2. Stop-flow injection.
3. Sample valve injection.

Injection with Solvent Flow

This method of injection is identical to that used in gas chromatography; a typical septum injection system for liquid chromatography is shown in Fig. 3.12. The rubber disk can be wrapped in PTFE tape to reduce the swelling

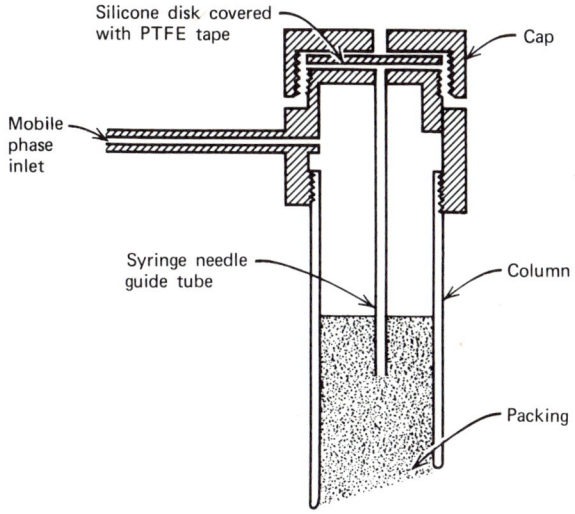

Fig. 3.12. Injection head for use in liquid chromatography.

effect of the solvent and the extraction of material from the septum that can result in detector instability. It is also important to ensure that the metal cap that secures the flat rubber septum has a flat conical seat, to provide lateral pressure on the rubber disk and thus seal the hole caused by the introduction of the syringe needle. This method of sample injection is the simplest and most convenient but, unfortunately, at pressures above the 1000 psi the system tends to leak (particularly with mobile phases of low viscosity) or needs very frequent septum replacement.

The factors limiting the life of a septum are the surface area pierced by the needle, the applied force on the unsupported area, and the pressure drop across the septum when a leak has developed.

In order to prolong the life of a septum as much as possible it is essential that the same hole be used every time a needle is inserted. This can be achieved by placing a metal disk on the low-pressure side of the septum having a central hole just large enought to accommodate the syringe needle. The disk performs another important function also: it supports the flexible septum against the high pressure in the column. The hole should be as small as possible to limit the forces acting on the unsupported area and keep the area of rubber septum perforated by the needle to a minimum. Calculated thrusts for various operating pressures and two sizes of hole are given in Table 3.1.

Table 3.1 Pressure Drop and Thrust Across Septum Pad

Pressure drop across pad	bar (ga)	69	138	206	69	138	206
	psig	1000	2000	3000	1000	2000	3000
Thrust on unsupported area of pad	lbf	0.3	0.6	0.6	0.6	1.2	1.9

It can be seen that the actual force on the unsupported area is very small. Since the internal diameter of the syringe needle is considerably smaller than the diameter of the hole, the force acting on the syring is even less than that indicated in Table 3.1. Accordingly it is considered practical to use an ordinary Hamilton syringe for sample injection.

The pressure drop across the septum, and hence the forces acting on the septum, may be reduced in the system developed by Pearce and Thomas [6], who employ a second septum pad, provided, like the first pad, with a backing disk. Such an injection device is shown diagrammatically in Fig. 3.13. The

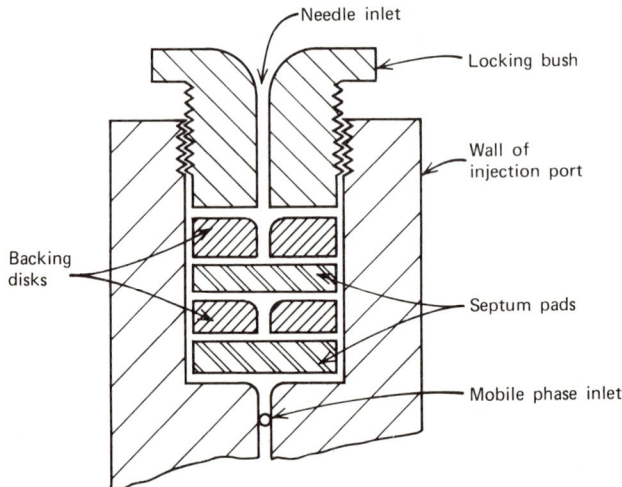

Fig. 3.13. High-pressure injection system.

septum pads, which are identical with those used in ordinary gas chromatography, are made of silicone rubber and are 0.928 cm diameter and 0.312 cm thick. Each septum pad is backed by a stainless steel disk of the same diameter and thickness. Holes of 0.051 cm diameter are drilled axially through each disk, and those holes are just large enough for 0.045-cm-diameter standard hypodermic syringe needles to be inserted without undue difficulty.

The septa and disks are located in a cylindrical chamber 0.932 cm in diameter and approximately 1.27 cm deep drilled out of a block of metal such as stainless steel. The disks and septa are retained in place by a locking nut that has a central hole of the same diameter as those in the disks. Easier insertion of the hypodermic syringe needle is ensured by chamfering the holes in the upper faces of the locking nut and backing discs, as shown in Fig. 3.13.

The components of the port are assembled as follows. The locking nut, backing disks, and septa are aligned and assembled in order, on a syringe needle. This assembly is inserted into the body of the tee, the locking nut is tightened by hand to compress the septa, and the needle is withdrawn. This technique ensures correct allignment of the holes in the disk. The injection port is pressurized with carrier fluid, and soap solution is used to detect any gas leaks. These are easily cured by tightening the locking nut slightly. When the port is gastight, surplus soap solution is wiped away.

It is advisable to use a Hamilton Kel-F plunger guide if Hamilton 10-μl syringes are employed. This guide serves a twofold purpose: first, it prevents bending of the small diameter plunger; second, a metal bush fitted to the free end of one of the guide rods acts as a stop and limits the outward movement of the plunger in the barrel.

At pressures below 180 bar (2500 psi), samples are injected by means of a syringe through the injection port just as in conventional gas chromatography. Above this pressure the plunger guide is positioned and secured. The syringe needle is pushed through the septa until its tip rests on the column packing. At the higher pressures a small pair of pliers are useful for inserting the needle through the injection port. The sample is injected in the usual manner. Syringes operating at pressures up to 6000 psi, such as that manufactured by Hamilton, are now beginning to become available, and this will simplify the procedures for injection at high pressures.

Leaks that develop over a period of time can be taken up by tightening the locking nut slightly. Eventually after prolonged use these leaks cannot be cured in this way and it then becomes necessary to replace the septa.

Although the authors found it possible to use an injection port containing only one septum pad at quite high pressures, a double septum is to be preferred, especially at the higher pressures, since far less attention is required to maintain it leak-proof. A double septum also has an appreciably longer life than a single septum, as shown in Table 3.2. (Note that the double septum was used at much higher pressures than the single septum.)

Table 3.2 Septum Life

Septum	Syringe	Ranges of Operating Pressures		Number of Injections Before Failure
		bar (ga)	psig	
Single	Hamilton 10 µl (glass)	41-152	600-2200	204
Double	Hamilton 10 µl (glass)	100-206	1600-3000	249
Double	SGE 1 µl (steel in steel)	124-206	1800-3000	87

The use of the guide tube was discussed in Chapter II and is necessary to ensure that the sample is distributed in the center of the cross section of the column and to allow the radical distribution of the charge to occur as rapidly as possible. This is particularly important when column diameters exceed 2 mm. With columns containing irregular packing material, it is best to inject the sample into a layer of plain glass beads that rest on the top of the column packing, because the radial distribution of the sample is more rapid in the larger geometrically regular glass beads.

Stop-Flow Injection

This method of injection is the simplest and most commonly used in liquid chormatography. The injection system itself is identical in form to that used for serum cap injection except that the metal cap carries no central hole for the hypodermic needle and for stop-flow injection the rubber septum is replaced by a PTFE disk. To inject the sample on to the column, the flow is stopped, and after a few seconds period that allows the column pressure to fall to atmospheric, the cap is removed. The sample is then injected into the column packing in the usual way with a suitable hypodermic syringe. After injection the cap is replaced, firmly tightened to obtain a good seal between the PTFE and the seating and the column flow turned on again. Because the diffusivity of those substances separated in liquid chromatography is very small, the stop-flow method of injection does not produce band dispersion to any significant extent. This method of injection is simple and reliable and because of this, it must be recommended as the best method of sample injection. The present trend is to place a porous metal disk on top of the column bed and to inject onto this frit. Such a system may improve the repeatability of the injection with respect to column efficiency particularly for high efficiency columns made from microparticulate packings. It can be particularly recommended for those workers inexperienced in liquid chromatography.

Sample Valve Injection

The sample valve method of injection is most useful for repetitive analyses where the sample size and column operating conditions can be kept constant. There are a number of suitable sample valves available, all of which operate on more or less the same principle. An example of a sample valve manufactured by Chromatronix Inc. is shown in Fig. 3.14. This valve provides repetitive injection

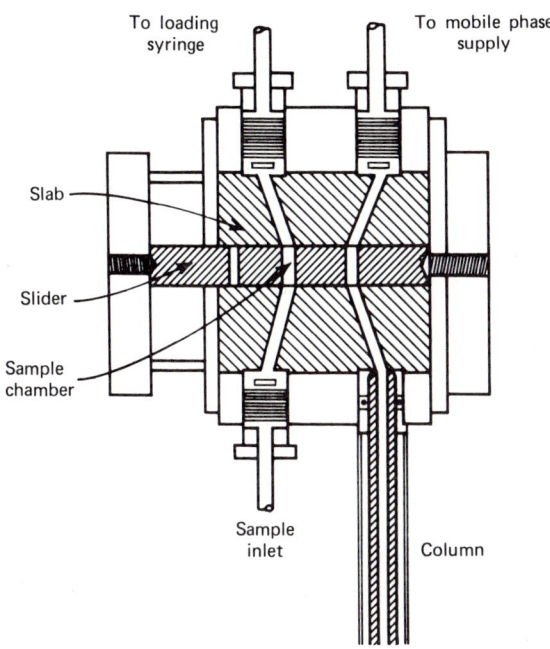

Fig. 3.14. Sample injection valve.

of samples into columns using any solvent without stopping the flow or decreasing the pressure. The only material that comes in contact with the mobile phase is PTFE and Kel-F. The Kel-F slider containing appropriate holes moves between two PTFE slabs. The column and connecting tube fittings make a seal directly against the slabs and press them against the slider, forming a leak-free seal for pressures up to 500 psi. The sample is drawn into a chamber using a syringe attached to the sampling tube and by moving the slider, the sample contained, in the sample chamber is placed in line with the solvent stream, and is carried into the column. The valve is available with various sample volumes from 2 to 100 μl. A photograph of the assembled valve, sample syringe, and column is shown in Fig. 3.15. Another valve manufactured by Valco Instruments of similar form can operate at pressures up to 7000 psi. This valve consists of a ceramic filled PTFE

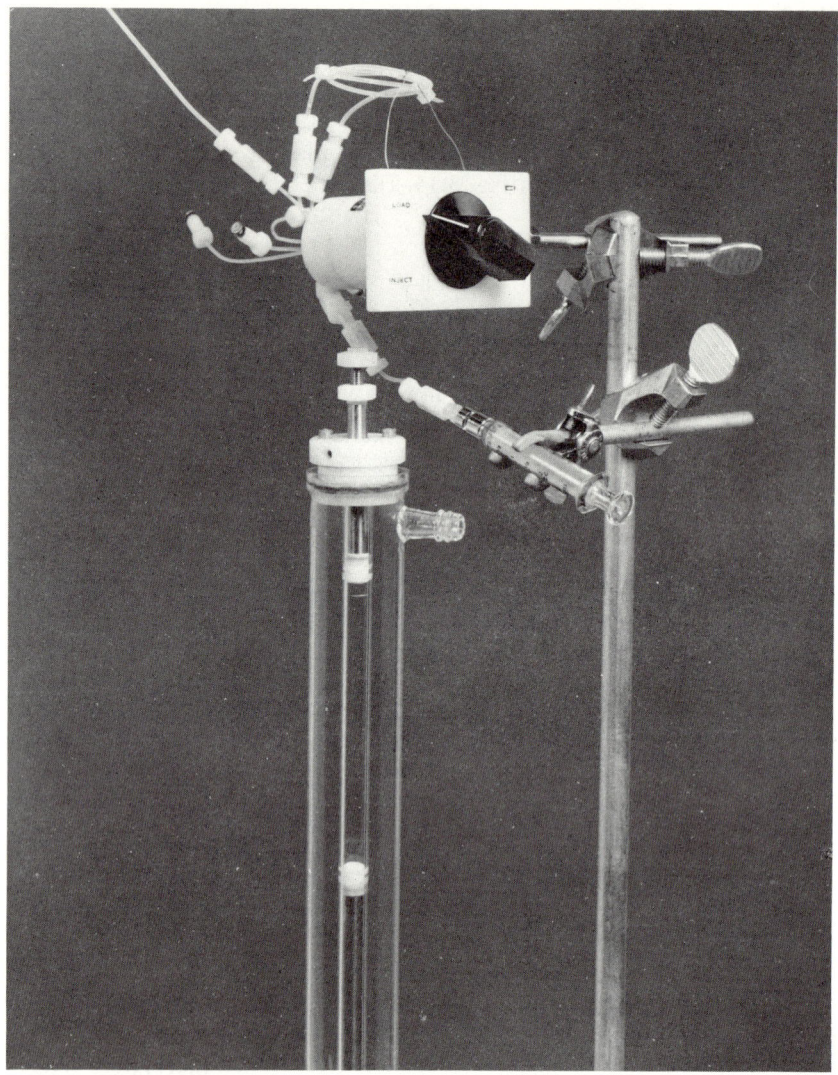

Fig. 3.15. The Chromatronix sample valve and column system.

cone that is compressed into a stainless steel seating, and the valve connections are engraved on the ceramic PTFE cones. These valves can have capacities up to 3 µl or can be fitted with interchangeable sample loops having volumes up to 100 µl. This valve can operate at a temperature of 175°C and thus can be situated in the column oven. Sample valves are very simple to operate but it should be emphasized that all samples must be free of solid contaminents and care must be taken to prevent any column packing entering the valve. If solid material does enter the valve, it is very liable to ruin the valve seating, which will result at high pressures in serious and continuous leaks.

6 LIQUID CHROMATOGRAPHY COLUMNS

Liquid chromatography columns are normally made from glass or stainless steel. Column diameters can range from fractions of a millimeter to a centimeter or even greater and column lengths from a few centimeters to many meters. The basis of choice for the dimensions of the column will be dealt with elsewhere, but it is obvious that the column dimensions will be conditioned by the nature of the sample and the quantity of material to be separated. Stainless steel columns can be made up in the laboratory from normal stainless steel tubing of appropriate length and fitted with standard stainless steel couplings. In a similar manner glass columns can be also fabricated, but the wall thickness of the tubing must be chosen to be compatible with the inlet pressure to be employed.

Generally, straight columns are to be preferred to coiled columns because they are easier to pack and tend to produce higher efficiencies. However, if the column lengths exceed 3 ft, straight columns can be awkward to use and cannot be fitted into a thermostatic oven. For this reason it is useful to use a coiled column. The effect of the column shape on the HEPT produced has been examined by Scott, Blackburn, and Wilkins [7] and Barth, Dallmeir, and Karger [8].

Karger and his workers showed that the effect of coiling the column on the HEPT depended on the column radius (r_0) and the coil radius (R_0). They show that aluminum colums 0.95 mm in radius packed with Corasil (27-35 μ) and coated in situ with 3,3-oxydipropionitrile, showed significant efficiency losses when the radius of the coil was less than 13 cm. The same column coiled to a radius of 1 cm showed a sixfold decrease in efficiency. It was also shown that columns having a radius of less than 0.38 mm showed no loss of efficiency even when formed into columns 1 cm in radius. Conversely, large-diameter columns having a radius of 2.4 mm when coiled to a radius of 2.4 cm resulted in a 14-fold decrease in efficiency. Generally, it can be said that providing the radius of the coil is sufficiently large relative to the column radius there will be little or no loss in efficiency. Alternatively, square-shaped columns can be constructed [7] where the corners of the square coil are crimped to contain the packing.

One particularly useful "do it yourself" column construction kit for the fabrication of chromatographic apparatus is manufactured by Chromatronix Inc., the design of their column system is shown in Fig. 3.16. The inlet tube, which is made of PTFE and fitted with a flange, is sealed against a backing disk

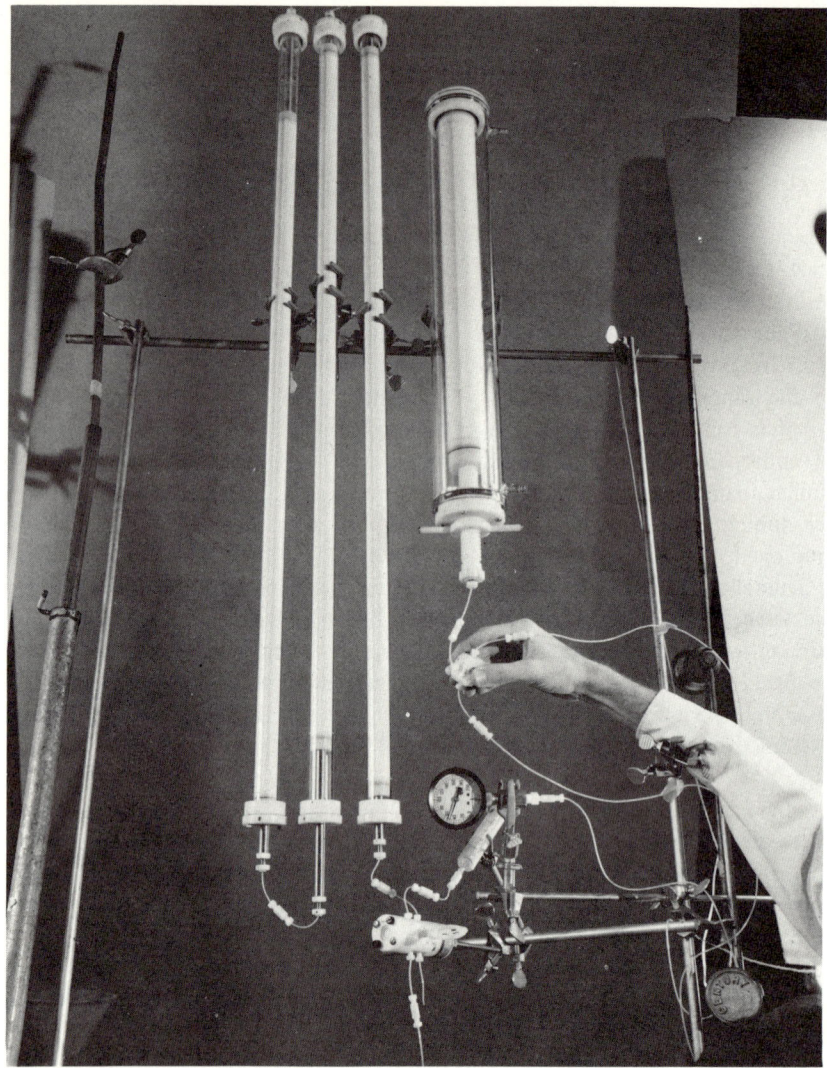

Fig. 3.16. Examples of different types of glass columns.

by screwing an inner rod and backing disk into a seal bushing. The high-pressure seal of the insert to the column wall is accomplished by pressing the PTFE skirt of the backing disk outward against the glass tube. This is achieved by the compression of a rubber O-ring (inside the skirt) with the outer rod. The)-ring supplies continuous force to the skirt despite any subsequent cold-flow of the

PTFE.

The bed supports are made of PTFE filter cloth stretched tightly across the flat face of a PTFE backing disk. A course PTFE mesh is sandwiched between the cloth and disk. Solvent entering the column goes through an orifice in the disk, then into the small spaces between the mesh filaments. These spaces allow solvent to flow radially outward and uniformly through the entire cross section of the cloth. The bed support presses against the bed top and makes it perfectly flat. In a like manner the outlet bed support collects the separated zones evenly to preserve resolution. The volume of the bed support (between the inlet-outlet tubing and the bed) is extremely small for instance, only 0.005 ml in the ½-in. column. There is, therefore, a minimum of sample dilution and zone mixing.

The cloths have a 40-μ pore size (about 400 mesh) and offer little resistance to gravity flow. They are easily replaced on the backing disk by the extra cloths that are supplied with the column.

Figure 3.16 is a photograph of columns of various sizes and demonstrate the versatility of this column system. However, it should be pointed out that this system has a maximum operating pressure of 500 psi and thus imposes limits on the speed and resolution that can be realized from chromatographic apparatus using such columns. Nevertheless, a high proportion of liquid chromatographic analyses can be successfully carried out with the aid of these column systems. In due course there will, without a doubt, be other manufacturers producing similar column systems and for the beginners in liquid chromatography such systems can minimize practical problems involved in setting up an effective liquid chromatograph.

A wide variety of columns are now available prepacked ready for use, many with guaranteed performance and usually supplied with copies of test chromatograms.

The prepacked columns are constructed to fit almost all commercially available chromatographs and can be obtained in a wide range of sizes and packings. Probably the most convenient size is the 25-cm-long 4-mm-I.D. column which, if packed with 10-μsilica gel, should provide in excess of 5000 theoretical plates and have a loading capacity of about 100 μg. Such columns provide adequate efficiency for most separations and at the same time can be used for semipreparative work if required.

Columns packed with the bonded phases presently manufactured will, however, provide significantly lower efficiencies and the loading capacity of such columns packed with these materials may only be a few micrograms.

Columns 1/8- and 1/16-in. I.D. are also available, the latter being usualy 1 m long and packed with Pellicular type packings; such columns, however, are only useful for analytical separations. The most common packing used in liquid chromatography is silica gel and the partical size normally employed is 5 or 10 μ. The characteristics of the different stationary phases will be discussed later, but it should be stated here that most stationary phases manufactured can be obtained in the form of prepacked columns and include a variety of bonded phases and microparticulate anion and cation exchangers, as well as the more common adsorbents.

7 COLUMN OVENS

The use of temperature as a variable in liquid chromatography has not been extensively investigated. It is, therefore, difficult to lay down specifications for column thermostats with any degree of certainty. Liquid chromatography separations have been carried out at temperatures ranging from −50°C to +250°C. Low-temperature applications of the technique are not common, but columns employed for certain gel filtration separations are sometimes thermostated at 100°C or above. Thus the oven should have an upper temperature limit of 250°C and for very sophisticated equipment, facilties for refrigeration to −50°C might be included. The accuracy of the temperature control should be with ±0.5°C, but again this accuracy is an assessment and a more accurate figure awaits the results of further work to determine the relative importance of temperature control in liquid chromatography. There are already indications that in some work temperature programing could be advantageous so the column oven might be fitted with a temperature programer that provides program rates from 0.1 to 5°C per minute. It should be pointed out that these oven conditions are to some extent extreme and most work in liquid chromatography today is carried out at room temperature with unthermostatted columns. There are basically two types of ovens, the air bath and the liquid bath. Owing to the high heat capacity of a liquid chromatogtaphic column relative to that of the gas chromatographic column, the thermostatting medium needs to have a high heat capacity itself. This is particularly so if the column is to be temperature programed or is greater than 3 mm is diameter. For this reason, it is suggested that liquid thermostatting baths are the best form of temperature control for liquid chromatography. This is suggested in the face of the fact that most commercial liquid chromatographs have air ovens. Another disadvantage of air ovens is the likeihood of explosion. Should any of the joints of the column system leak at high pressures, the oven can easily fill with an explosive mixture of solvent, vapor, and air and become ignited by the heater system. If the oven has any reasonable capacity this could result in extensive damage both to person and property. Air ovens should, therefore, be operated in an inert atmosphere or be fitted with vapor-sensing devices to provide warning when vapor is present. In the design of the oven sufficient space should be made available to carry dual columns, preparative scale columns, and also precolumns. Column ovens should only be considered as a useful adjunct to the chromatograph when the specific separations demand it. Operating liquid chromatographic columns at temperatures other than ambient is the extraordinary rather than the general case.

8 COMUMN DETECTOR CONNECTIONS

The subject of column detector connections has been discussed in detail in Chapter II and equations given that allow the calculation of the dimensions of any column detector connections such that band dispersion is kept to a minimum. There is little further to be said on this subject except that where calculation of the dimensions is deemed unnecessary, elimination of band dispersion can be achieved by crimping the connecting tube. Providing the laminar form of the flow patterns of the liquid in the tube is broken, this will result in convective mixing and reduce band dispersion. The laminar flow through a tube is easily destroyed by interfering with its regular geometry, and this can be achieved by crushing the tube almost flat every 2 mm along its length. With such a crimped tube, band dispersion is negligible. However, such a system suffers from the disadvantage that it is easily blocked by column bleed of support material, and for this reason a cylindrical tube of the correct dimensions is to be perferred to the crimping procedure.

9 LIQUID CHROMATOGRAPHY DETECTORS: CLASSIFICATION AND SPECIFICATIONS

The detector is a device that monitors the concentration of the solute in the mobile phase leaving the end of the column. Normally this monitoring process is continuous and, by using suitable electronic equipment, the signal from the detector is made to register on a potentiometric recorder, the output being a voltage/time analog of the elution curve. The efficient use of a liquid chromatographic column depends on the availability of versatile, high-sensitivity detectors. The problem of detection in liquid chromatography is far greater than in gas chromatography because low concentrations of solute in a liquid do not modify the overall physical characteristics of the liquid (e.g., density and dielectric constant) to the same extent that low concentrations of vapor in a gas modify the characteristics of the gas. With the advent of gas chromatography, considerable effort was put into the development of vapor detectors, and over a period of four or five years a large number of detecting devices were described, many of them very effective. In liquid chromatography, however, there are far fewer detectors available to choose from and even the best have severe limitations.

Broadly, there are two types of detector; one based on the measurement of a bulk property of the eluting liquid, for example refractive index. Detectors of the other type, which measure a physical property characteristic only of the solute, for example UV absorption detectors, are called solute property detectors.

Ideally a liquid chromatography detector should have the same characteristics as a gas chromatography detecor with a sensitivity of about 10^{-12}-10^{-11} g/ml and a linear dynamic range of six orders, characterized in gas chromatography by the flame ionization detector. The liquid detector should also be completely versatile and detect all the solutes while at the same time being independent of the characteristics of the mobile phase. This permits changes to be made in the composition of the mobile phase during development, for example, by gradient elution. It is unfortuante that no liquid chromatography detector so far devised nearly approaches the above specification. If the present liquid chromatography detectors are taken together, however, they do embrace most of the important characteristics of liquid detectors. For this reason present work in the field of liquid chromatography often requires a number of detectors of different types to be available, so that the appropriate one can be chosen for particular applications.

10 CALIBRATION OF DETECTORS: MEASUREMENT OF SENSITIVITY AND LINEARITY

There are two methods for determining the response of the detector. The first one, the incremental injection method, involves injection of known quantities of material onto a column, measuring the peaks, and calculating the sensitivity from the flow rate, peak width, and peak height. In a similar way the linearity of the detector can be measured by injecting a range of samples and plotting peak area against mass of sample injected. The second method it to use a logarithmic dilution procedure, where a known concentration of solute is diluted exponentially with time and the detector is made to plot the dilution curve directly on a potentiometric recorder. This system allows a very wide range of concentration to be examined and, because it is a dynamic system, gives the most precise results. Both systems for the examination of liquid detectors will be described.

The Incremental Method

This method employs the detector, together with its ancillary electronic apparatus and recorder, connected to a suitable column system. The column need not be packed with a stationary phase but may be conveniently packed with glass beads so that sufficient band spread is produced to allow a measureable peak to be eluted. Depending on the absolute sensitivity of the detector, suitalbe size samples of the chosen solute dissolved in the mobile phase are injected onto the column and the peak recorded. Duplicate injections should be made for each solute concentration and the sample size for successive calibrations should be increased by a factor of three until the sensitivity range of the detector is covered. The width of each peak is measured and, from the recorder chart speed and the mobile phase flow rate, the peak width in milliliters of mobile phase is calculated. The concentration at the peak maximum can be taken as twice the average concentration, and so the maximum concentration of the peak can be calculated from the following equation:

$$X_m = \frac{mC}{xQ}$$

where X_m is the concentration of solute in the mobile phase at the peak maximum in grams per milliliter, m is the mass of solute injected onto the column in grams, x is the peak width in cm at 0.6065 of the peak height (which equals one-half the base width), C is the chart speed in centimeters per minute, and Q is the flow rate in milliliters per minute.

This calculation is carried out for all peaks and a graph of solute concentration against recorded deflection in millivolts is constructed. The response factor of the detector is taken as numerically equal to the slope of the curve, excluding any nonlinear part of the curve, and the response is quoted in mV/g/ml.

The noise level of the detecting system should then be determined by increasing the amplifier sensitivity until the noise becomes approximately 10% of the full-scale deflection of the recorder. The noise must include both long- and short-term instabilities of the baseline and should be measured in millivolts. The minimum detectable concentration is taken as that concentration that would produce a signal equivalent to twice the noise level. Then the sensitivity is given by

$$\Delta x = \frac{2P}{D}$$

where Δx is the minimum detectable concentration (g/ml), P is the noise level in (mV), and D is the detector response, [mV/g/ml)]. This method of measuring and determining the performance of the liquid detector is satisfactory providing there is some concentration range over which it has a linear response. However, linearity is not always obtained from detectors, and a method is needed that will permit the extent of the nonlinearity to be determined and an equation given that provides a numerical value that can be directly associated with linearity. A method for obtaining a quantitative estimation of detector linearity was described by Fowliss and Scott [9], who assumed that the response of the detector to solute concentration could be expressed by the following equation:

$$D = Ac^n$$

where D is the signal produced by the detector, c is the concentration of solute in the eluent, n is the response index, and A is a constant.

If the detector has a linear response then $n = 1$ and thus a measure of the linearity of the detector is given by the proximity of n to unity. Figure 3.17

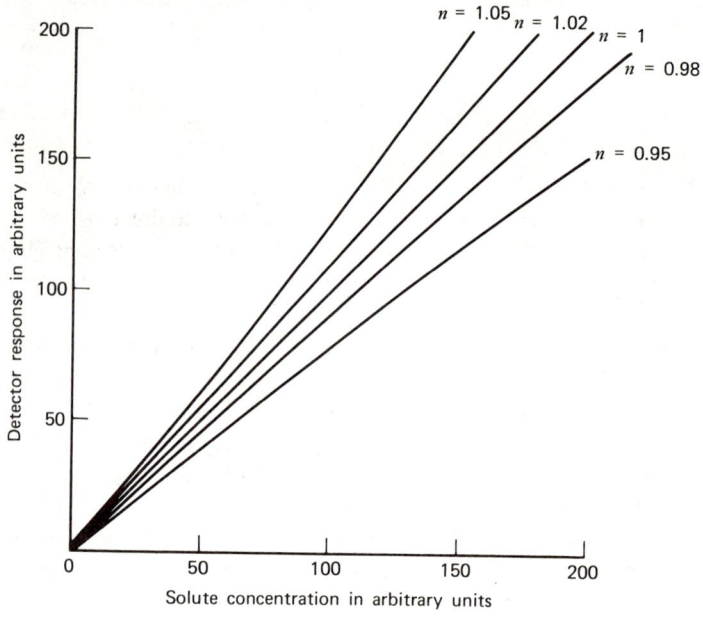

Fig. 3.17. Detector response curves for different response indices.

shows curves relating detector output against solute concentration for different values of n. It is seen that all the curves approximate closely to a straight line but the errors involved in assuming that they are linear are shown in Table 3.3. The results obtained from the analysis of a binary mixture of 5 parts of one compound and 95 parts of another are given for each value of n used in Fig.

Table 3.3 Results of the Analysis of a 5-95% Binary Mixture of Solutes Using Detectors Having Different Response Indices

Substance	% ^{W}/w Composition of Mixture				
	$n = 0.95$	$n = 0.98$	$n = 1.00$	$n = 1.02$	$n = 1.05$
1	5.75	5.51	5.00	4.73	4.34
2	94.25	94.49	95.00	95.27	95.65

3.17. It is seen from Table 3.3 that errors in the lower-level component can be as much as 15% for $n = 0.95$ and 13% for $n = 1.05$. For accurate analysis n must be between 0.98 and 1.02 if linearity is to be assumed. However, it should be pointed out that if n is known then a correction can be applied and thus it is possible to take into account any nonlinearity that may exist and permit accurate analysis.

Logarithmic Dilution Method

The logarithmic dilution method of detector calibration [10] provides a continuous flow of solvent containing solute, the concentration of which decreases logarithmically with time.

A known mass of solute is introduced into a well-stirred reservoir of volume v through which a flow of pure solvent continually passes. The resulting solution is thus continually diluted and the concentration of the solute in the exit flow from the reservoir is monitored by the detector under examination. A diagram of the dilution system is shown in Fig. 3.18. Let the vessel have a volume of V

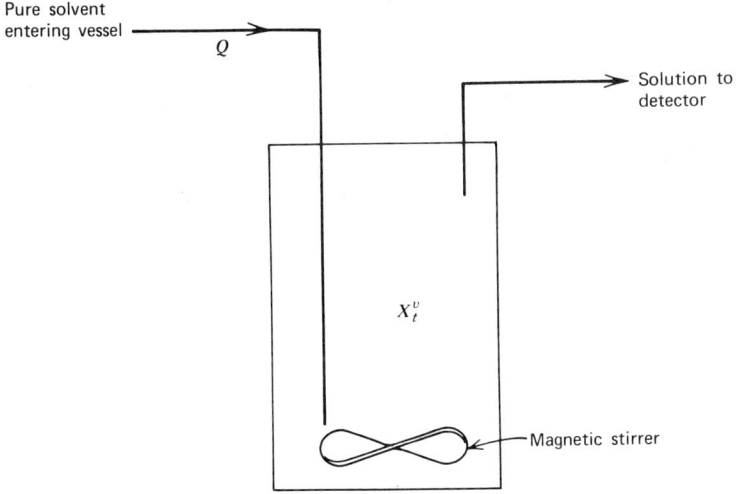

Fig. 3.18. The logarithmic dilution vessel.

and let the concentration of solute in the vessel be X_t after time t. Let the volume dv of pure solvent enter the vessel displacing a volume dv from the vessel. Now the mass of solute removed is $dm = X_t dv$. The removal of mass dm will result in a change in concentration of dX_t. Thus

$$V\, dX_t + X_t dv = 0 \qquad (3.7)$$

and

$$\frac{dX_t}{X_t} = \frac{-dv}{V}$$

or

$$\ln X_t = \frac{-v}{V} + K.$$

Now

$$v = Qt.$$

Thus

$$\ln X_t = \frac{-Qt}{V} + K. \tag{3.8}$$

When $t \to 0$, $X_t = X_0$ where X_0 is the initial concentration of solute in the vessel. Thus

$$K = \ln X_0.$$

Hence (3.8) becomes

$$\ln X_t = \frac{-Qt}{V} + \ln X_0$$

or

$$X_t = X_0 e^{\frac{-Qt}{V}}. \tag{3.9}$$

Thus a graph drawn of $\ln X_t$ against t will have a slope of $\frac{-Q}{V}$ if the detector has a linear response.

If the actual slope is $-\phi$ then n, the response index, is given by

$$n = \frac{\phi V}{Q} \tag{3.10}$$

Using (3.9) and (3.10) it can be seen that the logarithmic dilution method can provide the same information as the incremental method of calibration. However, the logarithmic dilution method, although requiring special apparatus to be set up, gives the greater precision expected from a dynamic calibration method.

11 BULK PROPERTY DETECTORS

The first group, the bulk property detectors, can be shown to have a very limited scope in sensitivity. Consider a bulk property detector that monitors the density of the eluent leaving the column and let it detect a concentration of a dense material, such as carbon tetrachloride (S.G. 1.595), at a level of 1 μg/ml in heptane (S.G. 0.684). This situation will be most favorable for such a detector and the change in density of the eluent (Δd) due to the carbon tetrachloride will be given by the following equation

$$\Delta d = (1.595 - 0.684) \times 10^{-6} = 9.11 \times 10^{-7}.$$

Now the coefficient of cubical expansion of heptane is approximately

$$1.6 \times 10^{-3} \text{ per } ^\circ C.$$

Thus the temperature change ($\Delta \theta$) that would give a signal from the detector equivalent to that of carbon tetrachloride at a concentration of 1 μg/ml is given by

$$\Delta \theta = \frac{9.11 \times 10^{-7}}{1.6 \times 10^{-3}} \ ^\circ C = 5.7 \times 10^{-4} \ ^\circ C.$$

It follows that in order to detect carbon tetrachloride at a concentration of 1 μg/ml in heptane, the temperature control of the column eluent would have to be about $\pm 10^{-4}$ °C. A similar argument can be applied to detectors measuring dielectric constant, refractive index, and so on, and such temperature control is extremely difficult to maintain. Even heat changes resulting from the desorption of the solute from the stationary phase as it leaves the end of the column can result in changes of eluent temperature that are significantly greater than 10^{-4} °C. It follows that the maximum sensitivity of bulk property detectors is severly limited. The most common bulk properties detector in practical use is the refractive index detector.

The Refractive Index Detector

The originators of refractive index measurement for monitoring chromatographic eluents were Tiselius and Claesson [11]; since their work many papers have been published describing different methods of refractive index measurement for this purpose.

The traditional techniques used for measuring refractive index are the *angle of deviation* method and the *critical angle* method. In the former approach, the liquid is passed through a hollow prism and the angle of deviation of a monochromatic beam is measured and related to the refractive index. In the critical angle method, the angle at which radiation incident on the sample surface is totally reflected is determined and related to refractive index. These methods are accurate and reliable and form the basis of many commercially available refractive index detectors. An alternative approach by Conlon [12] utilizes the relationship between reflectance from an interface between two transparent media and their refractive indices as given by the Fresnel equation:

$$R = \frac{1}{2}\left[\frac{\sin^2(i-r)}{\sin^2(i+r)} + \frac{\tan^2(i-r)}{\tan^2(i+r)}\right]$$

where R is the ratio of reflected light intensity to incident light intensity, i is the angle of incidence, and r is the angle of refraction. Now $\frac{\sin i}{\sin r} = \frac{n_1}{n_2}$ where n_1 is the refractive index of medium 1 and n_2 is the refractive index of medium 2. Thus, if medium 2 represents the liquid eluted from the column, then any change in n_2 will result in a change in R and thus a measurement of R could determine changes in the value of n_2 due to solute being present in the eluent.

The measuring cell devised by Conlon is shown in Fig. 3.19a, which consists of a rod prism sealed into a tube through which the column eluent flows. The rod prism was made from a 6.8-mm glass rod, 8-10 cm long, bent to the correct optical angle (4) and an optically clear flat ground on the apex of the bend as shown in Fig. 3.19a. The optical flat was then sealed into a window of a suitable tube that acted as a flow-through cell. The photocell is placed in the arm of a Wheatstone bridge and the out-of-balance signal fed to a suitable recorder. Another photocell can be placed so as to receive light directly from the light source, and if this is situated in the reference arm of the Wheatstone bridge it can help to compensate for slight variations in the light source due to voltage fluctuations. A example of a chromatogram recorded from this detector is given in Fig. 3.19b and shows peaks for a series of nonionic surfactants eluted from a silica gel column. The author of the original paper does not state the mass of eluent in each peak, but it probably was about 50-100 mg. The disadvantages of this system are that it does not contain a reference cell to compensate for

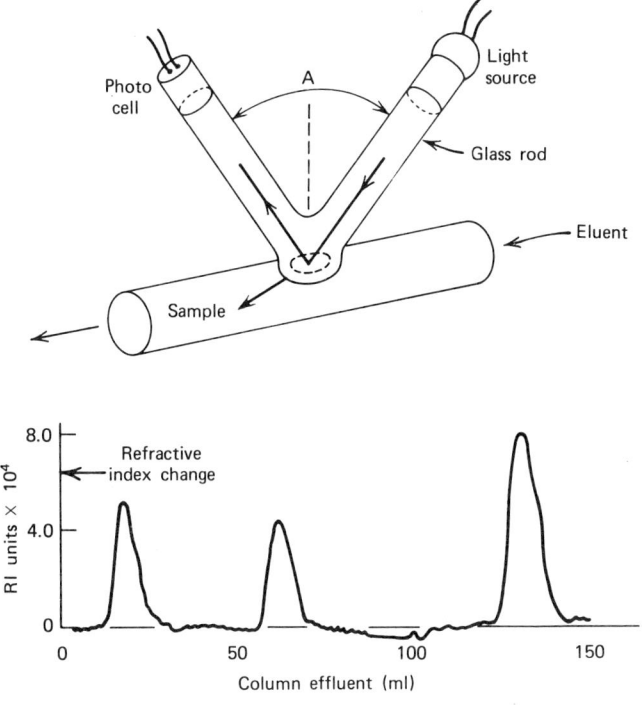

Fig. 3.19. (a) Rod prism attached to flow-through cell. (b) Chromatogram of nonionic surfactants from a silica gel column using the Fresnel type refractometer.

temperature changes or changes in solvent composition and, furthermore, the relatively large cell volume would be detrimental to the resolution of the substances chromatographed.

The limitation of all detecting devices is the so-called noise of the system. The sensitivity of any detecting system is at the maximum when the signal-to-noise ratio is 2 and thus, to improve a given detector, it is necessary to reduce the inherent noise.

Vandenheuvel and Sipos [13] improved the sensitivity of the refractive index detector by taking the model described by Zaukelies and Frost [14] and redesigning it to reduce the noise level. The general construction of the detector was strengthened to reduce the effect of mechanical vibration and the light source was operated from a 5-V supply instead of the normal 6-8 V so that the tungsten filament operated as a self-current-regulating device. Further improvement was achieved by using a stabilized voltage supply. To reduce thermal noise, the cells were thermostated to ± 0.002°C using a precision thermostat with a high thermal capacity and high-quality thermally stable components were used in all electronic circuits. The cell volumes were reduced to 50 μl and before entering the cell the eluent was passed through metal coils surrounded by the thermostating fluid to ensure thermal equilibrium. An example of a chromatogram obtained using this detector is shown in Fig. 3.20. It is seen that a very

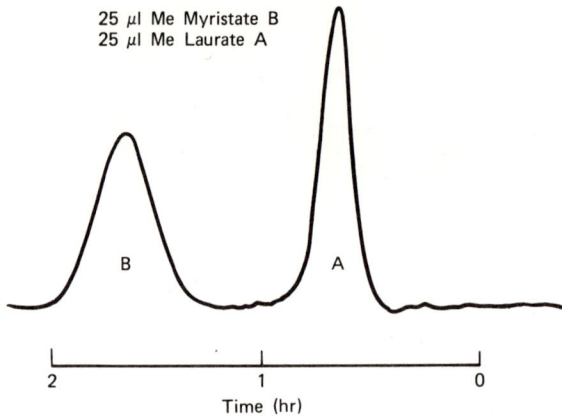

Fig. 3.20. Chromatogram obtained from reversed-phase partition chromatography of mixture of methyl laurate and methyl myristate showing peak deflection corresponding to less than 5×10^{-5} Δn for these major components using the improved refractometer detector.

stable base line is realized, but each of the peaks represents about 20 mg of sample, which indicates a relatively low sensitivity.

An effective commercial refractive index detector is that manufactured by Waters Associates Incorporated. A diagram of the basic detection system is shown in Fig. 3.21. A narrow beam of light, derived from a lamp and slit mask, passes through a collimating lens and the resulting parallel beam enters the twin cells. The cells have a oblique dividing wall through which the light passes undeflected providing both cells contain liquids having the same refractive index. The parallel beam then strikes a plane mirror, A, set at a slight angle to the normal, where it is reflected back through the cells and collimating lens producing a convergent beam that has been deflected by the mirror from the original incident beam path. The now convergent beam is again rendered parallel by a second collimating lens and the lower half of the beam passes to the photocell. C. The upper half of the beam is intercepted by a plane mirror, B, that deflects it through a right angle to a second photocell D. When both cells contain liquid of the same refractive index, the system is adjusted so that the ouputs from the photocells are equal. When solute enters the sensing cell it changes the refractive index of the medium relative to the reference cell and causes the beam to be deflected so that unequal amounts

11 BULK PROPERTY DETECTORS

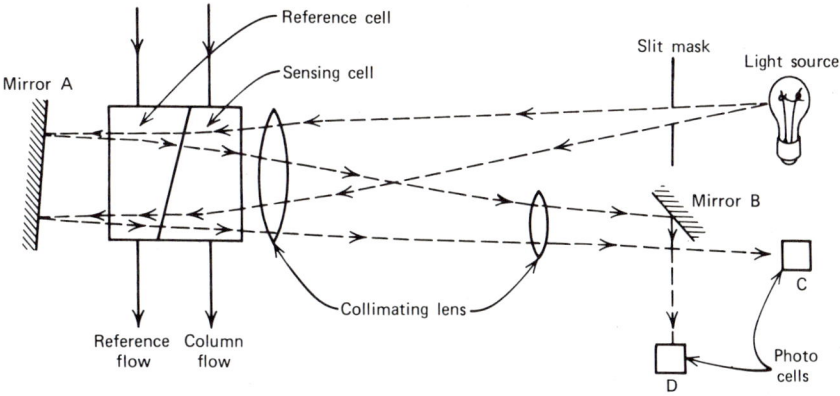

Fig. 3.21. Commercial refractive index detector for liquid chromatography.

of light fall on the two photocells. The differential output from the two cells is then fed to a potentiometric recorder. The volumes of the cells are 70μl, which, providing large diameter columns are being employed, will not severely affect column resolution. A chromatogram of a lipid separation obtained from the detector is shown in Fig. 3.22. The noise level is clearly

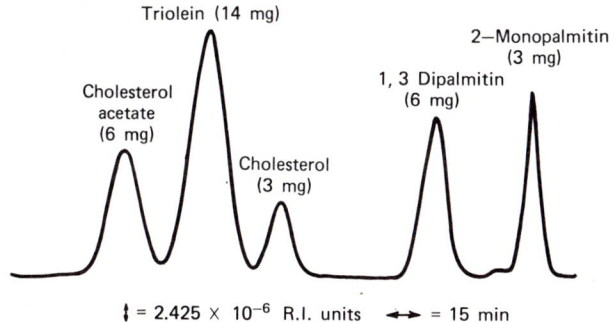

$\updownarrow$ = 2.425 × 10^{-6} R.I. units $\leftrightarrow$ = 15 min

Fig. 3.22. B-31 Factice 1447 × 10 mm. Solvent: 2% HO/acetone (v:v). Refractometric recording of a lipid class separation by factice chromatography, proceeding from right to left. The small vertical scalar representing approximately 2.5 units in the 6th decimal place of refractive index change indicating the sensitivity of this system of differential refractometry.

seen and the peak masses indicate that about 0.25-0.5 mg of solute can be easily detected. The cells are well thermostated but if temperature programming or gradient elution is employed the baseline tends to drift and the "noise level" increases so that the maximum sensitivity cannot be utilized.

Generally, refractive index detectors are best employed when the conditions of chromatographic development are constant and where individual components of the mixture are present in quantities greater than 0.1 mg. They are not recommended for use where gradient elution, temperature programming, or flow programming is to be used. Although with calibration the detector can be used for quantiative analysis, the refractive index method of detection does not give a wide linear dynamic range of response.

The Dielectric Constant Detector

The first paper to be published describing a liquid chromatography detector operating on the principle of dielectric constant measurement was that by Troitskii in 1940 [15]. His method, as with all dielectric constant detectors, measures the change in capacity of a condenser between the plates of which flows the mobile phase from the column. In some ways the dielectric constant detecting system is analogous to detection by refractive index measurement. For nonpolar solutes in nonpolar solvents the refractive index n is related to the dielectric constant k by the equation

$$K = n^2.$$

For semipolar solutes in nonpolar solvents this relationship is better replaced by

$$\frac{K-1}{K+2} = \frac{n^2-1}{n^2+2}.$$

However, for polar solutes in semipolar or polar solvents the analogy breaks down and there is no simple relationship between n and k in these circumstances.

The usual way of measuring the change in capacity of the condenser carrying the column eluant is to determine its impedance to a high-frequency alternating current. The condenser is made to form one arm of an A.C. bridge and the "out-of-balance" signal is amplified and rectified, the resulting D.C. output being fed to a suitable recorder. If the condenser plates are in direct physical contact with the mobile phase, and the mobile phase is conducting, there are two components of current passing through the condenser. There is a current that results from the capacity of the condenser which is called the capacity current, and the one that is conducted directly through the solution due to ions is called the resistive current. Now the capacity current will be out of phase with the voltage applied across the condenser while the resistive current will be in phase with the applied voltage. Thus the total out-of-balance current will be partly in phase and partly out of phase with the applied voltage. This situation reduces the bridge sensitivity and makes the initial balancing procedure difficult and tedious. To reduce this effect, the plates of the detecting condenser are often well insulated from the mobile phase, thus eliminating the resistive component of current.

Another method of measuring changes in dielectric constant is to make the condenser containing the dielectric part of an oscillator circuit. For example, if the capacity is connected in parallel with an inductance and placed in the grid circuit of a thermionic valve, provision being made for feedback, the valve will oscillate. The frequency of the oscillator will depend, among other things, on the capacity of the condenser. If the output of the valve is mixed, or heterodyned, with the output of the stable reference oscillator, the difference frequency can be filtered out to a rectifying circuit and the D. C. output fed to a recorder. Solid-state circuitry could be used instead of thermionic valves; with a crystal controlled oscillator, this would significantly imporve the electrical stability of the device. This alternative method of measuring changes in dielectric constant is extremely sensitive. However, it is not the electrical measuring system that determines the overall detector sensitivity, but the limitations inherent in bulk property detectors that have already been discussed.

A typical dielectric constant detector was described by Grant [16] and a diagram of the detector cell is shown in Fig. 3.23a. It consists of a flattened glass bulb containing two thin platinum electrodes, 1 cm^2 in area and 4 mm apart. The plates are immersed in, and in contact with, the mobile phase and are fused to the wall of the bulb to prevent vibration giving rise to detector noise. The cell holder is made of 1-in.-diameter brass tube and is connected to one plate of the condenser and to earth to provide electrical stability. The capacity measuring unit and detector cell derived its power from a stabilized power supply unit, and a relay unit was provided to protect the recorder from overload. A chromatogram of a mixture of amino acids obtained with this detector is shown in Fig. 3.23b. The minimum mass of amino acids that could be detected was about 2 mg and the peaks shown in Fig. 2.23b represent masses ranging between 3 and 5 mg. The volume of the cell was fairly large (2-3 ml) and this would obviously impair the resolution obtained from smaller columns. The author states, however, that the volume of the detector could be significantly reduced.

Johansson and Karrman [17] described a dielectric constant system that used a different type of cell in which the plates were kept out of contact with the mobile phase. As well as improving the performance of the measuring unit in the manner previously described, the isolation of the detector plates prevents the eluted solutes from being decomposed or adsorbed on exposed metal surfaces in the cell. A diagram of the detector cell having a volume of about 2-3 ml is shown in Fig. 3.24a. It consists of a straight glass tube that can be connected to the column by means of a ground glass spherical joint. The plates A and B are semicylindrical in shape, 3 mm radius and 70 mm long. To reduce wall effects the glass tube was chosen to have walls as thin as possible (0.3 mm thick). The

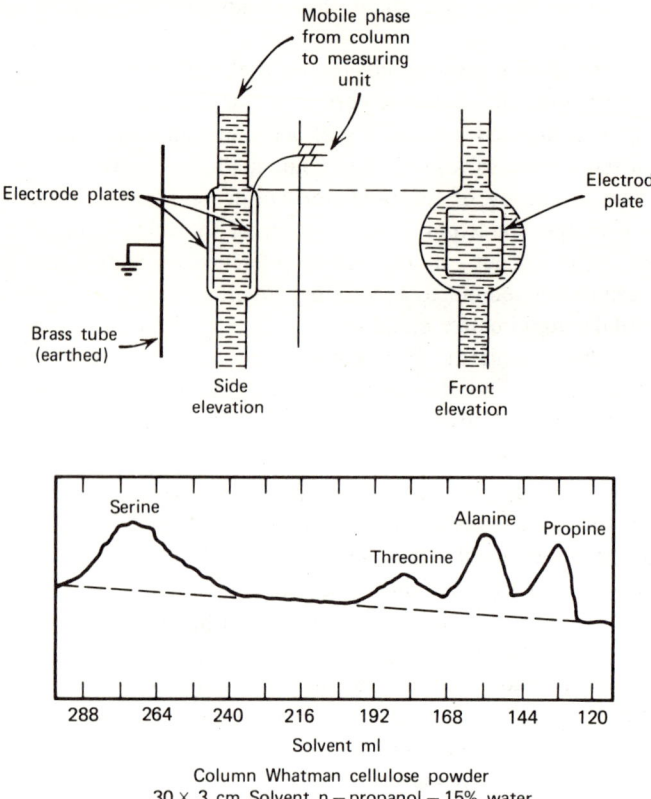

Fig. 3.23. (a) The dielectric constant detector. (b) Chromatogram from dielectric detector.

glass tube and semicylindrical plates were enclosed in a brass tube that was connected to one of the plates and earthed to improve electrical stability. A chromatogram obtained with this detector is shown in Fig. 3.24b and depicts the separation of three organic acids using the reversed phase system of Howard and Martin [18]. It is seen that the sensitivity of the detector is about the same as that realized by Grant with his apparatus. Johansson and Karrman suggest that the response of their detector is a function of both the eluant conductivity and the dielectric constant. This is very difficult to believe because the "resistive component" of the signal current must be negligible due to the liquid being separated from the electrode plates by the glass wall of the tube. Further, as the mobilities of ions in solution are relatively small, the detector must be responding exclusively to changes in dielectric constant. The authors' suggestion that the increased response of the detector is due to conductivity increasing with increased dissociation of the acids is a dubious one, since the

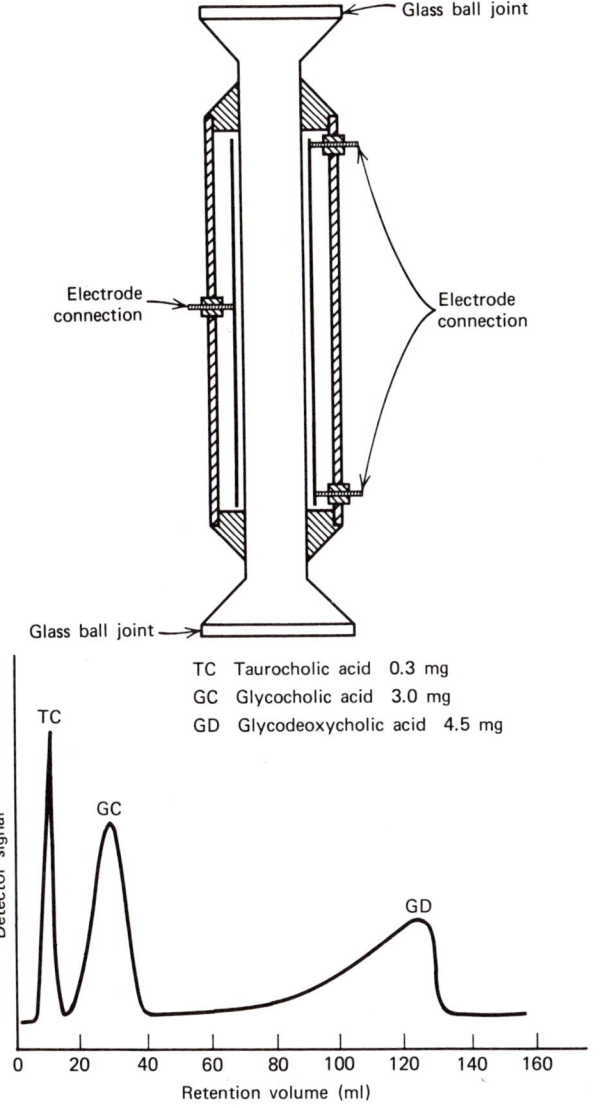

Fig. 3.24. (*a*) Dielectric constant detector. (*b*) Chromatogram of bile acids using the dielectric constant detector.

extent of dissociation of the acid could equally well correlate with an increase dielectric constant. However, irrespective of the true explanation for the detector response, it works effectively and reliably. Again, however, the volume of the cell (4 ml) would be too large for use with small-diameter columns.

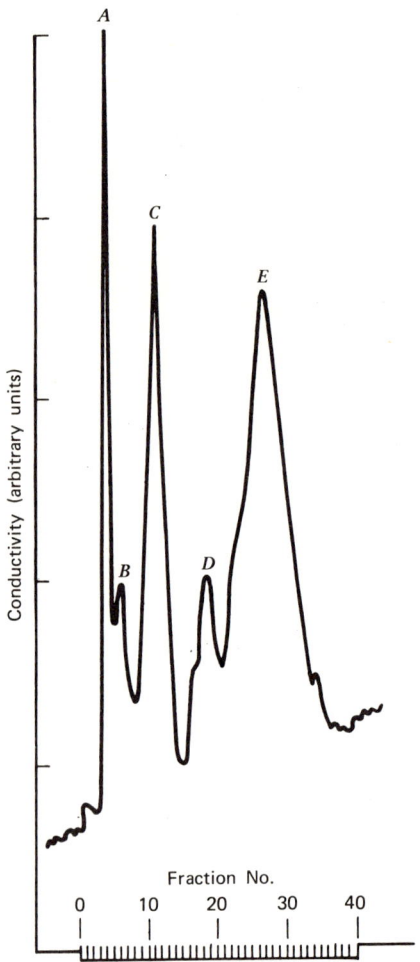

Fig. 3.25. Fractionation of a penicillin mixture. Tracing from conductivity-recorder chart. A, nonpenicillin acids; B, n-heptyl-; C, n-amyl-; D, n-pent-2-enyl-; E, benzylpenicillin.

Generally the same comments apply to dielectric constant detectors as to refractive index detectors. They are best employed where chromatographic conditions are kept constant and they are not suitable for use with gradient elution or temperature programming or where flow programming is used. If carefully calibrated they can be used for quantitative analysis but they do not have a linear response over a significant concentration range.

The Electrical Conductivity Detector

The change in electrical conductivity of the eluant from a chromatographic column can be an extremely sensitive method for monitoring inorganic ions [19] but does not have a linear response. By the use of a suitable measuring circuit, however, it can be made nearly linear over a concentration range of about one order. The method is confined to detecting those substances that ionize, which include organic acids, sulphonates, etc. The first effective conductivity detecting system was that described by Martin and Randall [20] in 1950 and consisted to two horizontal plantinum electrodes sealed into the end of the column and forming one arm of an A.C. bridge. The "out-of-balance" signal from the bridge was amplified, recitified, and fed to a recording milliammeter. A chromatogram of a penicillin mixture obtained using this detector is shown in Fig. 3.25. Since the paper by Martin and Randall, many workers have developed this type of detecting system, designing more sophisticated conductivity cells [21, 22] and more stable and sensitive bridge circuits and amplifiers [23, 24]. This method of detecting has been used, for example in the analysis of polysaccharide hydrolysates [25] and inorganic ions [26].

Conductivity detection is probably the most effective method of monitoring eluted substances that ionize. For such substances it can be very sensitive, if not linear in response, and can tolerate gradient elution providing the mobile phase does not change its conductivity. Providing also the cell is correctly designed, conductivity detectors can, to some extent, also tolerate flow programming and their dependance on temperature stability is not nearly so critical.

Other Bulk Property Detectors

A novel method of detection suggested by Simon, Clerc, and Dohner [27] was based on vapor phase osmometry. A drop of solution and a drop of solvent, when suspended in an atmosphere saturated with the solvent vapor, are in equilibrium with their environment at different temperatures. This temperature difference is due to the supression of the partial pressure of the solvent vapor over the solution relative to the pure solvent. For ideal solutions the temperature difference is linearly related to the molar concentration of the solute in the solvent. The authors suggested an apparatus of the form shown in Fig. 3.26. Pure solvent and column eluant are allowed to continuously drop onto each of two thermistor beads, capped with platinum gauze. The excess solvent and eluant flows over the outside of the thermisters into a reservoir at the base and then overflows to waste. The thermistors are enclosed in a well-thermostated copper cylinder to maintain strict temperature control. Each bead forms one arm of a Wheatstone bridge, and the "out-of-balance" signal is fed to a suitable recorder. The autors demonstrate that the time constant of the system can be made sufficiently small for use in liquid chromatography and give an example of a peak that contains only 32.5 μg of azobenzene, However, this mass was contained in only 0.05 ml of solvent, which represents a solute concentration of about 1 mg/ml. It is obvious that this detection system will not tolerate flow programing of gradient elution, but is has a linear and predictable response,

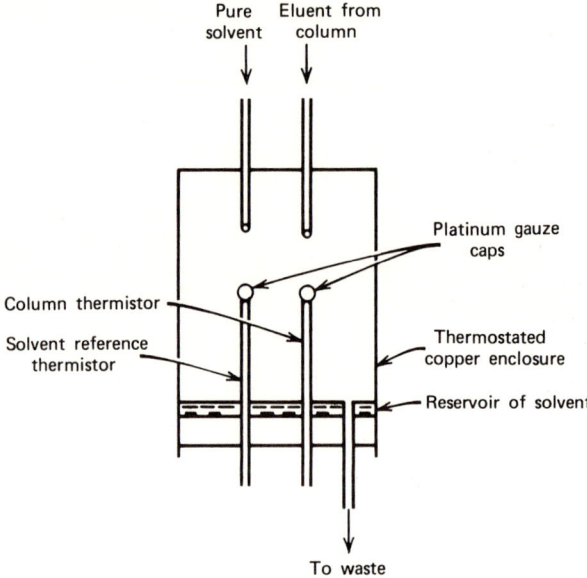

Fig. 3.26. Vapor phase osmometer detector.

Providing the molecular weight of the solutes are known and a reference substance of known mass is included in the mixture, it could be used effectively for quantitative analysis. In its present form it is relatively insensitive as a liquid chromatography detector.

Another bulk property detector, based on the mass detector [35] (to be discussed later) is the density detector developed by Fornstedt and Porath [28]. This detector takes the form of a float hung from a recording balance and suspended in the column eluant. Any change in density of the eluant increases the upthrust on the float and produces a step in the mass record. In order to achieve a reasonable sensitivity, the float must have a volume of about 1 ml and thus the volume of eluant required to suspend the float must also be about 1 ml or more. This represents a large detector volume, which would make it useless for small-diameter columns with efficiencies of more than 200 theoretical plates per foot. However, for large columns separating 100-mg samples under constant column conditions it could be a useful detector.

A further bulk property method of detection, described by Ohzeki et al. [29], uses the change in thermal conductivity and specific heat of the eluant, during the elution of a solute, as a means of detection. The detector cell consists of a thermistor bead, coated with an acrylic resin to protect it from the eluant, situated in a short length of glass tube through which the column eluant flows. The thermistor forms one arm of a Wheatstone bridge and the heat generated by the bridge current in the thermistor is conducted away by the eluant. The equilibrium temperature of the bead, and thus its resistance, is dependent on the specific heat and thermal conductivity of the eluant. When solute is present, producing a change in the thermal properties of the eluant, the equilibrium temperature of the thermistor, and thus its resistance, changes and the "out-of-balance" signal is fed to a recorder. The authors show chromatograms that indicate that 25 μg of suitable substance can easily be detected, but the detector has a poor linear response.

12 SOLUTE PROPERTY DETECTORS

Solute property detectors determine the presence of the solute in the solvent by measuring some physical property of the solute that the solvent does not possess or has to a markedly smaller degree. Alternatively, the solvent is discarded by some suitable means and the residual solute detected free of solvent. The most sensitive and versatile detectors are to be found in this class. Unfortunately, both these attributes are not found in any one detector. However, solute property detectors are the most popular in use and many papers have been published describing their construction including various modifications of established designs. The treatement of solute property detectors in this book will, for obvious reasons, only cover descritions of those detectors with the best performance or those that have particularly novel features.

The UV Absorption Detector

The UV absorption detector consists of a small cell through which the eluant from the column flows. It is fitted with suitable windows through which UV light can pass onto a photoelectric cell. The shape of the cell has taken several forms. The early type of cell was in the form of a Z, the UV light being passed down the center limb, the column eluant passing continually through the cell. However, such cells are sensitive to flow changes and in some designs have been replaced by split flow cells where the eluant entered the center of the cell and flowed in both directions to exits at either end. This significantly reduces the effect of flow on the light absorption and with well-designed cells would permit flow programming. The wavelength of the UV light is chosen such that that the solvent absorbs little or none of the light, whereas the solutes absorbed strongly and are detected by reduced output from the photo cell. In some instruments there are facilities for scanning the cell with several frequencies of UV light, absorption of each frequency being measured and the output fed to a multipoint recorder. This system results in a multiple chromatogram, each curve representing absorption at a specific wavelength. The relationship between transmitted UV light through the cell and solute concentration is given by Beer's law:

$$I_T = I_0 \, e^{-klc}$$

or

$$\ln I_T = \ln I_0 - klc$$

where I_T is the intensity of the transmitted light, I_0 is the intensity of the incident light, l is the path length of the absorption cell, c is the concentration of solute, and k is the extinction coefficient.

If the output of the photocell is linearly related to the intensity of the light falling on it, then a logarithmic amplifier must be used if the signal recorded is to be linearly related to the solute concentration. Such a system was used by Kirkland [30] ; a diagram of his apparatus is shown in Fig. 3.27a. The design is a modified form of the DuPont Double-Beam Photometric Analyser. The light source was a low-pressure mercury tube and filter, transmitting radiation from the 254-mμ mercury line only. The energy passing through the filter is divided into two beams, A and B, by a semitransparent mirror set at 45° to the direction of radiation. Ninety percent of the energy passes through the mirror to form the sample beam A. Ten percent of the energy is reflected off the mirror to form beam B. This beam is reflected off another 45° mirror to form the reference beam C. Beam C passes through a reference cell containing the mobile phase. The sample beam A is focused by a lens onto the micro flow-through cell containg the sample. The radiation transmitted through each cell passes through additional optical filters to minimize stray light and then strikes phototube D or E. The light intensity reaching phototube E is normally constant because it is passed through the solvent reference solution. Light intensity changes reaching phototube D create a voltage output change that is related to the absorbance changes in the sample cell.

The logarithmic amplifier associated with each phototube produces a voltage proportional to the negative logarithm of the phototube current. Thus the difference between voltages produced by two logarithmic amplifiers is directly proportional to the concentration of solute in the sample cell.

The lenses used in the sample beam are made from quartz to allow the transmission of the maximum amount of light energy through the sample. A diagram of a sample cell is shown in Fig. 3.27b. This design is for an old type of detector but serves to illustrate the basic principals very well. The dimensions of the cavities for the two cells were 1.5 mm I.D. by 10 mm and 0.79 mm I.D. by 10 mm. The total volume of these cells, including a 10-cm length of 0.02-in. tubing, were 20 and 7.5 μl respectively. This detector incorporated in the DuPont Liquid Chromatograph has a wavelength range of 380-650 mμ, using a quartz-iodine light source and 254 mμ using a standard low-pressure mercury light source.

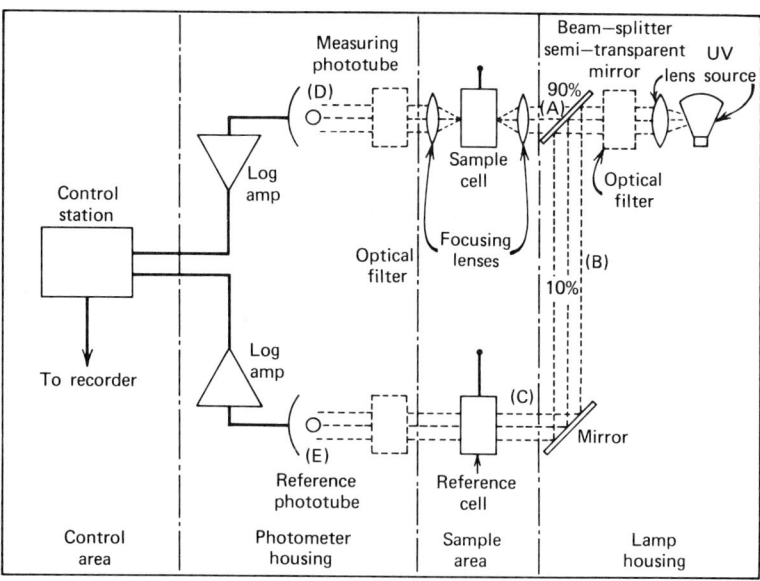

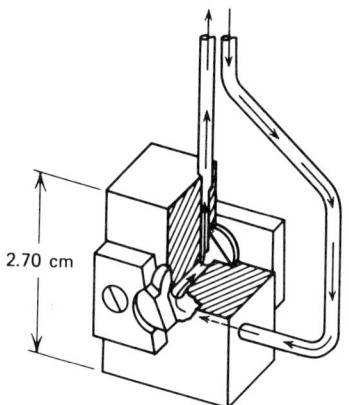

Fig. 3.27. (a) Schematic of the dual-beam UV photometric detector. (b) Micro flow-through sample cell.

A chromatogram of urea herbicides obtained from this detector is shown in Fig. 3.28. The detector has a linear response over a concentration range of about three orders, but depends on the solutes detected having a significant absorption at the light frequency employed. Horvath and Lipsky [31] devised a very effective micro absorption cell having a detector volume of only 4 μl, which they used in the analysis of thyroid extracts and nulceic acid residues.

The UV absorption detector is at present probably the most commonly used

158   LIQUID CHROMATOGRAPHY APPARATUS

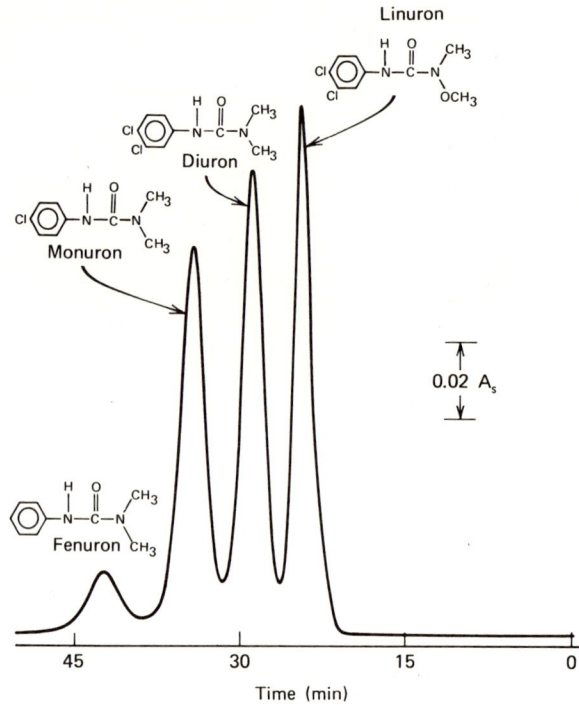

Fig. 3.28. Separation of substituted urea herbicides. Sample, 3 μl of 0.33 mg/ml solution of each; sensitivity, 0.2 absorbance, full scale; carrier, *n*-butyl ether; flow rate, 0.26 cc/min.

detector in liquid chromatography. The reason for this is that it is relatively simple and has a relatively high sensitivity. Another feature of this detector that makes it popular is that it can be made with cells that can fit a standard UV spectrometer and, since these instruments are fairly common in laboratories, it makes such a detector a readily available system. It is without doubt the most sensitive liquid chromatography detector we have available at present. The disadvantages, of course, are that it is specific and furthermore would be sensitive to gradient elution if the materials used in the gradient system absorbed in the UV. It is, however, relatively insensitive to temperature programming and flow programming.

Solute Transport Detectors

Solute property detecting systems in liquid chromatography have to function by measuring a property of the solute that is not common to the solvent. The restriction obviously limits the choice of solute property that can be used for detection and precludes a number of attractive possibilities for high sensitivity detection. The solute transport detectors attempt to overcome this limitation by taking a continuous sample of the column eluant, onto a suitable transport medium, removing the mobile phase by evaporation, and leaving the solute coated on the carrier for subsequent detection. There is now a completely free choice as to which solute property is measured to provide the detecting function. A number of detectors have been devised, based on this principle, and the most successful detection system will be described.

The Moving Wire Type

The moving wire detector [32] functions by allowing the column eluant to pass over a moving wire situated directly under the column exit. This results in a thin film of the column eluant adhering to the wire, which then passes into as evaporator to remove the mobile phase. The wire containing solute only then passes to a pyrolyzer where the involatile solute is pyrolyzed. The pyrolysis products are then swept by a gas stream into an argon or flame ionization detector, where they are detected, and the output from the detector is fed to a suitable recorder. A diagram of the Pye Unicam moving wire F.I.D. detector is shown in Fig. 3.29a.

Basically it consists of three ovens in series, through which a 0.005-in. (0.13-mm) diameter stainless steel wire continuously moves. The glassware in the three ovens is constructred of Supremex alumino-silicate glass and, to prevent excessive waste of argon gas from the oven ends, constrictions have been incorporated at both ends of the glass tubing. In the case of the pyrolysis oven the constrictions also minimize the loss of pyrolysis products, which would result in a loss of detector sensitivity. The maximum recommended temperature of operation of the glass is 730°C. The first oven acts as a cleaning device to remove any contaminants, such as lubricants, from the wire by pyrolysis in a stream of nitrogen. The wire is then coated with the column eluant at the coating block and the solvent removed at the evaporator oven in another stream of nitrogen. The wire, now coated with the compounds under investigation, passes to the pyrolysis oven where the materials are removed from the wire at a temperature of 600-700°C. The pyrolysis products are then conveyed to a gas chromatographic flame ionization detector in a further stream of nitrogen. The temperature of the detector is such that any condensation effects are reduced to a minimum. The detector output is fed into an amplifier and the signal displayed on a potentiometric recorder. An example of a chromatogram of a reaction mixture from the acetonization of sorbose is shown in Fig. 3.30. The detector has a linear response over about two orders (response index 0.98-1.02) and a maximum sensitivity of about 4 μg/ml. If the flame detector is replaced by an argon detector, the nitrogen purge gases have to be replaced by argon. This reduces slightly the linear dynamic range of the system but does not change the sensitivity. The system can be used with all types of chromatographic development; gradient elution, temperature programming, flow programming do not

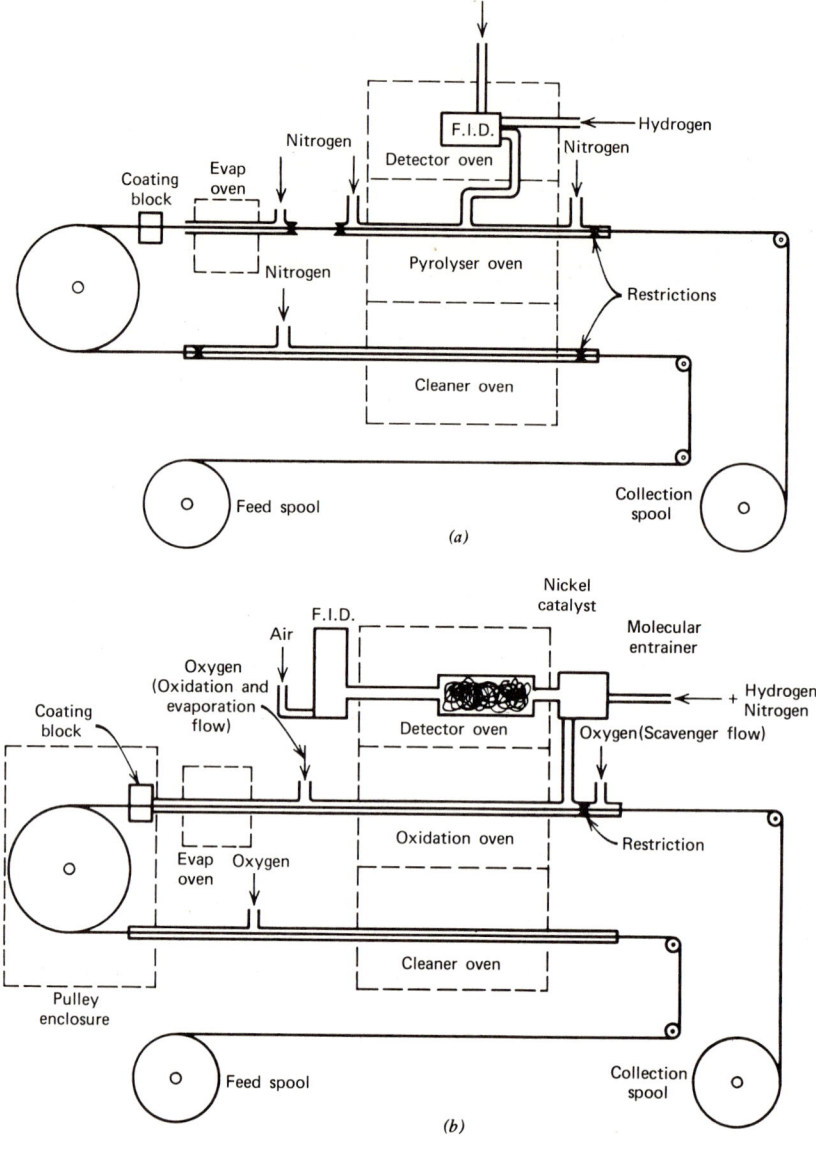

Fig. 3.29. (*a*) Normal wire detector system. (*b*) Modified wire detector system.

affect the detector performance in any way. However, the detector, in the form so far described, is difficult to set up and maintain at its maximum sensitivity.

These problems were overcome by modifying the Pye Unicam Detector in the manner described by Scott and Lawrence [33]. The modified form is shown in

12 SOLUTE PROPERTY DETECTORS

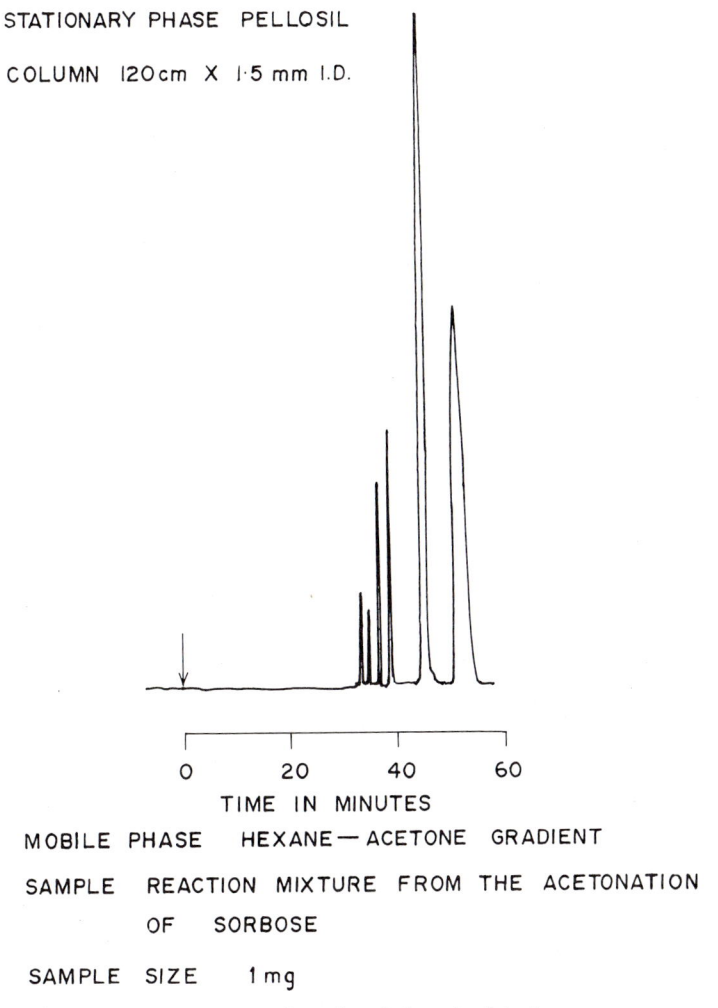

Fig. 3.30. Chromatogram from the moving wire detector.

Fig. 3.29b. The modification depends on the oxidation of the solute deposited on the wire to carbon dioxide and water as opposed to pyrolyzing them to low moelcular weight fragments. The carbon dioxide is then mixed with hydrogen and reduced to methane by a nickel catalyst. The methane is then detected by the F.I.D. detector. The F.I.D. detector was modified by enlarging the hydrogen lines in the detector bodywork to reduce the flow impedance of the detector system and permit the satisfactory operation of the molecular entrainer. The normal nitrogen inlet is connected to a 2-in. length of 3/16-in.-diameter thin-walled stainless steel tube, which is then connected to the molecular entrainer. The 3/16-in.-tube is closed with a loose pad of quartz wool and filled with about 2 g of nickel catalyst. The nickel catalyst is prepared by absorbing a saturated solution of nickel nitrate on 20/40 BS mesh brickdust, decomposing the nitrate to the oxide by heating to 500°C for 3 hr, and reducing the oxide to metallic nickel in a steam of hydrogen at 250°C. The catalyst tube is closed with another loose wad of quartz wool. The moleculatr entrainer is a device developed by W. G. Pye for use in one of their gas chromatographs. Essentially it consists of a jet and a Venturi and is placed in line with the hydrogen flow to the detector. The passage of hydrogen from the jet to the Venturi results in a pressure drop around the hydrogen stream. The reduced pressure side limb of the molecular entrainer is connected to the side limb of the oxidation tube by means of a silicone rubber sleeve. It is seen that the two tube system in the normal wire detector is replaced by a single tube. The oxygen if fed in that the center of the tube providing both the evaporator flow and the oxidation flow. The oxidation tube contains only one restriction at the end of the system. This simplifies the alignment of the wire in the tube and, since the restriction is subsequent to the oxidizing section of the tube, the position of the wire in the restriction is not critical. Oxygen is used for the cleaner flow to oxidize the material absorbed on the wire and clean it prior to coating. All restrictions in the cleaner tube are also removed to eliminate difficulties in wire positioning.

This modified moving wire system has a linear response over three orders of concentration range (response index 0.97-1.04) and a fourfold increase in sensitivity (1.1 μg ml) over the original pyrolysis system. Moreover, the sensitivity is maintained with materials of higher oxygen content such as carbohydrates and peptides, and change in sensitivity being accounted for by change in carbon content. The oxidation system is easy to operate and more reliable than the pyrolyzer system since its transport system is much simplified. Because complete combustion is easily achieved, the detector has a predictable response that can be determined from the carbon content of the solute.

The response factors obtained from this detector for a mixture of water soluble oxygenated materials are shown in Table 3.4. The response factors were obtained by taking the peak area and dividing by the mass of solute injected, which gave the peak area per microgram of each substance. Each value was then corrected for carbon content by dividing by the percentage carbon in the solute. The same response factor should then be given for each compound and the response factors obtained are shown in the last column using glucose as a reference substance and equal to unity. It is seen with the wide range of solute types that the response factors are indeed fairly predictable.

The moving wire detector is the most sophisticated form of the solute transport detectors that are at present manufactured. It is extensively used in Europe

12 SOLUTE PROPERTY DETECTORS 163

Table 3.4 Response of the Detector to Different Substances Using Water as the Solvent

Substance	Conc. of Solute (μg/μl)	% w/w Carbon in Solute	Response Factor (Glucose = 1)
Glucose	262	0.400	1.00
Citric acid	353	0.375	0.98
Sucrose	383	0.421	0.95
Glycine	98.5	0.319	1.04
Proline	95.7	0.458	1.01
Alanine	93.1	0.404	1.03

but to a lesser extent in the United States.

The advantage of this type of detector is that its performance is completely independnent of the method of development and it can tolerate gradient elution, temperature programming, and flow programming.

The Moving Chain Type

An alternative form of solute transport detector is the chain detector developed by Haati et al. [34]. Haati employed a loop of gold chain in the place of the continuous wire. The loop was supported on a series of pullies that caused the loop to rotate continuously. The chain first passed through the column eluant and then to an evaporator to remove the mobile phase. The chain then either passes to a pyrolyzer unit from which the pyrolysis products are swept into a flame ionization detector or the chain passes directly through the flame of the detector itself. In the latter case the detector collects the ions as they are produced in the flame during combustion of the solute and at the same time serves to clean the chain in preparation for subsequent coating.

Because the chain will pick up more eluent than the wire, the response to a given solute will be greater. However, the effective sensitivity, which depends on the signal to noise ratio of the system, is only about the same or less. This is due to local concentrations of solute accumulating at the points of contact of the chain links.

There have been several commercially available detectors of this type developed; a diagram of one once manufactured by Barber-Colman us shown in Fig. 3.31. It is seen that the basic system is far simpler than the moving wire detector and far more compact. A record of a series of peaks from a similar type of detector manufactured by Hitachi is shown in Fig. 3.32. The chromatogram gives an idea of the excessive peak noise produced by the chain system. Some time and effort spent in developing this detector might well result in a considerable improvement in its performance.

Gold chain
Synchronous motor
Air inlet
Evaporation tunnel
Pulley
Gauze jet
Chromatographic column
Flame
Ionization detector

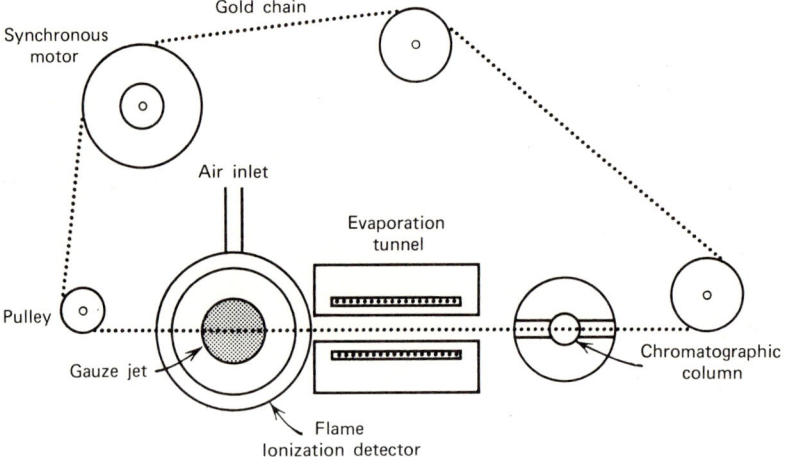

Fig. 3.31. The chain detector.

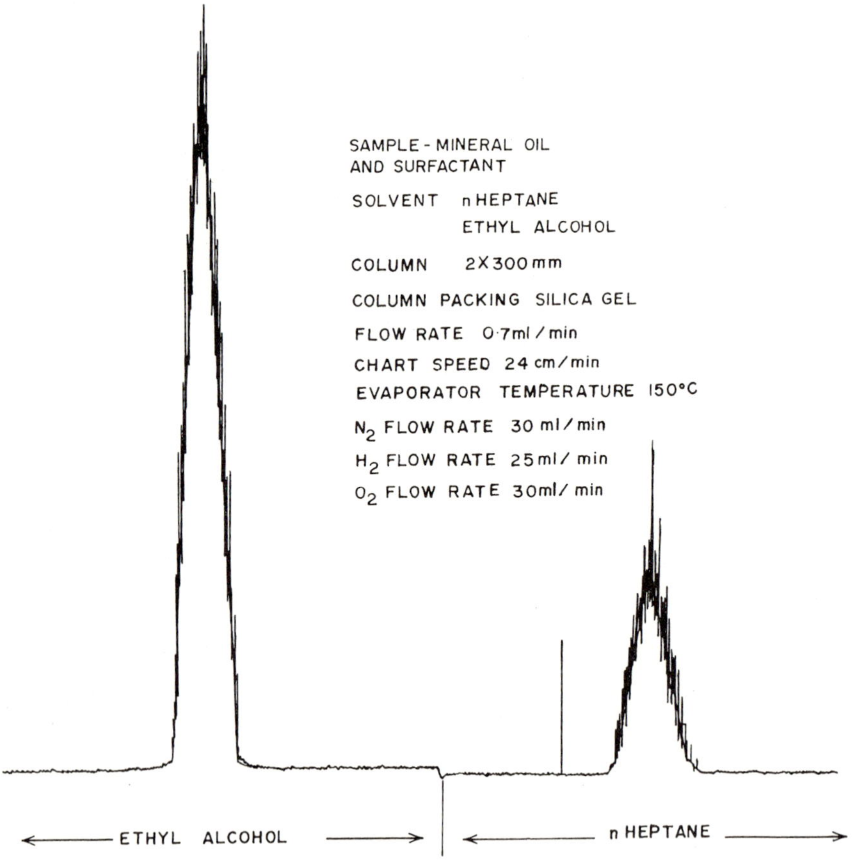

SAMPLE - MINERAL OIL
AND SURFACTANT
SOLVENT n HEPTANE
ETHYL ALCOHOL
COLUMN 2X300 mm
COLUMN PACKING SILICA GEL
FLOW RATE 0·7ml / min
CHART SPEED 24 cm/min
EVAPORATOR TEMPERATURE 150°C
N2 FLOW RATE 30 ml / min
H2 FLOW RATE 25 ml / min
O2 FLOW RATE 30ml / min
ETHYL ALCOHOL
n HEPTANE

Fig. 3.32. Chromatogram of mineral oil and surfactant using the chain detector.

The Mass Detector

This detector [35] functions by allowing the column eluent to drop continuously into a heated container suspended from a recording balance. Providing the temperature of the container is high enough, the mobile phase will flash evaporate leaving the eluted solute as a residue in the container. The trace recorded by the balance will thus be an integram, relating mass of solute eluted, with time. The elution of each component will be represented by a step on the integram. A diagram of the detecting system is shown in Fig. 3.33. The evaporation pot is made from aluminium and weighs about 100 g. It is constructed with a sloping internal base to prevent solute being thrown virtically from the pot during the flash evaporation of the mobile phase. The boiling point of the

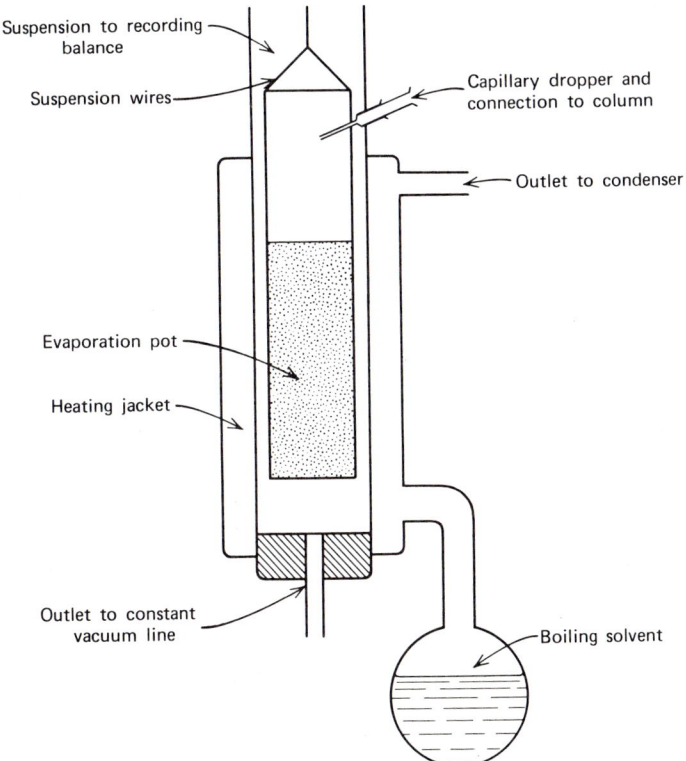

Fig. 3.33. Diagram of evaporation pot and vapor jacket used with the mass detector.

solvent in the vapor jacket is chosen to be 50-100°C above the boiling point of the mobile phase to ensure adequate heat for flash evaporation. A constant-suction vacuum line is connected to the internal tube of the jacket. This satisfies the dual role of withdrawing mobile phase vapor from the system and exerting a constant downward force on the evaporation pot. This constant force increases the apparent weight of the pot, accentuates the inertia of the balance, and increases the stability of the signal obtained.

An example of an integram obtained from this detector is shown in Fig. 3.34.

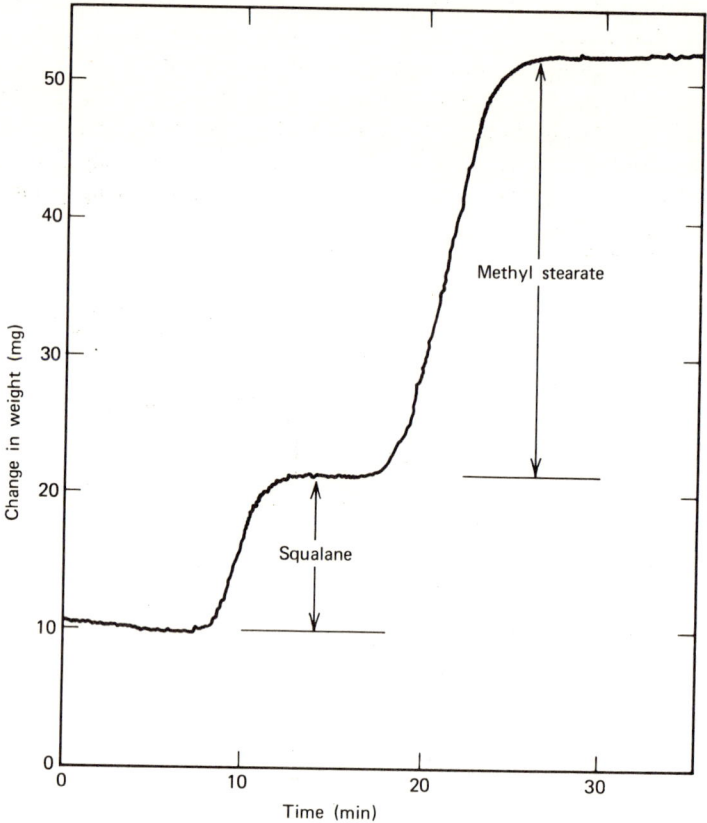

Fig. 3.34. The separation of squalane and methyl stearate on silica gel using the mass detector.

It is seen that it has a relatively low sensitivity but at the same time is a very stable system. The level portions between the "peaks" on the integram are unambiguous and it is obvious that accurate quantitative analysis could be obtained with such a detector providing the solutes are completely involatile. Although this detector has an absolute response, its use is severely limited due to its relative insensitivity.

12 SOLUTE PROPERTY DETECTORS

The Heat of Adsorption Detector

The heat of adsorption detector, devised by Claxton [36], consists of a small plug of absorbent, usually silica gel, through which the chromatographic eluent passes after leaving the column. Embedded in the silica gel or at a point downstream but close to it, is either a thermocouple or a thermistor that continuously measures the temperature of the absorbent or mobile phase. When an eluted solute comes into contact with the silica gel, heat is evolved as the solute is adsorbed, and the temperature rises. As the solute is subsequently desorbed from the silica gel, heat will be adsorbed causing the temperature to fall. The output from the temperature measuring device will thus first record an increase in temperature and then a decrease in temperature relative to its surroundings. Thus if the temperature is continuously recorded, an S-shaped curve will result.

The response of the detector can be determined theoretically in a similar way to the treatment by Scott [37], in his development of an equation that described the change in temperature at a point in a gas chromatographic column as a solute passes down it.

The theoretical response of the detector determined by this method [38] is given by the following equation

$$\theta_v = \psi \; e^{-\phi v} \int_0^v e^{\phi v} \left[X_0 \frac{e^{-v} v^n}{n!} - \frac{X_0 e^{-v/C_\alpha}}{C_\alpha} \int_0^v e^{-v/C_\alpha} \frac{e^{-v} v^n}{n!} \right] dv$$

where θ_v is the temperature of the detecting cell, ψ is a constant, ϕ is a constant representing the heat loss factor, C_α is detector plate capacity-column plate capacity ratio, v is the flow of mobile phase through the detector in plate volumes of the attached column, and n is the efficiency of the attached column.

With the aid of the computer the relative values of θ for $v = 74$ to 160 were calculated for a column having an efficiency of 100 theoretical plates and for C_α taking values of 0.25, 0.5, 1, 2, and 4 and for ϕ taking values of 0.01, 0.05, 0.25, and 1.25, respectively. The 20 curves shown in Fig. 3.35 represent the shape of the θ versus v_α curves and the integral θ versus v_α curves for the different values of C_α and ϕ and are all normalized to the same peak height. The curves shown cover the practical range of heat loss factors for the detector cell and demonstrate the effect of changes in detector capacity-plate capacity ratios that would result from different detection cell designs detecting a peak of constant width. The curves for different values of C_α would also represent the effect on peak shape of solutes of different retention and thus different peak widths passing through a detecting cell of fixed dimension but having an adsorbent different from the stationary phase of the column. Examination of Fig. 3.35 shows that the major effect on peak shape is the ratio of cell capacity to column plate capacity C_α.

168 LIQUID CHROMATOGRAPHY APPARATUS

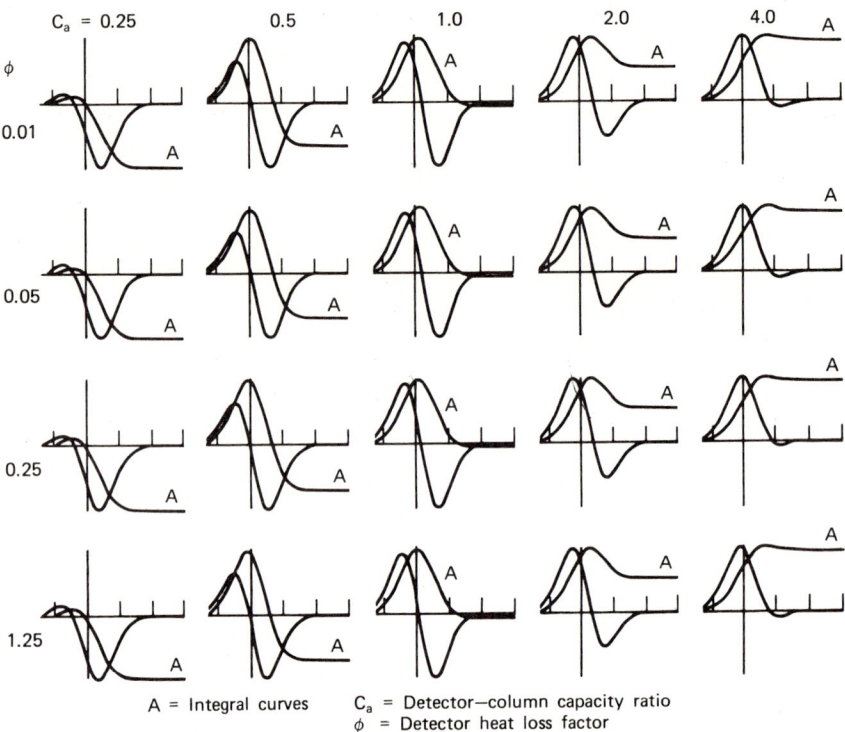

A = Integral curves
C_a = Detector–column capacity ratio
ϕ = Detector heat loss factor

Fig. 3.35. Temperature curves and integral temperature curves from the head of adsorption detector (theoretical).

It is seen that when the capacity of the detector cell is less then the plate capacity of the column ($C_\alpha < 1$) the negative part of the signal is predominant whereas when the detector cell capacity exceeds that of the column ($C_\alpha > 1$) the positive part of the signal predominates. For this reason the integral of the detector signal for $C_\alpha > 1$ rises to a peak but does not return to the baseline and for $C_\alpha < 1$ the integral curve first rises and then falls below the baseline and does not return. Only when $C_\alpha = 1$ does the detector signal simulate the differential form of the Gaussian curve and its integral describe the true elution curve. These conclusions are in general agreement with those of Smuts et al. [39].

The effect of the heat loss factor ϕ on the peak shape is small, but the magnitude of the detector signal varies inversely as ϕ. It should also be noted that for low values of ϕ where the maximum sensitivity is realized the peak maximum is displaced but, for large values of ϕ the maximum of the integral curve for $C_\alpha = 1$ is almost coincident with the maximum of the true elution curve.

It follows that for the detector to be effective and useful, C_α must at all times be unity and thus the detector must have the same plate capacity as the column for all *solutes*. This means that the detector must employ the same adsorbent, the same geometry, and be packed to have the same plate height as the column. It is obvious that to accomplish this, the column must be the detecting cell and the temperature sensing element be placed in the column packing itself.

If the end of the column is used as the detecting cell the temperature response is given by the following equation:

$$\theta_n = \frac{\alpha}{\beta-1} \sum_{r=0}^{r=n} \left(\frac{\gamma-1}{\beta-1}\right)^r \frac{dX}{dv} (n-r)$$

where θ_n is the temperature of the n th plate of the column, α is a constant, γ is the heat loss factor of the cell due to convection, β is the heat loss factor of the cell due to conduction, n is the number of theoretical plates in the column, $X = \frac{X_0 e^{-v} v^n}{n!}$, and v is the volume flow of mobile phase in column plate volume.

It is seen from the equation that α only affects the magnitude of the curve while γ and β affect the shape of the curve and its magnitude. To determine the shape of the theoretical curves describing the change in θ with v it is necessary to determine a practical range of values for β and γ.

The values taken for β and γ are as follows ($z = \beta - \gamma$):

$k' = 4$		$k' = 8$	
γ	1.50	γ	2.70
z	0.01	z	0.01
β	1.51	β	2.71
γ	1.50	γ	2.70
z	0.02	z	0.02
β	1.52	β	2.72
γ	1.50	γ	2.70
z	0.08	z	0.08
β	1.58	β	2.78
γ	1.50	γ	2.70
z	0.08	z	0.08
β	1.58	β	2.78
γ	1.50	γ	2.70
z	0.20	z	0.20
β	1.70	β	2.90

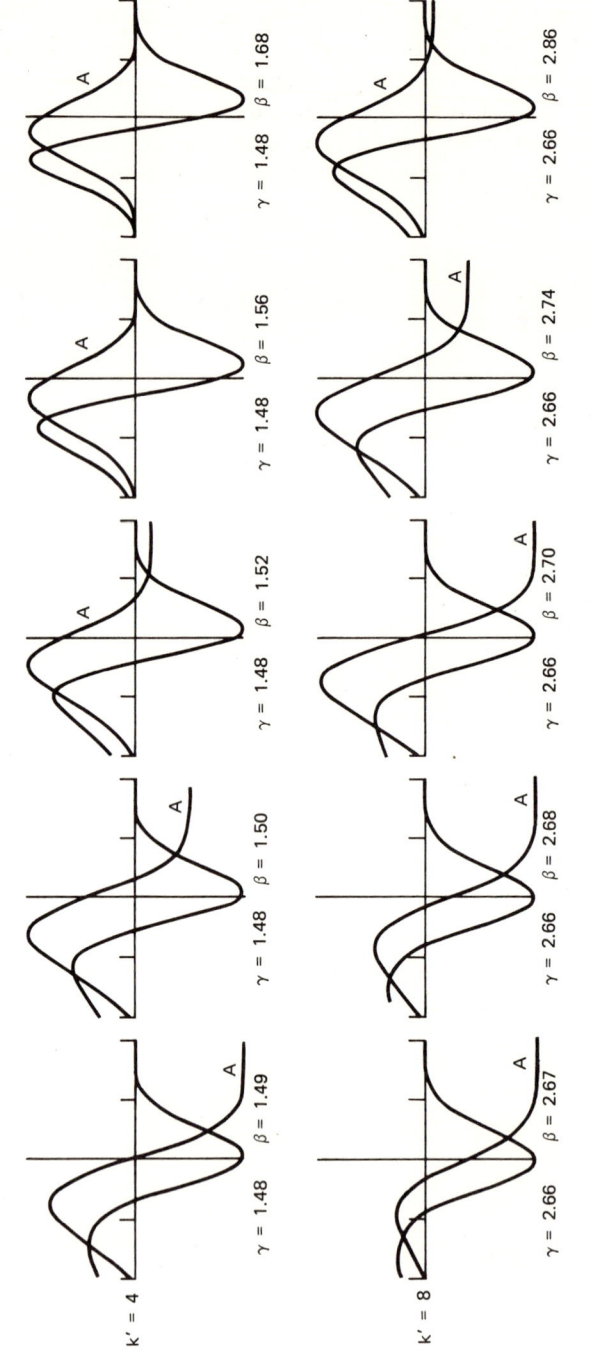

Fig. 3.36. Temperature and integral temperature curves from the heat of adsorption detector with the sensor situated in the column packing (theoretical).

12 SOLUTE PROPERTY DETECTORS

Using the computer the temperature and the integral of the temperature curves were calculated. Values of β were chosen to cover a practical range of conditions used in liquid chromatography based on the approximate value for z of 0.02. The curves obtained are shown in Fig. 3.36; it should be noted that G and B on the figure refer to γ and β, respectively. All curves are normalized to the same peak height.

It is seen from Figure 3.37 that, for a solute eluted at a k' value of 4, the value

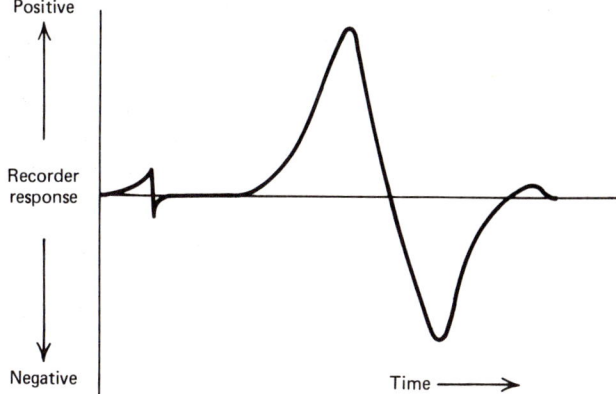

Fig. 3.37. Recorder trace of a peak from the heat of adsorption detector.

of z must be 0.08 if the differential form of the Gaussian curve is to be realized. Further, for a solute eluted at a k' value of 8 the value of z must be increased to 0.2 if the integral of the temperature curve is to take the form of the true elution curve of the solute. These values for z are considerably greater than those calculated for the column. The value of z can be increased by improving the radial transfer of heat across the column. In this treatment, z is only considered to be controlled by the thermal conductivity of the column contents but the radial mass transfer described by Knox [9] would also relate to the transfer of heat. However, even assuming the radial transfer of heat by the Knox principle to be equivalent to the heat transferred by conduction, this would only double the value of z. A practical value of 0.04 for z would only render the in-column detection method satisfactory for solutes eluted up to a k' value of 2 or 3.

From the theoretical treatment the following was concluded.

1. The output from a heat of adsorption detector will not provide the differential form of the elution curve of a solute unless the capacity of the detecting cell (as defined) is identical with the plate capacity of the column for all solutes. This condition can only be achieved in practice if the detector has

the same cross-sectional geometry as the column, contains the same absorbent, and is packed to give exactly the same efficiency (i.e., the same HETP). Thus each detector would have to be constructed to suit each particular column.

2. If the column itself is used as the detector by inserting a sensing element into the packing, then the detecting cell will have the same capacity as the column for all solutes. However, under such circumstances there can be no heat exchange between the last plate of the column and the detecting cell. In order to eliminate the effect of heat generated in the column prior to the detecting cell, which would be convected to the cell and distort its normal temperature profile, there must be adequate radial heat loss from the column. For normal packed columns it appears impossible, in practice, to achieve adequate radial heat loss that would provide a satisfactory signal from the sensing element. At best, an in-column heat of adsorption detector could only operate satisfactorily for solutes eluted at k' values below 2 or 3.

Bearing in mind other limitations of the detector such as its inability to cope with gradient elution and temperature program methods of development and that the adsorbent has to be changed from time to time, the practical difficulties of construction make the heat of adsorption detector not a viable system for detection in liquid chromatography.

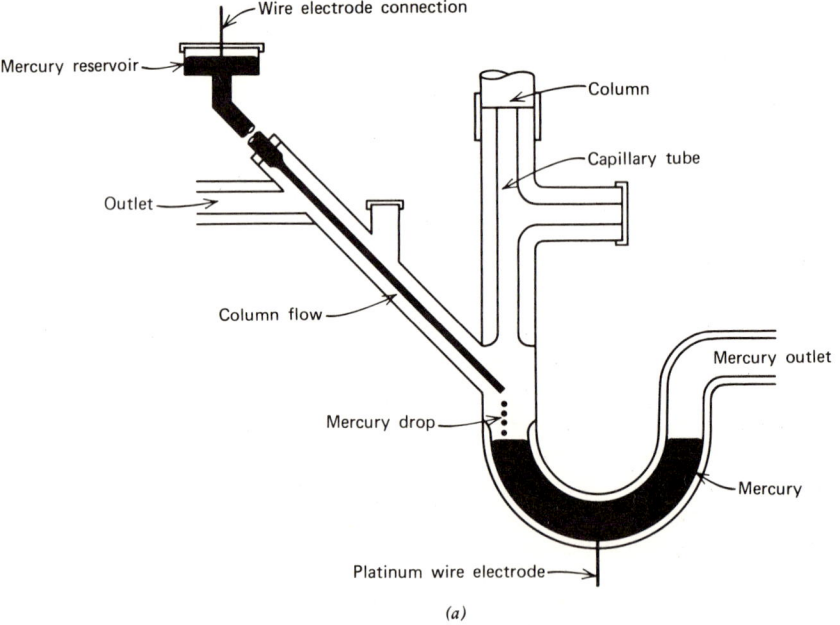

Fig. 3.38. (a) Diagram of the continuous flow dropping mercury cell. (b) The silicone-carbon membrane cell. (c) silicone-carbon membrane cell connected to a chromatographic column.

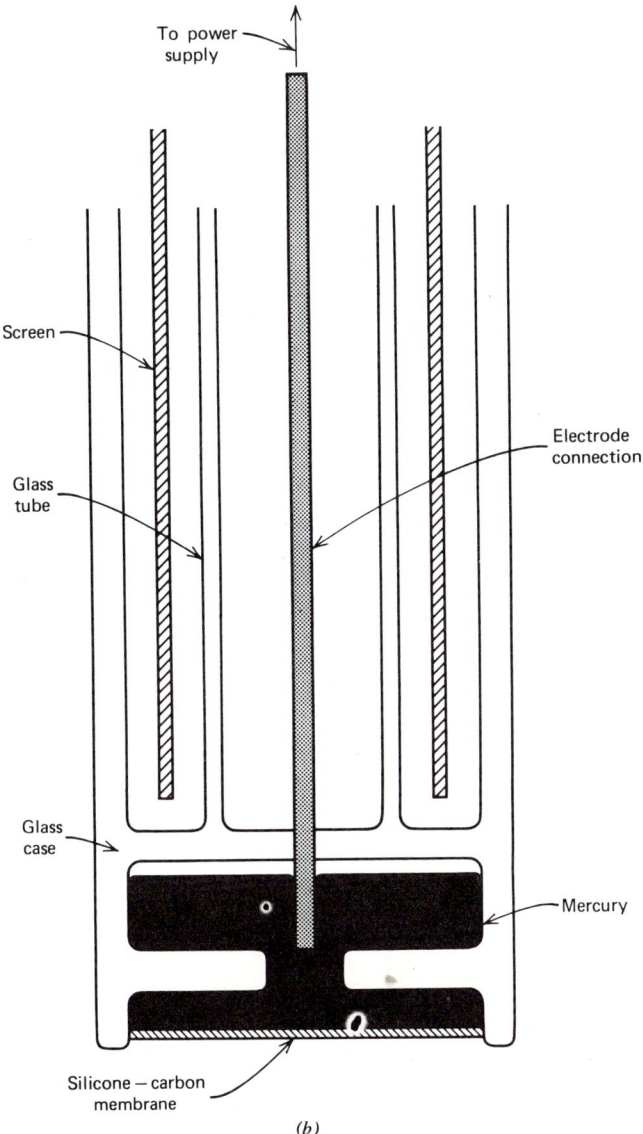

Fig. 3.38. (continued)

The Polarographic Detector

The polargraph was first used as a detecting system in liquid chromatography in 1952 by Kenula [40] and since then a number of workers have investigated the technique as a possible high-sensitivity type of detector; so far, it has only

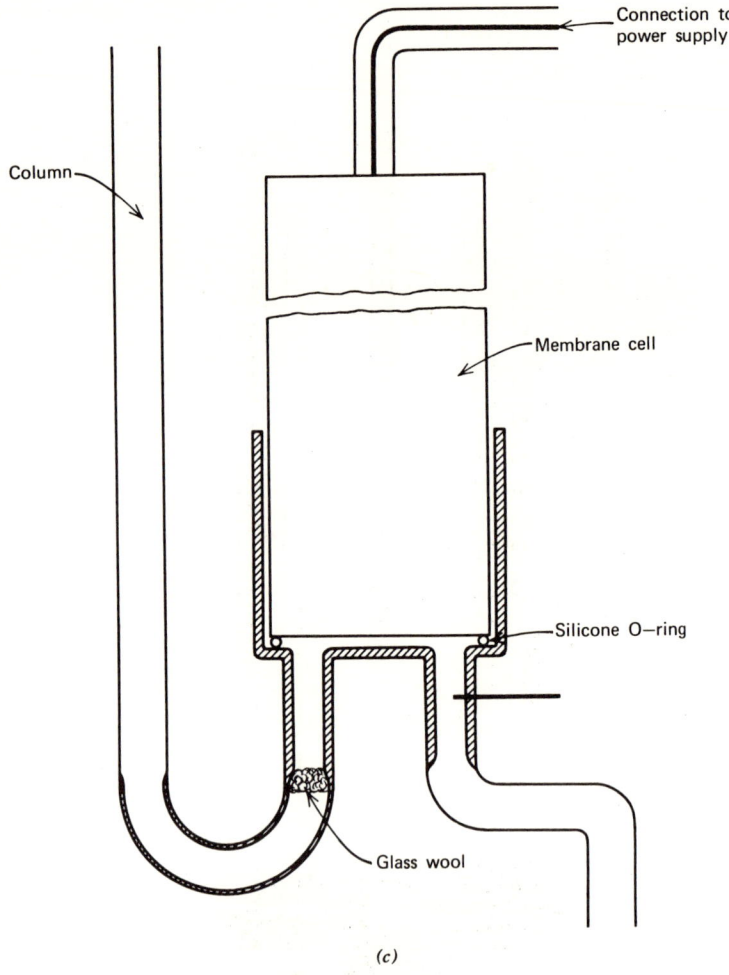

Fig. 3.38. (continued)

been shown to be effective in detecting inorganic ions of substances that exhibit electrochemical characteristics. The properties of the polarographic detector have been extensively investigated by Koen et al. [41] and Joynes and Maggs [42], who examined both the mercury dropping electrode and the silicone-carbon membrane electrode. The apparatus used by Joynes and Maggs is shown in Fig. 38a together with the silicone-carbon electrode, Fig. 38b and the method of column attachment in Fig. 3.38c. The eluant from the column passes over the jet providing the mercury drops and out to waste. The mercury accumulates as a pool at the base of the apparatus and flows out through a separate exit to that for the column eluant. Electrical connections were made to the mercury reservoir and the exit mercury pool. It was found that the dropping electrode suffered from two main disadvantages. First, the noise from the dropping mercury could only be reduced to an acceptable level by heavy electrical damping that would result in serious band broadening if used with efficient chromatographic columns. Second, traces of dissolved oxygen in the column eluant rendered the detector insensitive to the solute it was to detect. The alternative electrode having a carbon impregnated silicone membrane exhibited several advantages over the dropping electrode. Owing to its simple design it was more robust and showed a smaller tendency to form an oxide film on the surface. It also had a lower standing current and furthermore was found to be able to tolerate oxygen concentrations in the eluant 100 times greater than the dropping electrode could tolerate without effecting its performance. Both the electrode systems were examined by Joynes and Maggs using inorganic ions, 2-6-dinitrophenol, p-nitrobenzoic acid, and m-nitroaniline. Their results indicate that the carbon-impregnated silicone membrane electrode has a detection limit of 1.6×10^{-9} mol/liter and a linear dynamic range of about 10^5. It would appear, however, that at present this detector would only find application in specific areas. A chromatogram obtained by Kenula [40] using the polarographic detecting system is shown in Fig. 3.39.

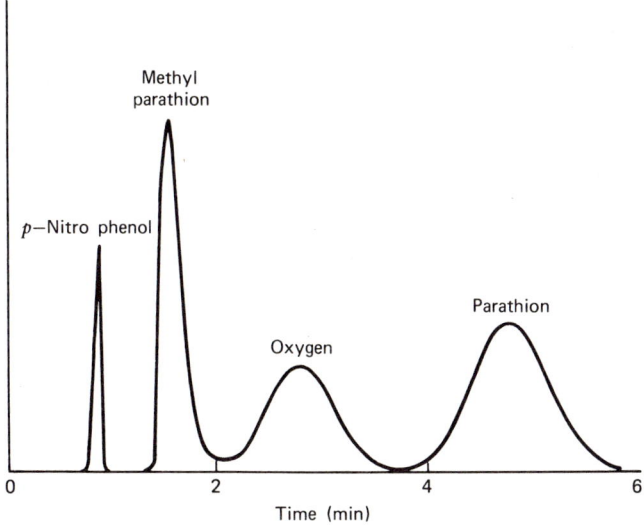

Fig. 3.39. Chromatogram of insecticide mixture using the polarographic detector.

The Electrochemical Detector

The electrochemical detector designed by Kissinger et al. [43] is an interesting developemnt of the polarographic detector.

The detector is a working electrode, set in the flow stream emerging from the column. The electrode is held at a fixed, predetermined potential to oxidize (or reduce) electroactive components passing by. A conventional potentiostatic system is used with an auxiliary and reference electrode. The electrode arrangement is shown in Fig. 3.40. The detector consists of two ½-in.-thick, 1¼-in.-square Lucite blocks separated by a thin (0.005 in.) Teflon spacer, S. The spacer, shown separately in Fig. 3.40, C, has a slit, 2 mm wide and 75 mm long, cut in it. One of the blocks (the left side in Fig. 3.40a) has two drill bores cut at the approximate indicated angles to provide an inlet (In) and outlet (Out) flow stream from the column through the Teflon slit region. These holes are drilled slightly oversized and standard 1/16-in. O.D. Teflon tubind is inserted. These were sliced flush at the inner face of the block and sealed with paraffin wax.

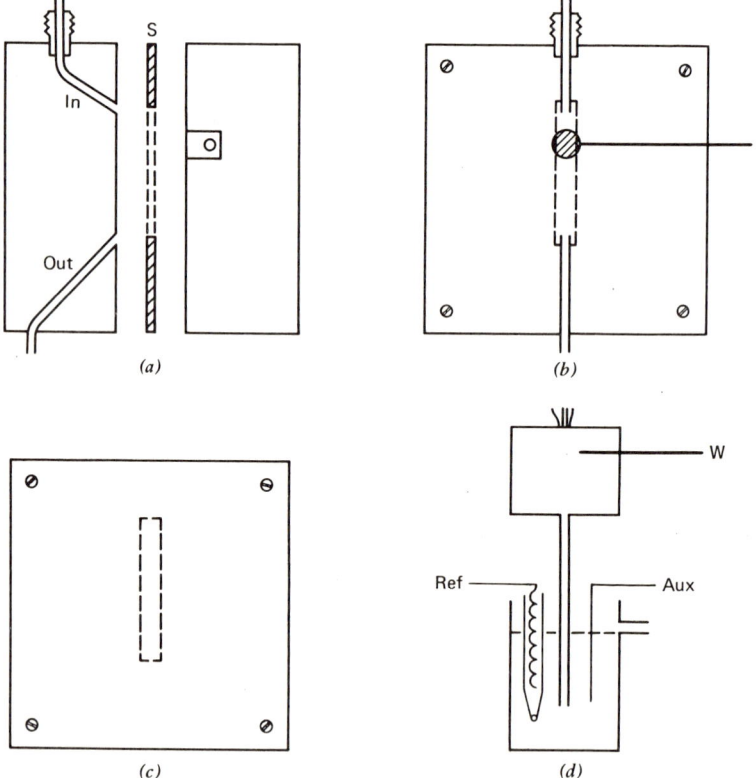

Fig. 3.40. The electrochemical detector.

12 SOLUTE PROPERTY DETECTORS

The upper end of the inlet tube is connected to a standard Chromatronix 1/16-in. fitting that is recessed and secured with epoxy cement to the top of the block.

The opposite block (right side fo Fig. 3.40a) contains a ¼-in.-deep recessed well whose diameter is about 0.10 in., into which a copper wire contact is sealed (the copper wire is shown as a solid black dot emerging perpendicular to the page in Fig. 3.40a). The well is packed flush to the rim with carbon paste, which forms the working electrode surface.

If the two blocks and Teflon spacer of Fig. 3.40a are pressed together and rotated counterclockwise 90°, they appear as indicated in Fig. 3.40b. Four drill holes in the corners of the blocks and spacer allow the unit to be aligned and tightly "sandwiched" together with 1¼-in. machine screws. The detector is then attatched to the end of the column via the standard fitting on top. The dead volume between the column and detector electrode is about 8 μl and the active volume of the detector per se is only about 0.7 μl.

Fig 3.40d shows the outlet tube connection to a simple overflow vessel. The auxiliary electrode (a platinum wire) and the reference electrode (a Ag/AgCl wire set in a small capillary salt bridge of 3 M NaCl) dip into the overflow volume and provide the three-electrode cell.

The detector measures the limiting current (at fixed potential) at the carbon paste electrode under forced convection conditions by means of a suitable amplifier. An example of a chromatogram obtained from this device is shown in Fig. 3.41.

The limitations of this type of detector are fairly obvious. First, it can only sence electroactive components. Second, the eluant must be electrically conducting. The latter is not very restrictive since most aqueous electrolytes or buffers will suffice. For carbon paste electrodes the choice of solvents will be limited to aqueous media, or at most, 20-30% nonaqueous mixtures. According to the authors similar thin-layer configurations utilizing metal films on glass, pyrolytic graphite, and mercury electrodes show promise for work in totally nonaqueous media. The authors claim that the device can detect as little as 50-100 pg but do not give the sensitivity in terms of minimum detectable concentrations.

The Fluorometer Detector

The flurometer detector functions by measuring the fluorescence of the column eluant when excited by UV light. A good example of an effective detector of this type is that manufactured by Laboratory Data Control. This detector has twin flow-through cells, each having a capacity of about 10 μl. The sensitivity of this device is claimed to be 10^{-9} g/ml of quinine sulfate, and this can be measured with a background fluorescence equivalent to 1000 times that of the sample. A diagram of the optical system is shown on Fig. 3.42. The excitation lamp EL, a low-pressure hot mercury lamp with a phosphor coating emits

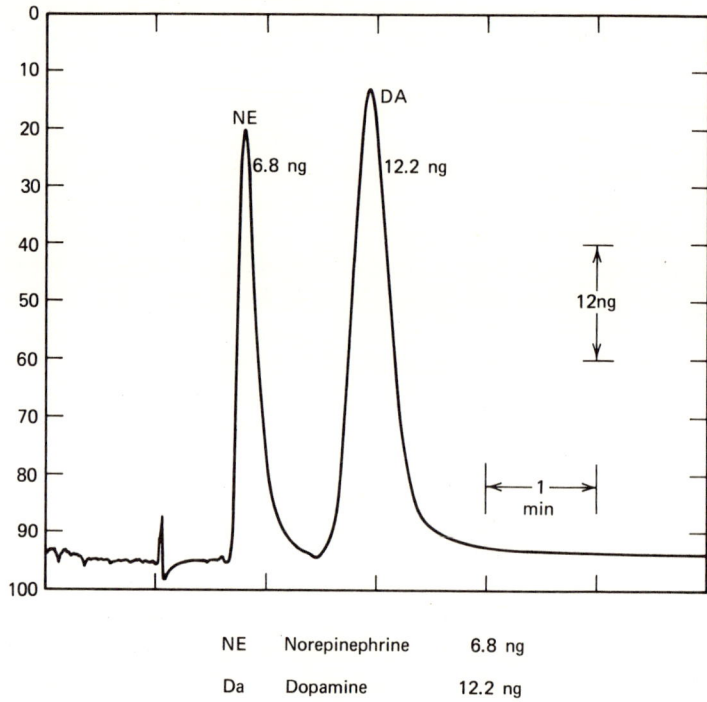

Fig. 3.41. Chromatogram from electrochemical detector.

NE   Norepinephrine   6.8 ng
Da   Dopamine   12.2 ng

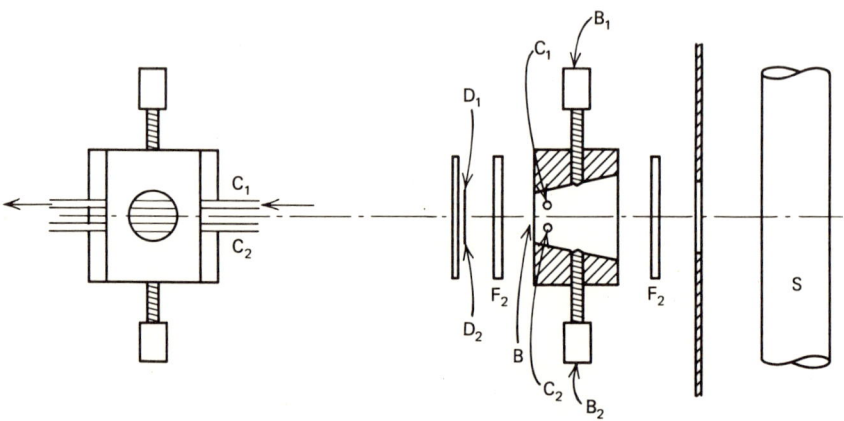

Fig. 3.42. Cell assembly of fluoromonitor.

UV light at about 360 nm. The light from the source passes through a visible light filter Fl, a fixed aperture A, and thence to the large-diameter end of a cone condenser. Cone condenser CC has a highly reflecting internal surface and is bifurcated into two sections by a metal plate whose surfaces are also highly reflecting. Therefore, excitation light that enters both sections of CC impinges on the corresponding flow cell chambers C1 and C2 by multiple reflections as well as by direct transmission. Fluorescence emission from cell chambers C1 and C2 passes through UV blocking secondary filter F2 and impinges on the corresponding photosensitive elements of dual photocell D. The purpose of the bifurcated cone condenser is to increase both the intensity of the excitation on the cell chambers and the efficiency of gathering emitted light, while at the same time dividing the optical system into two "beams." The system in effect is simulating the well-established Tyndale cone effect. In order to compensate for background fluorescence and to establish an initial illumination level on the dual photocell D, two blank adjusting thumbscrews B1 and B2 are provided. Fluorescent beads on the ends of these screws are excitable by the UV light and emit in the visible region to which the dual photocell is responsive. Optical coupling between each bead and the corresponding beam depends on the depth to which each adjusting screw is turned into its threaded bore.

With the MODE switch set to position 1 a large fixed aperture is fully exposed for maximum passage of UV excitation and thus the highest sensitivity. In position 2, smaller aperature reduces the excitation and in position 3, a fluorescent screen is placed over the aperature. A lamp monitoring photocell "views" the excitation lamp and senses its brightness and provides the signal that compensates for changes in lamp brightness caused by temperature changes, lamp aging, and so on. The initial illumination level on the photocell is set during manufacture.

This detector can be extremely sensitive, but its use is obviously limited to those substances with fluorescence and restricts the choice of solvents that can be used to those that do not fluoresce. It should be pointed out that minute traces of fluorescence materials commonly present in solvents can cause significant problems for this type of detector. This system has been very effectively used in the detection of amino acids when employed with fluorescing reagents described by Samejima et al. (44). An example of a chromatogram obtained using this type of detector in the separation of peptides is shown in Fig. 3.43.

The Spray Impact Detector

The spray impact detector was devised by Mowery and Juvet [45] using a well-established effect known as early as 1892 [46] and later investigated by Christiansen [47] and Loeb [48].

If a stream of the effluent from a liquid chromatography column is directed onto a conducting target electrode producing a spray, charged droplets are formed and potentials well over 2000 V may be produced at the electrode surface. As a result of this effect, when trace quantitites of organic or inorganic compounds are eluted from the column, a large change in electrode potential

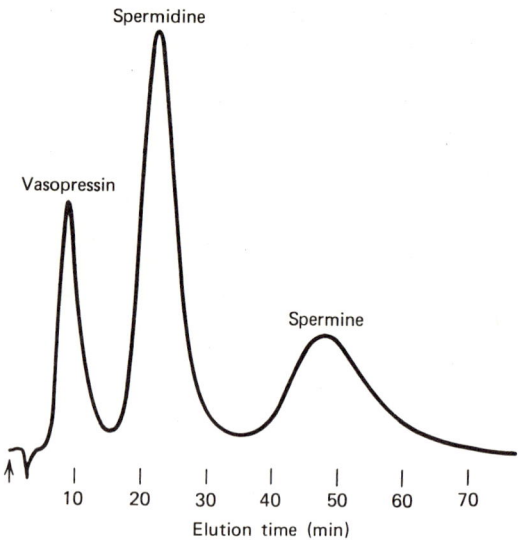

Fig. 3.43. Chromatogram from the fluorometric detector.

(and current) is observed. For example, 1.5×10^{-7} g of nitrophenol is sufficient to cause a 400-V change in the electrode potential. The operation of this detector is quite simple although the mechanism of operation is complex and not yet entirely understood. Its mode of operation depends upon producing an asymmetric charge separation of the mobile phase and monitoring the changes in current or voltage levels as solutes are eluted.

The spray impact detector consists of two major units, a pneumatic aspirator and a target electrode as shown in Fig. 3.44. A stainless steel inner capillary tube of 0.39 mm I.D. and 0.71 mm O.D. and an outer stainless steel concentric nozzle with a 0.83-mm interior diameter were used in the aspirator unit. Apertures of these diameters are suitable for operating the column conveniently at mobile phase flow rates of 2-7 ml/min and air flows of 2000-7000 ml/min.

The column effluent runs into the inner capillary tubing where the air flow from the surrounding aperture ruptures the liquid jet into droplets. The spray of droplets is then directed onto a target electrode, which acts as an impact wall and shatters the drolets into a fine spray. Spacing between the aspirator tip and target is about 5 mm at the flow rates typically used. A mobile phase flow rate of 3 ml/min and an air flow rate of 4000 ml/min generates a baseline current of about 3×10^{-8} amp with the aspirator described and is near optimum in signal-to-noise ratio when distilled water is used as the mobile phase in reverse-phase liquid chromatography.

The target electrode is a rod of conducting material isolated from electrical ground and connected to an appropriate amplifier. The targets themselves took

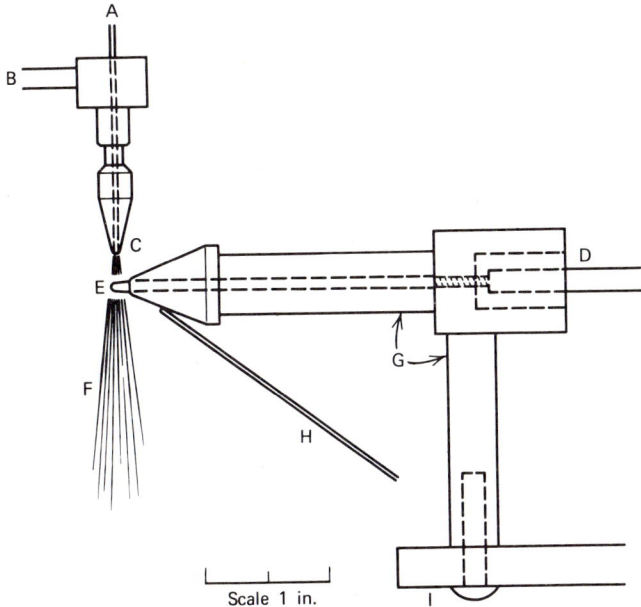

Fig. 3.44. The spray impact detector.

two forms: glassy carbon and gold plated platinum mounted in a PTFE body.

The PTFE body is supported by a high-resistance laminated plastic mounting to further decrease electrical leakage to ground.

When the spray impacts on the target, the target becomes highly negatively charged while the spent spray from the target is positiviely charged. This charge separation is monitored as current flow in the electrometer, and changes occur as components elute from the column.

The author claims an overall sensitivity ranging from 10^{-7} to 10^{-10} gm/sec, which at a flow rate of 2 ml/min is equal to a sensitivity of 3×10^{-6} to 3×10^{-9} g/ml. The device has a linear dynamic range of about 4 orders and thus appears to be suitable for quantitative work. A chromatogram obtained from this detector showing the separation of some fatty acids is shown in Fig. 3.45. This detector is an interesting new development and may have very useful potential in the field of liquid chromatography.

13 CHOICE OF DETECTOR

There is, at present, no liquid chromatography detector that possesses all the necessary attributes that are required for a completely versatile liquid chromatograph. The most generally sensitive detector is the UV detector and the most

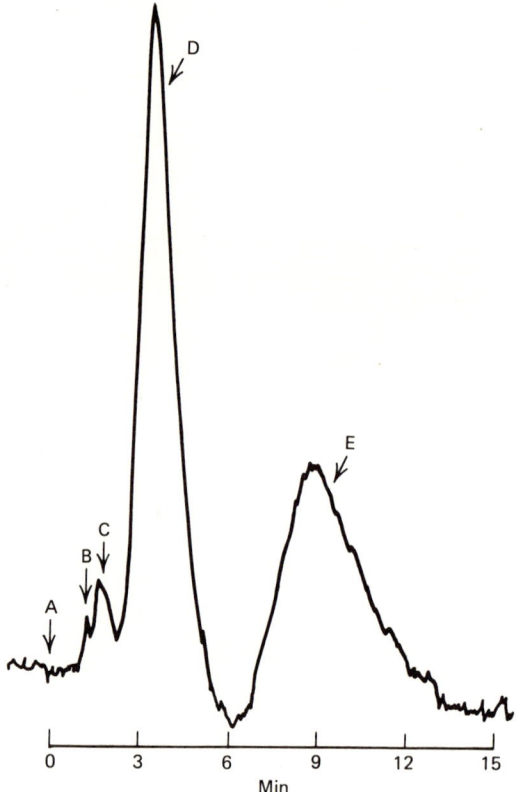

Fig. 3.45. Chromatogram from the spray impact detector.

versatile is probably the wire transport detector. Again it should be stressed that at the present state of the development of the technique it is necessary to have a number of different detectors available, if the liquid chromatograph is to be able to cope with a wide range of separation problems. A list of commercially available detectors are given in Table 3.5 together with their characteristics, advantages, disadvantages, and areas of applications. The detectors are listed in order of priority for those entering the field of liquid chromatographic analysis.

14 ANCILLARY EQUIPMENT

The necessary ancillary equipment for use with the liquid chromatograph is minimal if we exclude such devices as fraction collectors and automatic integrators. There are a number of fraction collectors available on the market with complete operating instructions and equipment for measuring peak areas can be

Table 3.5 Detector Characteristics

Detector	Sensitivity (Average)(g/ml)	Advantages	Disadvantages	Areas of Application
UV detector with dual wavelength option	10^{-7}	High sensitivity, stable to small changes in temperature and column flow rate, can be used for gradient elution providing solvents are chosen that do not adsorb in the UV.	Will only detect substances that adsorb in the UV, restricts choice of solvents for gradient elution.	Detects all substances that absorb in the UV.
Moving wire detector fitted with the methane converter	10^{-6}	Detects all involatile substances containing carbon, provides complete freedom of choice of solvents for gradient elution providing the solvents are volatile. Insensitive to column flow changes and thus permits flow programming and is insensitive to temperature changes of eluant. Predictable response.	Expensive, bulky; owing to the involved mechanical system for wire transport, requires more maintenance than other detectors. Only uses about 1% of the columns eluant available for detector, volatility of solute must be significantly less than solvent.	Detects all involatile substance: containing carbon.
Fluorometric detector with florophore reagent	10^{-8}	High sensitivity, stable to small changes in temperature and column flow rate.	Limited to the detection of fluorophores, requires separate pump and reaction system when used with fluorophore reagent.	All fluorophores, primary amines with fluorophore reagent, amino acids, and peptides.

Table 3.5 Detector Characteristics (continued)

Detector	Sensitivity (Average)/(g/ml)	Advantages	Disadvantages	Areas of Application
Conductivity detector	10^{-8}	High sensitivity, stable to small changes in temperature and column flow rate.	Limited to the detection of ions only	Detection of ions.
Refractometer detector	10^{-5}	General-purpose detector.	Sensitive to very small temperature changes, and the presence of dissolved gases, cannot be used with gradient elution, low sensitivity.	Detects all substances that have a refractive index significantly different to that of the solvent.

Microsyringes

For the effective use of the modern chromatograph a minimum of one each of 1-, 10-, and 100-μl microsyringes should be available. It is advisable in fact to have two each of these devices since they are liable to break or become damaged during use. The 1-μl microsyringes can be used at inlet pressures of up to 3000 psi, but the 10-μl microsyringes are limited to about 1000 psi. If the inlet pressure is too high for the syringe, back pressure on the syring piston can make injection difficult, produce leak of sample past the piston and sometimes result in fracture of the glass syringe itself. In preparative liquid chromatography where columns of 1 cm or more in diameter are to be used, then a 500-μl or 1-ml syringe may be needed.

Flow-Rate Measuring Apparatus

The simpliest method of measuring column flow is with a stopwatch and a small calibrated measuring cylinder. However, for very small flow rates of highly volatile solvents significant losses can occur from evaporation.

A commercially available flow meter manufactured by Laboratory Data Control Ltd. overcomes the problems of solvent volatility. The flow monitor consists of a precision-bore glass metering tube with two photodetectors precisely positioned and clamped onto the tube. The distance (volume) between the photodetectors sensing areas is accurately fixed during manufacture by calibration. The normal flow range for liquid chromatography is 9-110 ml/hr, and for this range of flow rates the metered volume is 0.5 ml. A specially constructed valve is used to inject a gas bubble of precise volume into the fluid stream to be measured. The time required for the bubble to traverse the metering tube, between the two photodetectors, is measured electronically as the bubble passes each of the photodetectors in turn. Electronically accurate to within 0.1%, the bubble transit time is indicated on a lighted digital display that can then be used as a flow index in conjunction with a calibration chart which is supplied with the instrument. The photodetectors are each capable of "seeing" a fraction of a millimeter movement of a bubble; however, bubbles significantly smaller than the bore of the metering tube will not trigger the unit.

Automatic flow meters of this type described above, although expensive, can be an extremely useful addition to the overall chromatography equipment.

Compression Fittings and Unions

There are many varities of compression fittings and unions available for connecting the various parts of a liquid chromatograph together. However, it should be emphasized that any unions used should be inspected to ensure that

186   LIQUID CHROMATOGRAPHY APPARATUS

all seals and union seats are clean and not marked or scored. The most common source of poor column performance, detector instability, and poor repeatability of analysis is leakage occuring at union connections. Good quality unions should always be used and it should not be assumed that a new union will be free from defect. It should also be emphasized that in order to minimize extra band spreading effects, mimimum dead volume unions should be employed when connecting the column to the detector.

APPENDIX: A COMPUTER PROGRAM FOR CALCULATING GRADIENTS USING THE INCREMETAL METHOD OF GRADIENT ELUTION

The program is written to accommodate eight mixtures of two solvents each made up to have a different volume ratio composition. The volume ratio of each mixture is entered as values in lines 210 to 280 (0 to Y) of the program. Lines 140 to 200 allow the time of introduction of each solvent mixture to the mixing vessel to be chosen over a period of 50 minutes by assigning specific times to A, B, C, through G.

The column flow rate Q is entered at line 110 and the volume of the dilution vessel at line 130. Line 120 allows a choice of outputs by allotting certain values of Q1 as follows:

120 LETQ1 = 0   A graph of solvent concentration versus time will be produced.

120 LETQ1 = 1   A graph of solvent concentration versus time together with a graph of log solvent concentration versus time will be produced.

120 LETQ1 = 2   A graph of solvent concentration versus time together with the exponent of solvent concentration versus time will be produced.

120 LETQ1 = 3   This will provide the same output as to Q1 = 1 but will include a table giving numerical values for the solvent concentrations, log solvent concentrations, and the exponents of the solvent concentration for each minute of the solvent program.

In the example given solvent volume ratios of 0.1 to 0.8 have been allotted to concentrations of O to Y, but these could equally well have been a series of buffer solutions having pH values of 1 to 8 or 3 to 10, and so on. Initially the effect of V should be eliminated by making V very small, for example 0.01 ml, but not zero as the program will not operate for $V = 0$. Assigning different times to the constants A to G, a stepped program will be produced. When the form of the stepped program is correct, then V is increased progressively until a smooth solvent program of the desired form is obtained. Examples of the curves obtained for the linear, logarithmic, and expoential gradient elution programs

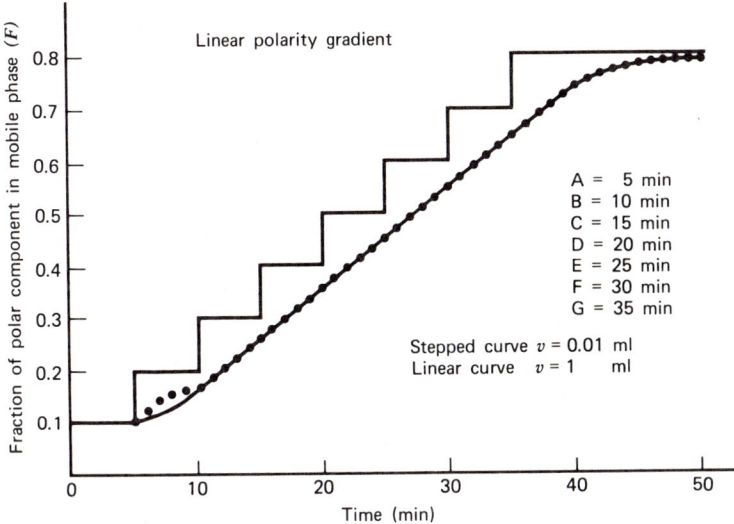

Fig. 3.46. Graph of fraction of polar component in mobile phase supply against time (calculated from the computer program).

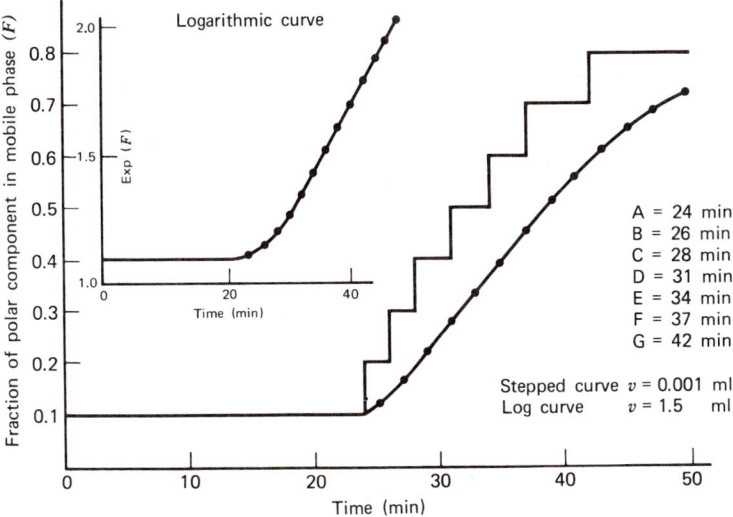

Fig. 3.47. Graph of fraction of polar component in the mobile phase against time (calculated from the computer program).

are shown in Figs. 3.46-3.48. Table 3.6 shows the computer program to provide polarity profiles for the incremental method of mixing.

188 LIQUID CHROMATOGRAPHY APPARATUS

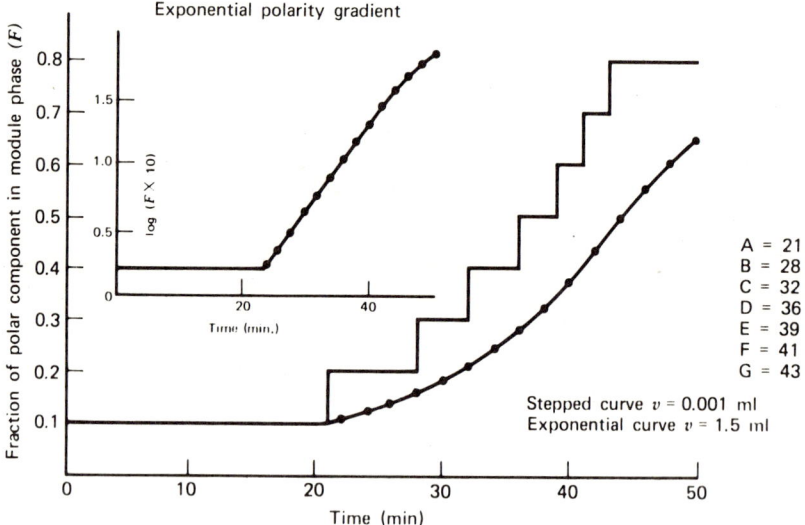

Fig. 3.48. Graph of fraction of polar component in the mobile phase against time (calculated from the computer program).

Table 3.6 Computer Program to Provide Polarity Profiles for the Incremental Method of Mixing

```
100 DIM X(50),Y(50)
110 LET Q=0.2
120 LET Q1=3
130 LET V=2
140 LET A=21
150 LET B=27
160 LET C=31
170 LET D=35
180 LET E=38
190 LET F=40
200 LET G=42
210 LET Ø=.1
220 LET P=.2
230 LET X=.3
240 LET R=.4
250 LET S=.5
260 LET T=.6
270 LET U=.7
```

Table 3.6 Computer Program to Provide Polarity Profiles for the Incremental Method of Mixing

```
280 LET Y=.8
290 FOR J=1 TØ 50 STEP 1
300 IF J<A THEN 600
310 LET W=Ø*EXP(-Q*(J-A)/V)
320 IF J>B THEN 350
330 LET W=W+P*(1-EXP(-Q*(J-A)/V))
340 GØ TØ 570
350 LET W=W+P*(1-EXP(-Q*(B-A)/V))*EXP(-Q*(J-B)/V)
360 IF J>C THEN 390
370 LET W=W+X*(1-EXP(-Q*(J-B)/V))
380 GØ TØ 570
390 LET W=W+X*(1-EXP(-Q*(C-B)/V))*EXP(-Q*(J-C)/V)
400 IF J>D THEN 430
410 LET W=W+R*(1-EXP(-Q*(J-C)/V))
420 GØ TØ 570
430 LET W=W+R*(1-EXP(-Q*(D-C)/V))*EXP(-Q*(J-D)/V)
440 IF J>E THEN 470
450 LET W=W+S*(1-EXP(-Q*(J-D)/V))
460 GØ TØ 570
470 LET W=W+S*(1-EXP(-Q*(E-D)/V))*EXP(-Q*(J-E)/V)
480 IF J>F THEN 510
490 LET W=W+T*(1-EXP(-Q*(J-E)/V))
500 GØ TØ 570
510 LET W=W+T*(1-EXP(-Q*(F-E)/V))*EXP(-Q*(J-F)/V)
520 IF J>G THEN 550
530 LET W=W+U*(1-EXP(-Q*(J-F)/V))
540 GØ TØ 570
550 LET W=W+U*(1-EXP(-Q*(G-F)/V))*EXP(-Q*(J-G)/V)
560 LET W=W+Y*(1-EXP(-Q*(J-G)/V))
570 LET X(J)=W
580 LET Y(J)=LØG(10*W)
590 GØ TØ 620
600 LET X(J)=Ø
610 LET Y(J)=LØG(10*Ø)
620 NEXT J
630 PRINT TAB(1);"TIME";TAB(32);"CØNCENTRATIØN";
640 PRINT
650 PRING TAB(4);"0";TAB(17);"0.2";TAB(31);"0.4";TAB(45);"0.6";
660 PRINT TAB(59);"0.8";TAB(71);"1.0";
```

Table 3.6 Computer Program to Provide Polarity Profiles for the Incremental Method of Mixing

```
670 PRINT
680 FØR J=1 TØ 50
690 LET Z=X(J)*70+4
700 PRINT TAB(1);J;TAB(Z);"X";
710 PRINT
720 NEXT J
730 IF Q1=0 THEN 880
740 IF Q1=2 THEN 780
750 PRINT TAB(9);"LØG CURVE";
760 PRINT
770 GØ TØ 800
780 PRINT TAB(9);"EXP CURVE";
790 PRINT
800 FØR J=1 TØ 50
810 IF Q1=2 THEN 840
820 LET Z=Y(J)*70/(50)+4
830 GØ TØ 850
840 LET Z=EXP(X(J))*70/EXP(X(50))+4
850 PRINT TAB (1);J;TAB(Z);"X";
860 PRINT
870 NEXT J
880 IF Q1<3 THEN 940
890 FØR J=1 TØ 50
900 LET A=EXP(X(J))
910 PRINT TAB(4);J;TAB(16);X(J);TAB(28);Y(J);TAB(40);A;
920 PRINT
930 NEXT J
940 END
```

REFERENCES

1. L. R. Synder and D. L. Saunders, *Advances in Chromatography*, A. Zlatkis, Ed., Preston Technical Publications, New York, **1969**, 289.
2. R. J. Maggs and T. E. Young, *Gas Chromatography, 1968*, C. L. A. Harbourne, Ed., Institute of Petroleum, London England, 1968.
3. R. P. W. Scott and J. G. Lawrence, *Advances in Chromatography*, A. Zlatkis, Ed., Preston Technical Publishers, New York, 1970, p. 389.
4. R. P. W. Scott and J. G. Lawrence, *J. Chromatogr. Sci.*, **8**, 619 (1970).

5. S. H. Byrne, J. A. Schmit, and P. E. Johnson, *J. Chrom. Sci.,* **9**, 592 (1971).
5a. B. Laverne and B. L. Karger, *Anal. Chem.,* **9**, 819A (1973).
6. B. Pearce and W. H. Thomas, *Anal Chem.,* in press.
7. R. P. W. Scott, D. W. J. Blackburn, and T. Wilkins, *J. Gas Chromatogr.,* 183 (1967).
8. H. Barth, E. Dallmeir, and B. L. Karger, *Anal. Chem.,* in press.
9. I. A. Fowliss and R. P. W. Scott, *J. Chromatogr.,* **11**, 1 (1963).
10. I. A. Fowliss, R. J. Maggs, and R. P. W. Scott, *J. Chromatogr.,* **15**, 471 (1964).
11. A. Tiselius and D. Claesson, *Arkiv Kemi Mineral Geol.* **15B** (No. 18), 1 (1942).
12. R. D. Conlon, *Rev. Sci. Instr.,* **34**, 1418 (1961).
13. F. A. Vandenheuvel and E. Sipos, *Anal. Chem.,* **33** (No. 2), 286 (1961).
14. D. Zaukelies and A. A. Frost, *Anal. Chem.,* **21**, 743 (1949).
15. G. V. Troitskii, *Biokhimiga,* **5**, 375 (1940).
16. R. A. Grant, *J. Appl. Chem.,* **8**, 136 (1959).
17. G. Johansson and K. J. Karrman, *Anal. Chem.,* **8**, 1397 (1958).
18. G. A. Howard and A. J. P. Martin, *Biochem. J.,* **46**, 532 (1950).
19. R. P. W. Scott, D. W. J. Blackburn, and T. J. Wilkins, *Gas Chromatogr.,* 183 (1967).
20. A. J. P. Martin and S. S. Randall, *Biochem. J.,* **49**, 293 (1951).
21. H. D. Harlan, *Chem. Anal.,* **54** (No. 3), 89 (1965).
22. C. I. Sjoberg, *Acta Chem. Scand.,* **8** (No. 7), 1161 (1954).
23. P. W. Avirzonis, F. Fritz, and J. C. Wriston, *Anal. Chem.,* **34**, 58 (1962).
24. Drake Birger, *Arkiv Kemi,* **4**, 401 (1952).
25. Drake Birger, Svenlgardell, *Arkiv Kemi,* **4**, 469 (1952).
26. A. M. Batich, *C. R. Acad. Sci. Paris,* **236**, 2055 (1953).
27. W. Simon, J. T. Clere, and R. E. Dohmer, *Microchem J.,* **10**, 495 (1966).
28. N. Fornstedt and J. Porath, *J. Chromatog.,* **42**, 376 (1969).
29. K. Ohzeki, T. Kambara, and K. Saitoh, *J. Chromatogr.,* **38**, 393 (1968).
30. J. J. Kirkland, *Anal. Chem.,* **40** (No. 2), 391 (1968).
31. C. G. Horvath and S. R. Lipsky, *Nature,* **211**, 748 (1965).
32. A. T. James, J. R. Ravenhill, and R. P. W. Scott, *Gas Chromatography 1964,* A. Goldup, Ed., The Institute of Petroleum, City, 1964.
33. R. P. W. Scott and J. G. Lawrence, *Anal. Chem.,* XX, XXX (19XX).
34. E. Haahti, T. Nikkari, and X. Karkainen, *Gas Cromatography 1964*, A. Goldup, Ed., The Institute of Petroleum, City, 1964, p. 190.
35. R. P. W. Scott and J. G. Lawrence, *Anal. Chem.,* **39**, 830 (1967).
36. G. Claxton, *J. Chromatogr.,* **2**, 136 (1959).
37. R. P. W. Scott, *Anal. Chem.,* **35**, 481 (1963).
38. R. P. W. Scott, *J. Chromatogr. Sci.,* **11**, 351 (1973).
39. T. W. Smuts, P. W. Rechter, and V. Pretorius, *J. Chromatog. Sci.,* **9**, 457 (1971).
40. Kenula, *Roczniki Chem.,* **26**, 281 (1952).
41. J. G. Koen, J. F. K. Huber, H. Poppe, and G. denBoef, *J. Chromatogr. Sci.,* **8**, 192 (1970).

42. P. L. Joynes and R. J. Maggs, *Advances in Chromatography 1970*, A. Zletkis, Ed., Preston Technical Abstracts, City, 1970, p. 379.
43. P. T. Kissinger, et al., *Anal. Letters,* **6** (5), 465 (1973).
44. D. Samejima, W. Diaman, and S. Undenfriend, *Anal. Biochem.,* in press.
45. R. A. Mowery, Jr., and R. S. Juvet, Jr., *J. Chromatogr. Sci.,* **12**, 687 (1974).
46. P. Lenard, *Ann. Phys. (Leipzig), Ser. 3,* **46**, 584 (1892).
47. C. Christiansen, *Ann. Phys. (Leipzig), Ser. 4,* **40**, 107 (1913).
48. B. Loeb, *Static Electrofication,* Springer-Verlag, Berlin, 1958.

Chapter **IV**

STATIONARY AND MOBILE PHASES FOR LIQUID CHROMATOGRAPHY

1 **Methods of Assessing Stationary Phases 194**
   Experimental Procedure 195
   Presentation of Results 196
     Resolution 196
     Peak Capacity 197
     Scope 198
     Loading Capacity 201
     Column Permeability 204
   Reduced Plate Height Curves 204

2 **Adsorbents 207**
   Silica Gel 207
     Characteristics of Silica Gel Packings 212
   Alumina 216
   Bonded Phases (Modified Adsorbents) 216
   Carbon 221
   Polyamide 221
   Ion-Exchange Media 222
     Cation-Exchange Resins 222
     Anion-Exchange Resins 222

3 **Supports for Liquid-Liquid Chromatography 222**

4 **Choice of Stationary Phase 224**

5 **Choice of Phase System for Liquid-Liquid Chromatography 225**

6 **The Mobile Phase 227**
   Isocratic Period: Elution Development 230
   Gradient Process: The First Displacement Effect 230
   Gradient Process: The First Elution Effect 230
   Gradient Process: The Second Displacement Effect 231
   Gradient Process: The Second Elution Effect 232

7 **Chromatographic Properties of the Solvent Series for Use with Incremental Gradient Elution 232**

1 METHODS OF ASSESSING STATIONARY PHASES

The purpose of liquid chromatography is to provide an efficient means of separation and the separation is carried out in the column. Thus, the column packing and the mobile phase used with it form the crux of the separation system. If the combination of stationary phase and mobile phase does not provide both the required selsectivity and the necessary efficiency, no separation will be achieved regardless of the investment made in the chromatograph and ancillary equipment involved. It is therefore necessary to have a clear understanding of the nature and performance characteristics of the stationary phases available, to know how to choose the correct one for any given separation, and finally to have a rational procedure for detemining the best mobile phase to employ with the chosen stationary phase.

However, before identifying the parameters necessary to determine the efficacy of any given stationary phase, it is necessary to consider the attributes that would be provided by the column system as a whole. The desirable attributes from a chroamtographic system can be diagrammatically represented by a tetrahedron as shown in Fig. 4.1. Chromatography is a separation technique and

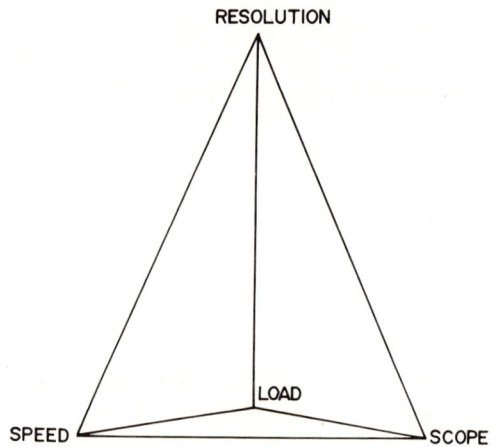

Fig. 4.1. The chromatography tetrahedron.

therefore, resolution is an a priori requisite expected from the column system. Secondary to resolution, for analytical systems, speed may be the next desirable property, and both resolution and speed are conditioned by the type of stationary phase employed. The scope of the chromatographic system, that is, its capacity for separating mixtures of wide polarity range, is usually achieved by modifying the mobile phase during development as with the technique of gradient elution. However, for practical or other reasons it may not be convenient to employ a gradient elution procedure so that the scope of the adsorbent may be important and should be known. The capability of the system to separate large quantitites of materials is dependent on the loading capacity of the stationary phase, that is, the mass of any solute that can be placed on the column wihtout causing band dispersion or peak asymmetry. Therefore, loading capacity is another important property of the stationary phase that should be known.

There are various ways by which a stationary phase can be assessed, and opinions differ at present as to which method is the best for comparing different stationary phase materials. The following procedure is the one adopted by the author for comparing different stataionary phases and examples will be given for a number of commercially available silica gels.

Experimental Procedure

The apparatus consisted of a Waters 6000 pump, column system fitted with an injection device and a LDC UV detector working at 254 nm. The standard column used was ¼ in. O.D., with an I.D. of 4.6 mm, 50 cm long, and made from 304 stainless steel. For supports having a particle size in excess of 20 μ the packing procedure used was the dry packing method. For supports having particle diameter less than 20 μ, the slurry packing method developed by Majors [1] was employed. After packing, three column volumes of each of the solvents used by Scott an Kucera [2] for incremental gradient elution were passed through the column and the column was then reconditioned for the mobile phase water using 10 column volumes of ethyl alcohol, acetone, ethyl acetate, 1.2-dichloroethane, and heptane, respectively. This ensured that all columns were activated in a reproducible manner. All solvents used were cleaned and dried by passing through active carbon, molecular sieve 5A, and silica gel.

The solvent used as the mobile phase was n-heptane and 1-μl volumes of approximately 0.02% solutions of the following solutes were injected onto the column for test purposes.

1. Carbon tetrachloride
2. Benzene
3. Naphthalene
4. Anthracene
5. Diphenylether
6. Anisole
7. Nitrobenzene
8. Nitropropane
9. Nitromethane

In some cases the concentration of the solute was adjusted to accommodate significant differences in extinction coefficients between the solutes. The specific solutes used for a given adsorbent were chosen to be appropriate to the scope of adsorption capacity of the adsorbent. Chromatograms were obtained for the appropriate solute mixture at a standard flow rate of 1.0 ml/min, which corresponded approximately to a linear mobile phase velocity of 1.5 mm/sec and the k' value and efficiency of each peak. The dead volume of each column was taken as the retention volume of n-nonane measured by using a LDC refractometer detector. The adsorbents investigated were as follows:

1. Corasil II (C2)
2. Pellosil HC
3. Vydac A (VA)
4. Porasil C (PORC)
5. Biosil A (BA)
6. Biosil HA (BHA)
7. Lichrosorb 30 μ (L30)
8. Lichrosorb 10 μ (L10)
9. Partisil 10 μ (R10)
10. Partisil 5 μ (R5).

Presentation of Results

Resolution

Resolution can best be determined by calculating the minimum α value (retention ratio) of two solutes that can be separated on the given stationary phase packed in the standard column [3].

From the plate theory the condition for resolution between solute A and solute B will be given by:

$$4 \sqrt{n} \; (v_m + K_A a_s) = n \; (K_B a_s - K_A a_s)$$

where v_m is the volume of mobile phase per plate, n is the efficiency of the peak of solute A ($v_m = V_m/n$), a_s is the effective surface area of adsorbent per plate ($a_s = A'_s/n$), and K_A and K_B are the distribution coefficients for solutes A and B.

Dividing throughout by v_m and noting that $K_A a_s/v_m = k'_A$ and $K_B a_s/v_m = k'_B$,

$$4 \sqrt{n} \, (1 + k'_A) = n(k'_B - k'_A).$$

Dividing throughout by k'_A,

$$4 \sqrt{n} \, (\frac{1}{k'_A} + 1) = n \, (\frac{k'_B}{k'_A} - 1) = n \, (\alpha - 1).$$

1 METHODS OF ASSESSING STATIONARY PHASES

Rearranging,

$$\alpha = \frac{4}{\sqrt{n}} \left(\frac{1}{k'_A} + 1\right) + 1.$$

The equation provides an expression for the minimum α value of a solute pair that can be resolved where the first peak is eluted at k'_A and an efficiency of n. Curves relating the minimum α value against the k' value of the eluted solvent for different adsorbents is shown in Fig. 4.2. The lowest α value indicates the adsorbent with the highest potential resolution.

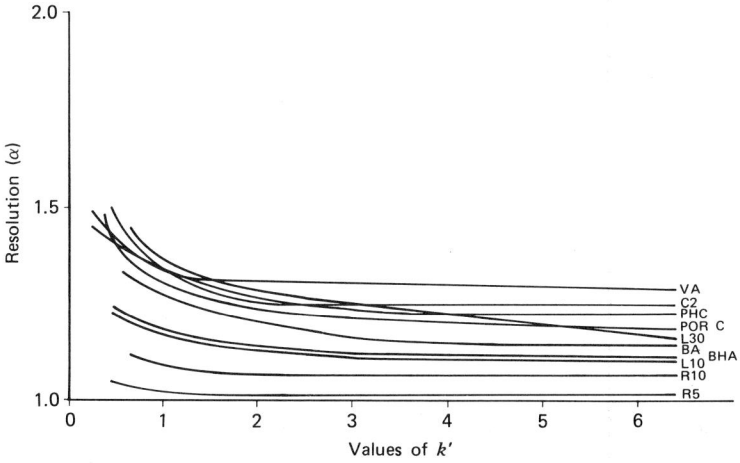

Fig. 4.2. Graph of potential resolution of the standard column packed with different adsorbents against k' of solutes.

Peak Capacity

Another important aspect of the adsorbent is its capability of separating multicomponent mixtures. Thus, although an adsorbent may be capable of producing adequate resolution for a given pair of solutes, it has also to provide adequate space in the chromatogram to elute discretely other components present in the mixture. This capability is termed peak capacity and can be calculated from the column efficiency and k' value of the solute using the equation developed in Chapter II.

$$\frac{k'}{1+k'} = \frac{4}{\sqrt{n}} \left\{ \frac{1 - \left(\frac{n-2\sqrt{n}}{n+2\sqrt{n}}\right)^r}{1 - \left(\frac{n-2\sqrt{n}}{n+2\sqrt{n}}\right)} - 0.5 \right\}$$

where r is the peak capacity and n is the efficiency of the last eluted peak. A more practical arrangement of this equation is as follows:

$$r = \frac{\log\left(1 - \left(\frac{\sqrt{n}}{4}\left(k'/(1+k')\right) + 0.5\right)(1-P)\right)}{\log P}$$

where

$$P = \left(\frac{n - 2\sqrt{n}}{n + 2\sqrt{n}}\right).$$

In fact, peak capacity is another way of expressing column efficiency and those adsorbents that provide high resolving power columns will also exhibit high peak capacity. If Fig. 4.3 the peak capacity is plotted against k' values for

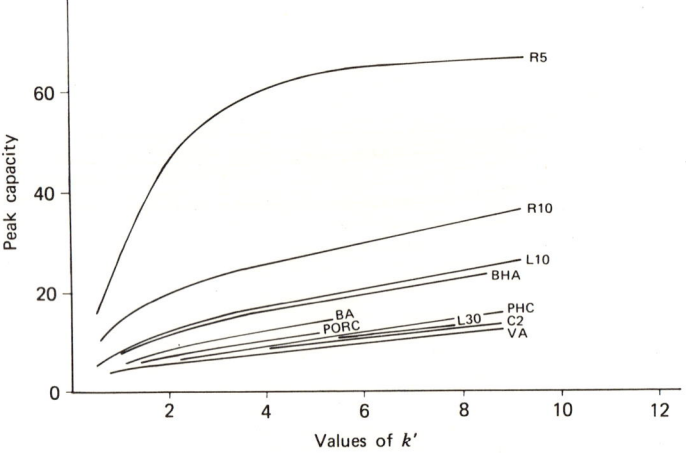

Fig. 4.3. Graph of peak capacity against k' for standard columns packed with different adsorbents.

the same series of absorbents. It can be seen comparing Figs. 4.2 and 4.3 that adsorbents with low α values in Fig. 4.2 show high peak capacities in Fig. 4.3.

Scope

In Fig. 4.4 the elution time for a series of solutes chromatographed on standard columns, operated under standard conditions, at a flow rate of 1 ml/min is plotted against the retention volumes of the respective solutes chromatographed on a standard Porasil column. The X axis then becomes an *arbitrary* polarity scale and indicates the polarity range of the solutes used. If a retention time of

1 METHODS OF ASSESSING STATIONARY PHASES   199

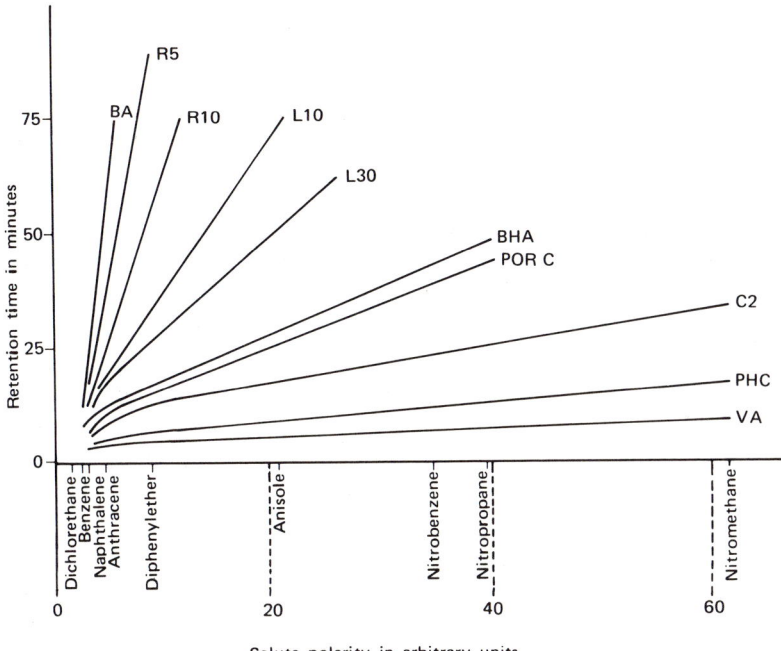

Fig. 4.4. Graph of retention time against solute polarity for different adsorbents.

25 min is considered as a maximum it is seen that the high surface area adsorbents (see Table 4.1) can only elute solutes up to a polarity of naphthalene,

Table 4.1 The Characteristics of Different Silica Gel Adsorbents

Name	Particle Size (μm)	Shape	Surface Area (m²/g)	Pore Size (Å)
Biosil A	20-44	I	200+	<100
Partisil 10μ	9	I	400+	40-50
Partisil 5 μ	6	I	400+	40-50
Lichrosorb 30 μ	30	I	200+	60
Lichrosorb 10μ	10	I	200+	60
Biosil HA	44	I	200+	<100
Porasil C	37-75	S	50-100	100-200
Vydac A	30-44	S	12	57
Pellosil HC	37-44	S	8	na
Corasil II	37-50	S	14	na

S, spherical; I, irregular; na, not available.

which represents a very small polarity change from that of the mobile phase heptane. Conversely, the low surface area of pellicular packings can cover a polarity range that extends further than nitromethane, which represents five solvent changes in the Scott and Kucera [2] series of solvents. The other adsorbents range themselves in a manner that would be expected from their respective surface areas, one exception being BioSil A, which has the greatest retention although it only has intermediate surface area. It is possible that the large pore size may contribute to this effect but the BioSil HA (a chemically deactivated adsorbent) with the same surface area also has a pore size less than 100 Å and gives a loading capacity commensurate with its surface area. Generally, the results indicate that the lower the surface area the wider the polarity range of solutes that can be chromatographed in a given time.

The importance of the scope of the adsorbent sould not be underestimated. An adsorbent with a low scope will only elute solutes in a reasonable time that have polarities close to that of the mobile phase. Conversely, adsorbents with a high scope will elute solutes that have polarities significantly higher than that of the mobile phase. Since many liquid chromatography separations are routine and repetitive in nature, it is highly desirable from the point of view of both simplicity and economy to operate such analyses under isocratic conditions of development. Thus for a simple wide polarity range mixture an adsorbent with a wide scope would permit rapid analysis under isocratic conditions. An example of the importance of having adsorbents of different scope is shown in Fig. 4.5.

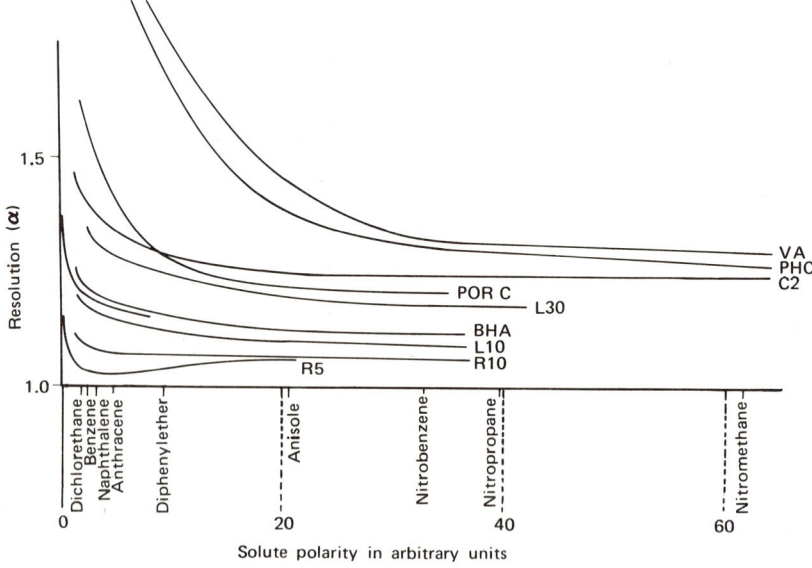

Fig. 4.5. Graph of potential resolution from the standard column packed with different adsorbents against solute polarity.

Here again it is seen that the high surface area adsorbents, although producing the highest resolution, have a relatively restricted polarity range and can only separate solutes that have polarities close to that of the mobile phase when used under isocratic development. The pellicular type material, however, when used under the same conditions can separate solute mixtures that have polarities significantly greater than the mobile phase.

It can be argued that the high surface area, low scope microparticulate packings could be used for wide polarity range samples by increasing the flow rate or using a shorter column. Both alternatives, however, would reduce both the column efficiency and thus the resolving power of the column, whereas a microparticulate column of the same length but higher scope (lower surface area) would maintain the same resolving power but provide the required separation in a shorter time. Another alternative would be to increase the polarity of the mobile phase and thus reduce analysis time. However, increasing the polarity of the mobile phase will generally reduce the retention times of all solutes, but separations will only be obtained for those solutes that have polarities close to that of the now more polar stationary phase. The effect of surface area on the scope of the adsorbent is similar to the effect of changing the stationary phase loading on a gas chromatographic column. When wide boiling range samples are separated on a GC column (compare wide polarity range of solutes in LC) under isothermal conditions the loading of the column is reduced to permit separation is a reasonable time (compare low surface area adsorbent in LC).

It would appear that a series of four microparticulate adsorbents having surface areas of approximately 15, 75, 150, and 350 m^2/g, respectively, would produce a comprehensive family of adsorbents, from which the appropriate packing could be selected, to suit most separation problems.

Loading Capacity

The definition of loading capacity must be, of necessity, somewhat arbitrary. If a column is loaded progressively with solute and the efficiency of the solute band measured then, first, the efficiency will remain constant, then at a particular load the efficiency will begin to fall.

In line with procedures in gas chromatography as discussed in Chapter II, the loading capacity will be taken as that mass of solute that causes column efficiency to fall by 10%. Again for comparative purposes the same solutes should be used for each adsorbent, but since the loading capacity increases with the k' value of the solute employed, the k' value of the solute on the specific adsorbent must be taken into account, when assessing the significance of the loading capacity as measured.

The loading capacity of each adsorbent, packed in a standard column, can be determined by chromatographing progressively larger charges of solutes and measuring the efficiency of each solute band for each mass of charge. A graph can then be plotted relating efficiencies to charge size, two examples of which

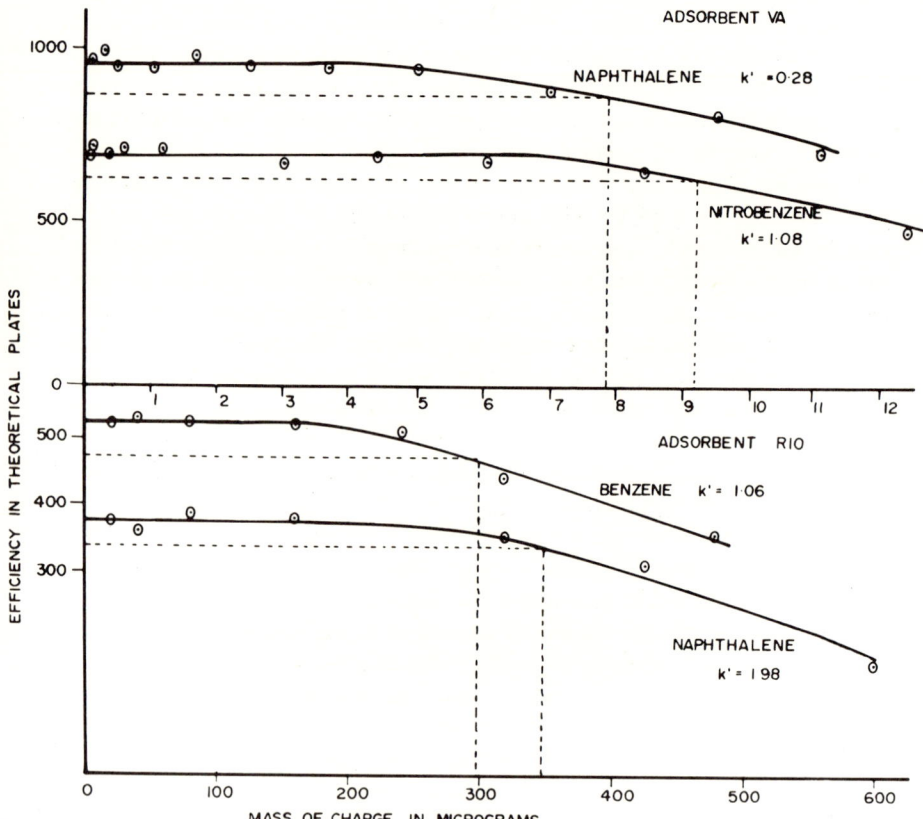

Fig. 4.6. Graphs of efficiency against mass injected from standard columns packed with two different adsorbents.

are given in Fig. 4.6. The loading capacity of the adsorbent is taken as that mass of solute that could be placed on the column to limit the fall in efficiency to 10%. Majors [1] used the change in the k' value of the solute with charge size as a means of determining the loading capcity of the adsorbent. However, excessive loading first affects the column efficiency before significant changes in k' value are observed. It follows that measurement of efficientcies will be a more sensitive assessment of loading capacity. The loading capacities of a number of adsorbents for two appropriate solutes together with the respective k' values are shown in Table 4.2.

It is seen that the loading capacity for the standard column appears to be a direct function of the surface area of the adsorbent and it is also interesting to

Table 4.2 Loading Capacity of Different LSC Adsorbents (μg)

Adsorbent	Biosil A		Partisil 10		Partisil 5		Lichrosorb 30		Lichrosorb 10	
k'	2.25	4.49	1.06	1.93	1.58	2.82	0.89	1.32	1.25	4.6
Solute	Benzene	Naphthal.	Benzene	Naphthal.	Benzene	Naphthal.	Benzene	Naphthal.	Naphthal.	DFE
−5	190	200	300	350	420	450	220	230	240	240

Adsorbent	Biosil HA		Porasil C		Vydac A		Pellosil HC		Corasil II	
k'	1.22	3.6	1.34	2.90	0.28	1.08	0.66	1.72	1.30	7.2
Solute	Anthracene	Anisole	DFE	Anisole	Naphthal.	NO$_2$ Benzene	Anisole	NO$_2$ Benzene	Naphthal.	Anisole
−5	200	200	60	90	8	9	10	13	19	60

Operating Conditions:
Column: 50 cm × 4.6 mm I.D., 304 Stainless Steel
Mobile phase: heptane
Detector: UV at 254 nm
Column temperature: 27°C
Flow rate: 1.0 ml/min.

note that the two standard Partisil columns that have surface areas in excess of 400 m²/g can still only handle a total charge of any one solute of 300 - 400 μg without significantly affecting the efficiency obtained.

Column Permeability

The permeability of a packing is important because it determines the pressure required from the pump to provide the necessary flow. The permeability of an adsorbent can be calculated from the pressure drop across the column measured at a given flow rate and/or using the following equation suggested by Halasz [4]:

$$\Delta P = \frac{v \eta L}{k^\circ}$$

where ΔP is the pressure drop across the column (atm. $\times$ 10^6), v is the linear velocity of the mobile phase (cm/sec), L is the column length (cm), η is the solvent viscosity (poises), K° is the permeability (cm²). Where the particle diameter is in excess of 20 μ, the permeability is given approximately by the equation

$$K^\circ = \frac{dp^2}{1000}$$

where dp is the particle diameter.

Some results for a number of commercially available packings are given in Table 4.3. It is felt by the author that at the present time pressure should not be considered a limiting factor in liquid chromatography. Five years ago 500 psi was considered high pressure in liquid chromatography. Two years ago 3000 psi was readily available, and today 6000 psi is commonly used as a column inlet pressure. Already pumps providing 20,000 psi are being designed and it would seem that as high pressures are demanded the pump engineers can provide them.

The Reduced Plate Height Curves

The normal HETP curves, as already discussed, allows the comparison between one column and another and the lower the HETP curves the higher the efficiency and the better the column. However, if one compares a column packed with 20-μ silica gel with one packed with 5-μ silica gel unless the 5-μ silica gel column was very poorly packed indeed it will always provide a more efficient column than the 20-μ silica gel. The HETP curves do not allow the comparison of the packing procedure, because the results are masked by the effect of the particle diameter. In order to compare the quality of the packing procedure it is necessary to eliminate the effect of particle diameter and for this

Table 4.3 Column Permeability $k°$

Adsorbent	Pressure Drop $\Delta P_{heptane}$ (p.s.i.)	Pressure Drop ΔP_{H_2O} [p.s.i]	$k°_{hept}10^{-8}$ (cm^2)	$k°_{H_2O}10^{-8}$ (cm^2)	$K°_{average}10^{-8}$ (cm^2)	Calculated Value for d_p (μ)	Value for dp given by Manufacturers (μ)
Biosil A	50	120	0.94	0.92	0.93	~30	20-44
Partisil 10	556	1250	0.081	0.088	0.085	~9	9
Partisil 5	1250	2760	0.036	0.040	0.038	~6	7
Lichrosorb 30	45	112	1.00	0.98	0.99	~32	30
Lichrosorb 10	560	1240	0.081	0.089	0.085	~9	10
Biosil HA	48	120	0.90	0.92	0.91	~30	44+
Porasil C	20	50	2.25	2.200	2.23	~47	37-75
Vydac A	35	90	1.29	1.22	1.26	~35	30-44
Pellosil HC	27	72	1.67	1.53	1.60	~40	37-44
Corasil II	25	65	1.80	1.70	1.75	~42	37-50

Operating Conditions:
Column: 50 cm × 4.6 mm I.D., 304 Stainless Steel
Mobile phase: Heptane or water
Column temperature: 27°C
Flow rate: 1.0 ml/min.

reason Giddings [5] introduced the reduced plate height curve, which is as follows:

$$\text{Reduced plate height} = h = H/dp,$$

$$\text{Reduced fluid velocity} = v = udp/Dm,$$

where H is plate height of the column, dp is the particle diameter of the support, u is the linear velocity of the mobile phase, and D_m is the diffusivity of the solute in the mobile phase.

Values for h are calculated from measured values of H and the particle diameter of the adsorbent. Values for v are obtained from the corresponding values of the mobile phase velocity u, the adsorbent particle diameter and known or calculated values for D_m [6]. Some values of Dm for different solutes in different solvents are as follows and can be used for calculating reduced plate height curves.

Solute	Solvent	$D_m \times 10^{-5}$ cm²/sec
Benzene	n-heptane	2.47
Toluene	n-heptane	3.72
Benzene	carbon tetrachloride	1.53
Nitrobenzene	carbon tetrachloride	1.00
Phenol	chloroform	2.00

Since values of h and v are large and are often determined over wide ranges of values for u, it is usual in reduced plate height curves to plot $\log h$ against $\log v$. A set of reduced plate height curves from Knox and Kennedy [7] are shown in Fig. 4.7. They are presented as bands incorporating the extreme values for solutes of differing k' values. The lower the curves the better the quality of the packing, and it is seen that the Zipax column provides a significantly better packing than Porasil. However, it must be stressed that these curves only relate to the quality of the packing; the Zipax column may or may not be more efficient than the Porasil column, but the results show that the Zipax packs more efficiently and more nearly realizes its potential efficiency than does the Porasil column. The full reduced height equation is as follows:

$$h = \frac{2\gamma}{v} + Av^n + Cv.$$

Under the conditions normally used in liquid chromatography the first term is negligible. Over a moderate range of reduced velocity values the equation approximates to $h = dv^n$ where $n < 1$. This equation for limited reduced velocity ranges is a reduced version of the empirical Snyder equation $H = Du^n$ previously

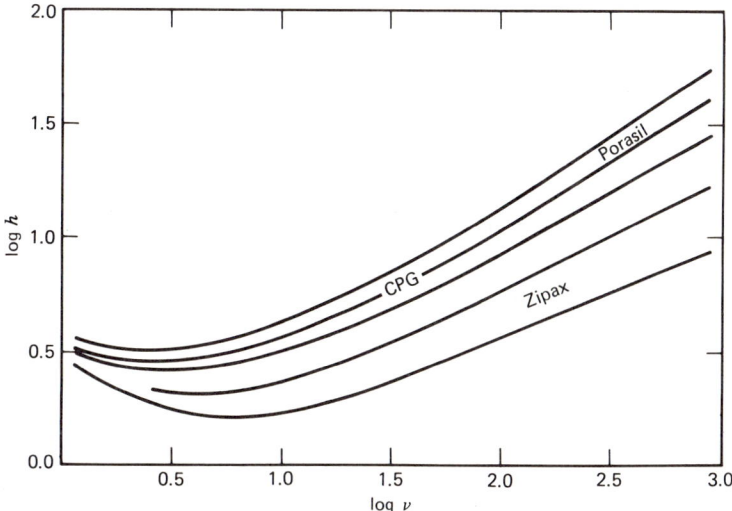

Fig. 4.7. Reduced plate height-velocity curves for Zipax, Porasil, and controlled pore glass chips: lower limits of bands for unretained solutes, upper limits for retained solutes.

discussed in Chapter II.

From plots of $\log h / \log v$ several conclusions can be drawn.

1. The lower the curves produced the better the column has been packed as stated previously.

2. The gradient of the curves for reduced velocities above 20 indicates the significance of the resistance to mass transfer in the stationary phase and static pockets of mobile phase. Thus for nonporous materials n should take a value of about 0.3 and for porous materials n will take larger values and increase with the reduced velocity v.

3. An increase in h for solutes having k' values in excess of unity also indicates the importance of slow transfer rates in stationary phase.

The use of the reduced plate height curve to identify the factors effecting dispersion, however, is no better, and perhaps less helpful, than the use of the normal HETP curves. The former, however, as previously stated does help in assessing the procedure used for packing whereas the normal HETP equation does not.

2 ADSORBENTS

Silica Gel

Silica gel is the most commonly used stationary phase in liquid chromatography. It is relatively inexpensive and in the correct form can provide columns of extremely high efficiency. Besides being used as a stationary phase per se, it is also used as the support material for the manurfacture of most bonded phases and thus, owing to the role it plays in the practice of liquid chromatography, some discussion of its formation, structure, and chromatographic characteristics is appropriate and pertinent.

Silica gel can be considered as a polycondensation product of orthosilicic acid. It can be prepared by precipitation from silicate solution with acids or by the hydrolysis of silicon derivatives such as silicon tetrachloride or ethyl silicate. X-ray examination has shown that silica gel is not crystalline and in fact consists of an agregate of elimentary particles about 100 Å in diameter. During the initial formation, a colloidal solution is formed that eventually coagulates to a hydrogel. On subsequent drying the polymerization continues, to produce a xerogel, which is a more dense polymer and is in fact the silica gel used in chromatography. The basic physical characteristics of silica gel that effect the chromatographic properties are the hydroxyl or water content (which for an unmodified silica gel is directly related to the surface area), the pore diameter, and pore volume. The macroparticle diameter also strongly influences the efficiency obtained from a column packed with silica gel, but this is not a basic property of the silica gel itself since the product can be ground or screened to any particle diameter that is required. The pore diameter is related to the primary or micro particle diameter, which is inversely related to the surface area. Thus, in general, the higher the surface area of the silica gel the smaller the primary particle and thus the smaller the pore diameter. The pore diameter affects the transfer rates of solutes through the particle and thus also affects the efficiency obtained from a column packed with silica gel to any given pore diameter. The surface area and pore diameter of any given silica gel are controlled both by the condition of precipitation as a hydrogel and by the condition under which it is further polymerized to the xerogel. The surface area does not seem to be strongly affected by changing the pH of formation between 1 and 5 but above a pH of 5, increase in pH of formation produces silica gels of progressively lower surface areas. Subsequent treatement of the silica gel at high pH also reduces the surface area and increases the pore diameter, as does heating the silica gel in contact with strong electrolytes over extended periods of time. Linsen [7a] devotes a chapter to the formation and physical and chemical properties of silica gel and provides a more detailed discussion than is appropriate to give here. The chemical nature of the silica gel surface has been, and still is, a subject of considerable controversy. It would appear that the nature of the surface can indeed be very complex consisting of hydroxyl groups, geminal hydroxyl groups containing two or three hydroxyl groups silanyl groups, and hydrogen bonded water of many different forms. The types of structure that can occur on the surface of the silica gel are depicted in Fig. 4.8. However, the actual chemical nature of the

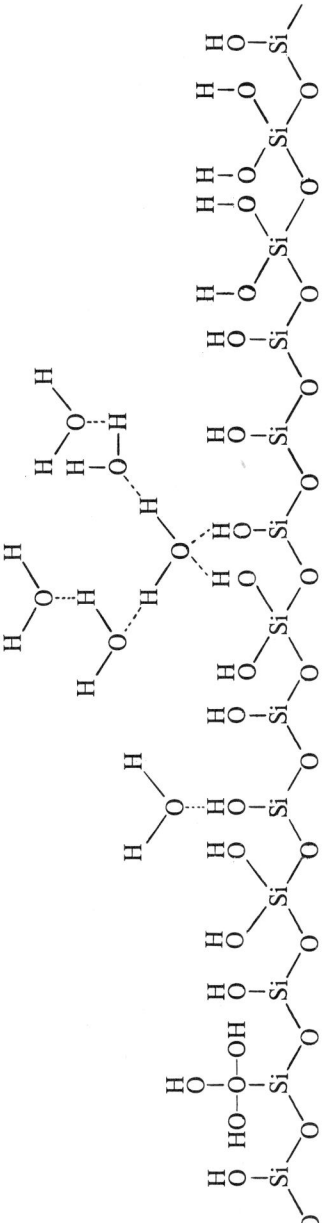

Fig. 4.8. Some possible structures of the surface of silica gel.

surface is not important from a chromatographic point of view, because it is the retentive and mass transfer characteristics of the silica gel that will affect the quality of the separations obtained from it.

It is usual to activate silica gel before use as column packing by heating it to about 200°C. Alternatively, the silica gel can be activated after packing by using a series of appropriate solvents. The effect of thermal treatment of silica gel on its chemical and physical properties will depend on the temperature of modification.

In Fig. 4.9 the residual loss on ignition expressed as water retained by a silica

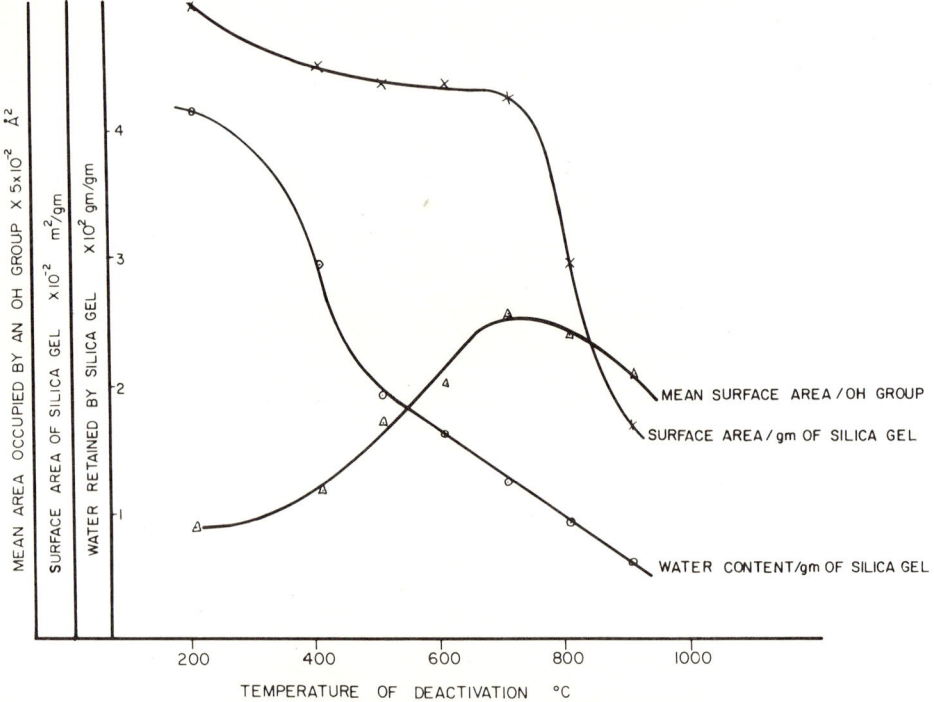

Fig. 4.9. Curves relating the change in physical properties of silica gel with deactivation temperature.

gel is plotted against modifying temperature. Included in the same figure are the graphs relating surface area per gram of silica gel and mean surface area per OH group, also plotted against the temperature. It is seen that between 200 and 500°C the water content of the silica gel falls rapidly; however, between 500 and 900°C the rate of loss of water is reduced, and subsequent to 500°C the water content falls linearly with the temperature.

Between 200 and 750°C there is only a slight change in the surface area per gram of the silica gel. However, between 750 and 900°C the surface area falls rapidly to about 35% of its initial value. As a result of the different rates of change of water content and surface area of the silica gel when heated between 200 and 900°C, the mean surface area per OH increases to a maximum between 700 and 800°C and then as a result of the rapid fall off in surface area is subsequently reduced. From the results it would appear that there is no obvious relationship between the change in surface area of the silica gel and its water content during the entire modification procedure.

Curves relating the corrected retention volume per gram of silica gel, modified at different temperatures, against the water content of silical gel for different solutes are shown in Fig. 4.10. It is seen that the curves show the character-

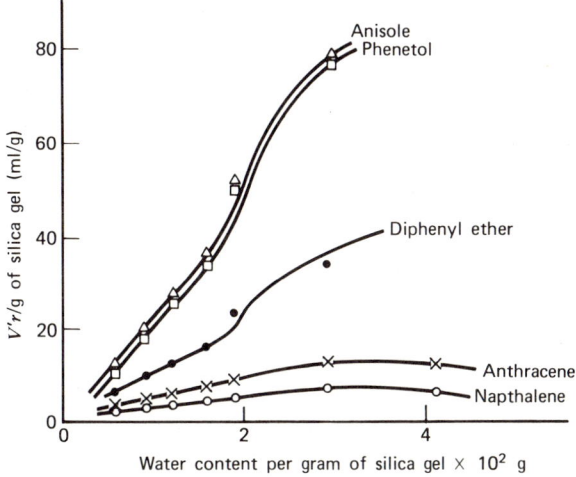

Fig. 4.10. Graphs of $V'r/g$ against water content of silica gel for different solutes.

istic increase in retention for water contents between 0.6 and 3% followed by a subsequent decrease for water contents betweeen 3 and 4%. Comparing these results with those shown in Fig. 4.9 it is seen that between 200 and 400°C the silica gel is activated with respect to solute retention, whereas above 400°C the silica gel is deactivated. It should also be noted that between water contents of 0.6 and 1.9%, representing silica gels deactivated between temperatures of 600 and 900°C, the relationship between the corrected retention volume per gram and water content of the silica gel is linear and that there is no intercept on the retention axis for zero water content of the silica gel. Thus, the residual siloxyl groups appear to contribute little or nothing to the retention of the solutes and thus display neither polarity or dispersive retention. It should be pointed out, however, that the mobile phase employed was heptane, which will exhibit only dispersive forces on the solutes, and these will be relatively large in nature. It is possible that some dispersive effect of the siloxyl groups might be shown if more polar solvents of significantly lower molecular size were employed as the mobile phase, for example, methanol with an alternative series of solutes. It also follows that the retention characteristics of silica gel in liquid chromatography are wholly dependent on the hydroxyl or water content of the silica gel. Furthermore, for silica gels deactivated at 600°C and above, the retention of a solute is linearly related to the water content and, therefore, indicates a chromatographic homogeneous surface with respect to the OH groups present.

It is interesting to note that silica gels deactivated between 600 and 900°C exhibit significant changes in surface area. This change in surface area, however, has no effect on the retention characteristics of the silica gel, since the linear correlation between water content and retention volumes indicates that this is solely related to hydroxyl content of the silica gel. It follows that surface area per se plays no part in the retention characteristics of the silica gel and only appears to do so for unmodified silica gel where the surface is saturated with hydroxyl groups and thus the surface area is related directly to the hydroxyl and water content.

Whether all the OH groups contained by silica gel deactivated between 600 and 900°C are of the same type is uncertain and irrevelent from the chromatographic point of view, since it has been shown that they all exhibit the same type of retentive forces on the range of solutes examined. The constant nature of the surface for the silica gels deactivated between 600 and 900°C is confirmed by the retention ratio of the solutes to naphthalene as shown in Table 4.4. It is seen that between 600 and 900°C the silica gels provide sensibly the same retention ratios for each solute. It would appear that silica gel deactivated over this range of temperature performs as an ideal polar bonded phase, with the active moiety being simply OH groups.

The stability of silica gel deactivated thermally, however, is open to question since it will rehydrate in the presence of water and thus, although the above results illustrate the importance of water content as opposed to surface area on solute retention, silica gels deactivated at high temperatures could only be used in practice over extended periods of time under anhydrous conditions and in the absence of alcohol.

Because retention on silica gel is dependent on the hydroxyl or water content of the silica gel, separations achieved by it are almost solely based on solute polarity and any selectivity based on dispersion forces can only be obtained by choosing the appropriate mobile phase.

Characteristics of Silica Gel Packing

Silica gel packings are supplied in various particle sizes ranging from 35-60 μ to 3-7 μ and even smaller. Silica gels having particle sizes above 20 μ can be dry

Table 4.4 Retention of Ratios of Solutes Chromatographed Relative to the V'_R of Naphthalene on Thermally Deactivated Silica Gel

Temperature of Deactivation (°C) Solute	400	500	600	700	800	900
			Retention Ratios			
Benzene	0.58	0.56	0.53	0.55	0.57	0.56
Toluene	0.71	0.70	0.70	0.67	0.71	0.70
Ethyl benzene	0.67	0.69	0.64	0.67	0.71	0.69
Propyl benzene	0.56	0.60	0.60	0.62	0.66	0.63
Butyl benzene	0.54	0.57	0.55	0.60	0.65	0.61
p-Xylene	0.75	0.83	0.83	0.79	0.82	0.75
m-Xylene	0.78	0.88	0.81	0.82	0.83	0.77
o-Xylene	0.89	0.97	0.92	0.91	0.90	0.84
Mesitylene	0.77	1.03	0.95	0.95	1.00	0.84
Naphthalene	1.00	1.00	1.00	1.00	1.00	1.00
Anthracene	1.88	1.83	1.84	1.78	1.85	1.83
Phenyl methyl ether	12.17	11.02	9.47	8.96	9.32	9.18
Phenyl ethyl ether	11.79	10.41	8.86	8.53	8.88	9.29
Diphenyl ether	5.45	4.98	4.25	4.16	4.60	4.44

packed, but silica gels of particle diameter of 20 μ and less have to be slurry packed to provide efficient columns. Most silica gels are irregular shaped particles and have a close size range; for example, a 10-μ silica gel would have 80% of its particles having diameters ranging between 8 and 12 μ. As shown by the Huber equation in Chapter II, the smaller the particle diameter (dp) the higher the efficiency obtained from the column but the lower the permeability. It can be theoretically predicted that packings having particle diameters of 1 or 3 μ would provide advantageous performance, but at present there is not an effective reproducible technique available to pack such particles. A microphotograph of 10-μ particle diameter silica gel is shown on the left of Fig. 4.11. It is seen that particles are irregular and that the range of diameters is relatively small. On page 215 Fig. 4.11 shows a microphotograph of 10-μ spherical silica gel. The silica gel has been produced by hydrolizing an emulsion of ethyl silicate. It has been claimed that such spherical particles pack more easily, provide higher column efficiencies with lower pressure drop across the column for a given flow rate. Such columns certainly tend to produce low pressure drops relative to the irregular particles but their advantage in column packing, and resulting efficiencies seem to be minimal and need further substantiation.

Silica gel is also available as a pellicular packing where the silica is coated on the surface of glass beads. The photomicrograph on the right in Fig. 4.11 is of the pellicular type packing first introduced by Horvath and Lipsky [8] and con-

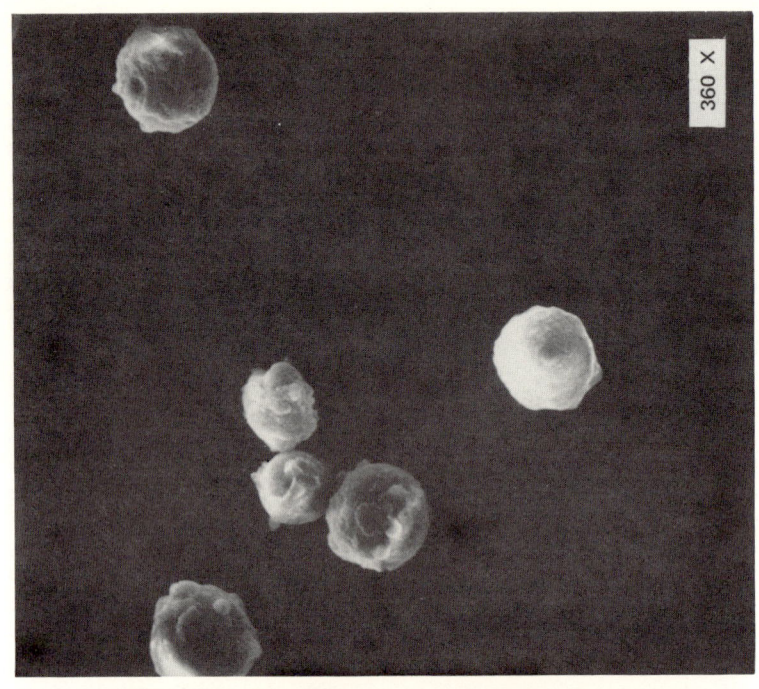

Fig. 4.11. Micrographs of different types of adsorbent particles.

214

Fig. 4.11. (continued)

sists of silica gel bonded to the surface of a glass bead. Such packing provide low pressure drops across the column; they permitted long columns to be used in the early days of liquid chromatographic development when high-pressure pumps were not available. The pellicular packing introduced by Horvath and Lipsky was the forerunner and perhaps the initiator of high-performance liquid chromatography. With the advent of high pressure pumps and the packing techniques introduced by Majors [1], the microparticulate silica gels eclipsed the pellicular packings by their vastly imporved efficiency and higher loading capacities. The use of pellicular packings is now largely confined to the separation of simple mixtures, particularly of wide polarity range. Nevertheless, the pellicular packings played a very important role in the development of liquid chromatography. Silica gel either in the form of microparticles or in the form of pellicular packings also form the basis of the so-called bonded phases, which will be discussed later.

Silica gel is probably the most versatile and useful of all the adsorbents used in liquid-solid chromatography. It has only a very small catalytic activity and provides very good selectivity for polar solute mixtures. A silica gel column, as with most liquid-solid columns can be readily reequilibrated after an analysis using gradient elution and can, with reasonable care, be used for analysis over a period of many months. Whenever liquid-solid chromatography has been selected as the most likely chromatographic system to be used, then silica gel should be the first adsorbent to examine as the most likely stationary phase to provide a satisfactory separation. Silica gel will provide poor separation ratios between solutes of the same polarity but different molecular weight but large separation ratios for solutes of different polarity. One disadvantage of silica gel is its slightly acidic characteristics; where substances that are labile under acid conditions are being separated, then decomposition can occur and the separation would be best carried out using a neutral polar bonded phase or an alkaline adsorbent such as alumina.

Alumina

Alumina is an adsorbent similar to silica gel insofar as it separates substances on a basis of polarity. However, it is in general not as commonly used as silica gel, largely because it has a high catalytic activity and has not yet been optimized to the same extent as silica gel for use in liquid chromatography.

Alumina is normally obtained by dehydration of the trihydroxide. There are many forms of alumina, each one denoted by a Greek letter suffix. The alumina normally used in liquid chromatography is γ alumina produced by heating aluminium hydroxides to temperatures not exceeding 600°C. γ alumina has the emperial formula $Al_2O_3 nH_2O$ where $0 < n < 0.6$. Much controversy exists with respect to the exact nature of the dehydration process and the precise structure of the surface. The surface area of alumina can vary from 1 to about 200 m^2/g, having water contents of 80 to about 26 g per 100 g, respectively, the surface area is conditioned largely by the aging process in its manufacture and by the subsequent temperature of activation. The use of alumina in liquid chromatography is not extensively reported in the literature and is usually avoided owing to its catalytic activity. It is, however, alkaline in nature and thus is often examined as an alternative to silica gel when substances that are labile in contact with acid media are being separated. Alumina is now available in the form of 10- and 5-μ-diameter particles, and for such materials the slurry packing method is usually employed. The microparticulate packing in a spherical form is manufactured by the Atomic Energy Estabilshment at Harwell England and is supplied by a number of distributors. It is claimed that the 10-μ spherical alumina can be dry packed to provide the efficiency normally expected from a slurry packing procedure.

Bonded Phases (Modified Adsorbents)

A bonded phase consists of a basic substrate, usually silica or sometimes alumina, to which an organic moiety is attached. When in contact with a mobile phase the resulting chromatographic system acts as a hybrid between absorption and partition chromatography. The concepts of the bonded phase was introduced to obtain a stationary phase that exhibited some of the characteristics of liquid-liquid chromatography while maintaining the stationary phase stability of the liquid-solid system. The bonding of organic material onto the surface of silica gel has been known for some time [9, 10]; one of the first papers describing the use of such materials in chromatography was published by Halasz and Sebestian [11], who esterified Poracil C with 3-hydroxypropionitrile and used it as a stationary phase in gas chromatography. Since that time the possibility of the use of bonded phases in liquid chromatography has been extensively studied and an excellent text on the subject, edited by Grushka [12], gives a good account of the present status of both the production and use of such materials as stationary phases for both gas and liquid chromatography. The need for such materials arises from the fact that the presently effective absorbents, for example, silica gel and alumina, retain solutes almost exclusively on a basis of polarity and thus the separation of mixtures containing substances of similar polarity but different molecular weight is extremely difficult. Very high column efficiencies are needed together with carfully choosen mobile phases since dispersive force selectivity can only be effected by the mobile phase in liquid-solid adsorption chromatography. Even under these conditions, however, minimum success is obtained and thus one of the first bonded phases to be made was a long-chain hydrocarbon linked to silica gel to provide separation based largely on dispersion forces. From a selectivity point of view, using an octadecyl bonded phase and a methanol water mobile phase, the bonded phase was highly satisfactory in that it provided selectivity between solutes of very similar polarity but different molecular weight. However, the columns produced from the early bonded phases had relatively poor efficiencies compared with the parent silica gel from which they were manufactured and, furthermore, required long periods for equilibration when the mobile phase composition was changed. Owing to the early difficulties experienced in the manufacture of bonded phases, it is of interest to consider their structure and how their structure might effect their ultimate chromatographic preformance.

Considering the surface of the bonded phase in the first instance, the ideal bonded phase should take the form shown in Fig. 4.12a. The organic moieties are attached to a silicon atom and thence to the substrate surface and are a sufficient distance apart to allow good contact with the solute molecule but at the same time they permit rapid mass transfer between the mobile phase and the bonded phase and provide columns of good efficiency. Normally bonded phases are produced by reacting the silica gel with the appropriate silanyl reagent. However, if all the OH groups present on the silica gel surface are reacted, then the organic moiety is too dense, resulting in poor efficiency from columns

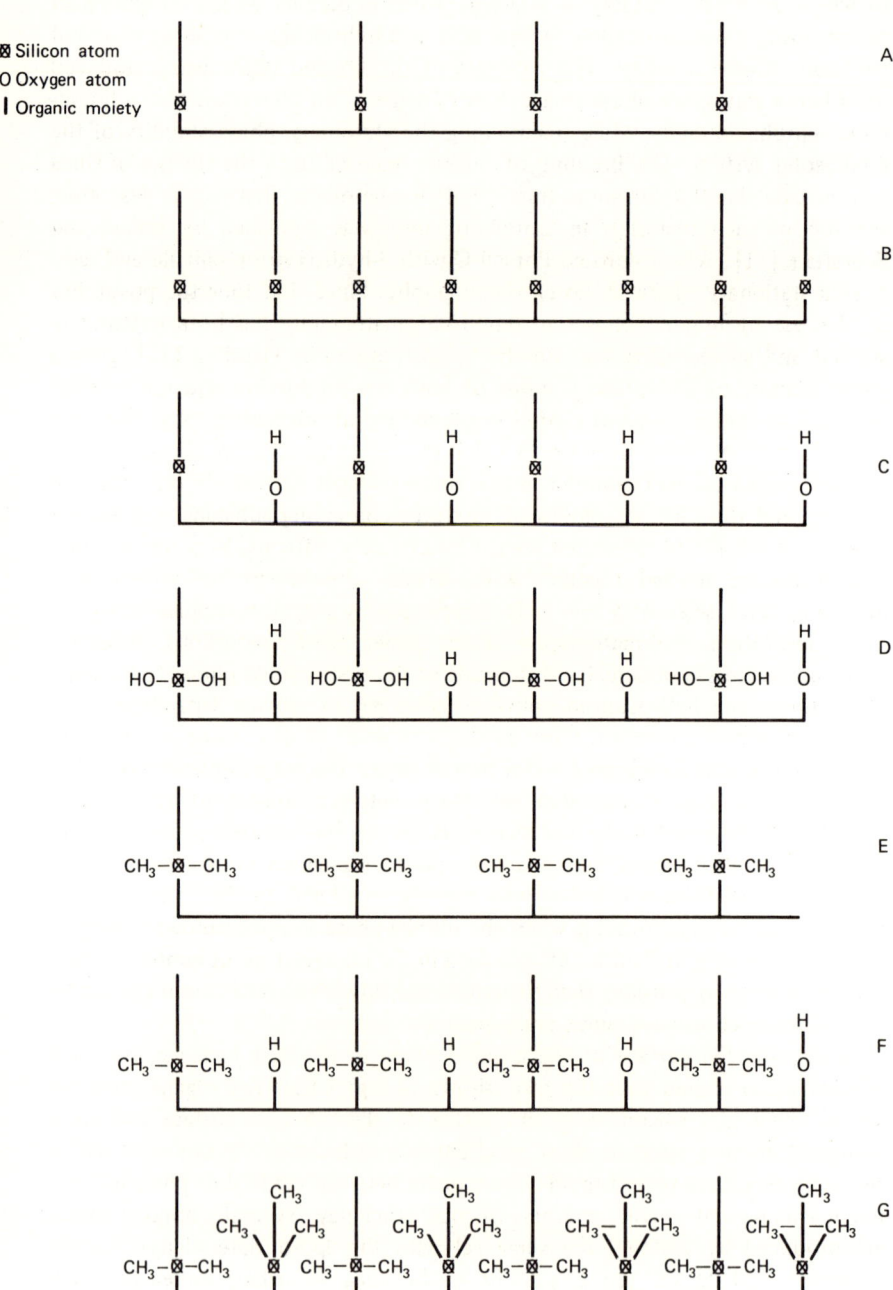

Fig. 4.12. Possible structures of the bonded phase surface.

packed with such material. This form of bonded phase is depicted in Fig. 4.12b. To avoid this problem the silanyl reaction is usually restricted, either by limiting the time of reaction or the quantity of reagent employed. However, this procedure results in a mixture of bonded material because the residual OH groups on the surface of the silica gel are still chromatographically active. Such a surface is shown in Fig. 4.12c. This is further aggravated if the trichloro silanyl reagents are used where the unreacted chlorine atoms hydrolyze to produce OH groups at the base of the bonded organic moiety as shown in Fig. 4.12d. The presence of such OH groups together with the residual OH groups on the surface of the silica gel not only modifies the characteristics of the bonded phase, which will vary with the amount of OH groups present, but also causes other problems when packed into the column. This is particularly so for semipolar and nonpolar bonded phases where the lyophilic polar surface resulting from the OH group present combined with the lyophobic nonpolar organic entity can result in packing difficulties, poor efficiency (relative to the efficiency obtained from the parent silica gel), packing instability (where the volume of the bed changes with solvent composition), and finally long equilibrium times when changing the mobile phase from one composition to another. Scott and Kucera [13] suggested that these problems could be eliminated by thermally deactivating the silica gel at a chosen temperature before reaction to provide the desired hydroxyl content and thus the necessary degree of bonding. An excess of the silanyl reagent could now be used and the reaction driven to completion since the degree of bonding would be controlled by the hydroxyl content of the silica gel. Furthermore, the use of the monochlorosilanyl reagents as opposed to the trichlorosilanyl reagents would eliminate the subsequent production of OH groups on the terminal silicon of the organic moiety. The surface resulting from such a reaction is shown in Fig. 4.12e. However, because aqueous solvents are to be used, then the remaining unreacted silica surface must be rehydrated at elevated temperatures and the OH groups produced between the bonded moieties deactivated with trimethylchlorosilane. These two stages of reaction are depicted in Figs. 4.12f and 4.12g. Again it should be emphasized that the monochloro reagent must be employed to eliminate the further production of OH from unreacted chlorine atoms remaining on the terminal silicon atom. Using thermally deactivated silical gel in this way could eliminate the undesirable effects of the lyophilic-lyophobic characteristics of the bonded material together with the practical difficulties that can arise from it. Furthermore, the use of thermally deactivated silica gel would allow a more reproducible bonded phase from the point of view of a well-controlled organic content to be produced.

The above discussion, however, only deals with the substrate surface, which in the case of silica gel and alumina takes the form of microscopic pores inside the particle of substrate. If these pores are roughly the same radius as the organic moiety, then the pores will be blocked and only the external surface will be available for solute interaction. Furthermore, the pores that are large enough to permit the entry of solute molecules will exhibit serious resistance to mass transfer effect, producing very poor column efficiencies. In Table 4.5 the surface

Table 4.5 Pore Radii and Surface Areas of Different Silica Gels and Chain Lengths of Some Characteristic Organic Moieties Used in Bonded Phases

Silica Sample	Surface Area (m^2/g)	Mean Pore Radius (Å)
1	752	13
2	697	8
3	675	23
4	670	10
5	625	16
6	555	9
7	450	31
8	414	14
9	345	45.1
10	330	52
11	205	113
Organic Moiety	Chain Lengths (Å)	
Octa decyl $-\text{Si}-\text{C}_{18}\text{H}_{37}$	22	
$-\text{Si}-\text{C}_6\text{H}_5$	5.5	
Cyanopropyl $-\text{Si}-(\text{CH}_2)_3\text{CN}$	6	

area and mean pore radii of a number of silica gels commonly used in liquid chromatography are given together with the approximate lengths of the commonly used organic moieties for bonded phases. The values given are only approximate because they are obtained by summarizing the various bond distances of the atoms in the molecule. They are sufficiently precise, however, to allow a comparison between the pore diameter and chain length of the organic moietites in order to deduce the effect of pore diameter on the subsequent chromatographic performance of the bonded phase. From Table 4.5 it is seen that unless the pore diameter is significantly greater tha 45 Å the octadecyl moiety will completely block the pore since the bonded material will be located all around the perimeter of each pore. However, the pore radii given are mean values and the standard deviation of the radii about the mean will be large. Furthermore, the pores themselves are not likely to be cylindrical in shape but will have narrow "bottlenecks" that can become blocked even by the small organic moieties. Thus, in practice, the surface area of the silica substrate may need to be as low as 100 m^2/g to provide a satisfactory stationary phase. It is interesting to note that the more successful commercially available stationary phases have been made from 200-m^2/g silica gel. It should be pointed out, however, that as the pore size is increased and the surface area reduced, the mechanical strength of the silica gel may be seriously impaired, which would make the material unsuitable for high pressures. At this time all commercially available bonded stationary phases have two distinct disadvantages.

1. They have significantly lower efficiencies then the parent silica gel substrates.
2. They have very low loading capacities.

Both factors indicate that the pores of the substrate have been blocked with the organic moiety and the bonded phase is probably acting as a pellicular type packing with only the external surface being effective. In spite of the present defects, which it is hoped will be eliminated by the time this book is in print, the present-day bonded phases play an important role in liquid chromatography separations but they have not yet been manufactured in a way that allows their full potential to be realized.

Carbon

Carbon has not been used to any significant extent in liquid chromatography although it has been shown to have some useful applications in gas chromatography. However, it has considerable potential as an adsorbent for liquid-solid chromatography since it is hydrophobic in character and, in conjunction with polar or semipolar mobile phases, it could effect separations on a basis of molecular weight by London's dispersion effects. Theoretically, carbon should allow the rapid elution of polar materials while retaining nonpolar substances preferentially. In this way carbon could act as a complementary adsorbent to silica gel. The problem is to obtain carbon in a suitable physical form that will provide linear adsorption isotherms and have rapid mass transfer properties. If carbon could be made into a suitable stationary phase for liquid chromatography it could replace the octadecyl type bonded phase for reversed-phase chromatography.

Polyamide

Polyamides coated as a thin layer on glass beads are available as a proprietory packing. Employing the proprietory packing Pellidon, Rabel [14] has reported its use in the separation of the *o, m,* and *p* phentidines, sulfonamides, nucleotides, and quinones. Rabel also showed that polyamides were an effective stationary phase for the separation of food preservatives and tryptophan metabolites.

However, polyamide is more likely to be effective as a bonded-phase type of material than as a coating on a bead. When the bonded-phase problems have been solved then a polyamide type bonded phase could have many useful applications.

Ion-Exchange Media

Although not within the scope of this book, ion-exchange media are similar in nature of bonded-phase material and should be mentioned here. There are basically two forms of ion-exchange resins that can be used, the anion-exchange resins and the cation-exchange resins. The column packing can take the form of solid resin beads or resin-coated glass beads. Both types of packings are manufactured specifically for ion-exchange chromatography, the coated glass beads providing columns of high efficiency but somewhat less retention.

Cation Exhcange Resins

These resins have been used for the separation of nucleic acid basis, basic and neutral peptides, amino acids, and carbohydrates. The pH range of the mobile phase buffer solutions usually range between 2 and 9.

Anion Exchange Resins

These resins are largely used for the separation of organic acid, base salts, and acidic peptides. The pH range of the mobile phase buffer solution can also be between 2 and 9.

Both resins are most effectively used in conjunction with appropriate gradient elution systems, and under the correct operating conditions the resin-coated beads can be made to give very rapid separations. Recently, ion-exchange bonded phases have been introduced where the ion-exchange radical is bonded to the surface of a microparticulate silica gel. These materials are showing great promise because, although they are exhibiting the same disadvantages as other bonded phases, their performance relative to the standard ion-exchange bead or the pellicular ion-exchange media is extremely good.

3 SUPPORTS FOR LIQUID-LIQUID CHROMATOGRAPHY

The most inert support for liquid chromatography is Celite or some other similar diatomaceous material. The particle size of the support must be chosen commensurate with the column inlet pressure that is available. The smaller the particle diameter, the higher the efficiency of the column produced but the greater the pressure drop across the column. Generally, the particle diameter should be as small as possible and have a narrow particle diameter range. Ideally, for high efficiencies the particle diameter should be 10-20 μ, but such supports in liquid-liquid chromatography are extremely difficult to pack; therefore it is better to use particle diameter of about 40-60 μ. In the packing procedure it is important to avoid the production of fines and, therefore, the harder particles of support should be used. One way of ensuring that the particles used have a satisfactory mechanical strength, suggested by Huber [15], is to grind the support in a pestle and mortar using a much coarser grade of support than that required for the column packing. During the grinding process the more friable particles are reduced to dust and when the material is sieved to the grade required, the particles that remain are those which have the highest mechanical strength. For very small particle diameters this sieving process can be very tedious and an automatic sieving apparatus should be employed if possible. An example of a very good sieve for this purpose is the Allen-Bradley Sonic Sifter manufactured by Fisher Scientific. After grading, the support particles will carry a coating of dust. This is removed by washing with methanol and removing the dust by decanting the supernatant liquid. If the support is to be used for liquid-liquid systems where the stationary phase will be polar, the support can be used directly after the methanol wash and subsequent drying. If the support is needed for reverse-phase chromatography however, then it must be carefully silanized. The best method for doing this is to spread the dry graded support in a thin layer on the surface of a Petri dish, which is then placed in a desiccator over trimethylchloro silane. The support should be allowed to stand in contact with the chlorosilane vapor for a least 48 h and preferable 72 h. For effective reversed-phase chromatography the support must be completely silanized or the stationary phase will be stripped from the support by the mobile phase when used in the column. It should be pointed out that the instability of the reversed-phase liquid-liquid chromatographic columns is often due to the incomplete silanization of the support. The support is then removed from the desiccator and washed with dry alcohol. The use of dry alcohol is also very important since, if the freshly silanized support is allowed to come in contact with water or water vapor, the active sites of the support that have been chlorinated as opposed to silanated will revert to hydroxyl groups. In the presence of alcohol, however, the chlorine atoms will be replaced by the C_2H_5O groups.

Coating the support is quite a different procedure in liquid chromatography from that of gas chromatography since the stationary phase chosen will have a low molecular weight to improve the column efficiency and, therefore, it will also be fairly volatile. For this reason the stationary phase should be added directly to a weighed quantity of the support, initially being spread in drops over the surface. The loading of stationary phase on the support will depend somewhat on the mixture to be separated, but for most separations a stationary phase loading of about 40-45% w/w is recommended; this is approximately equivalent to the optimum loading discussed in Chapter II. The support and the stationary phase is then well mixed by gently stirring and the material is then ready for packing into the column. It should be stressed that the support and stationary phase must be well mixed to obtain homogeneous layer of stationary phase over the total surface of the support.

Silica gel has also been used in a support for liquid-liquid chromatography and, in fact, was one of the first employed by Martin and Synge [16] in 1946. However, using silica gel as the support can introduce some surface interactions with the solutes separated and thus the system may not be entirely liquid-liquid chromatography. If silica is used as the support, it can be coated before packing or after packing in situ in the column.

For coating involatile stationary phases before packing, 10 g of silica gel can be coated by mixing it with about 30 ml of chloroform containing the stationary dissolved in it at a level of about 7% w/v. The mixture is well stirred and the solvent removed in a rotary evaporator at 50-60°C. The material is then dry packed into the column in the usual way. A coating procedure after packing entails the use of an approximately 30% solution of the stationary phase in a suitable solvent, which is then slowly passed through the column at a flow rate of about 0.2 ml/min. The solution is then replaced by nitrogen and the column pumped free of the coating solution and the residual solvent evaporated by heating the column to 60-70°C. When the column is free of solvent the mobile phase is passed through the column, care being taken to ensure that the mobile phase is saturated with the stationary phase to prevent stripping the stationary phase from the coated support. Columns coated in this way under carefully controlled conditions can be highly reproducible. The insitu coating of supports with stationary phases have also been shown [17] to provide improved separations relative to those obtained from columns packed with precoated material.

4 CHOICE OF STATIONARY PHASE

The choice of the correct chromatographic system to use for a given mixture of solutes cannot be made with certainty and in most instances can only be confirmed by experiment. However, a staisfactory system can be chosen with a greater degree of probability from the nature of history of the sample or by a few simple tests. If the likely chemical nature of the compound of the sample is known, then the phase system can be chosen from the areas of application given in the literature for such mixtures. If nothing is known about the chemical nature of the sample and nothing can be deduced from its history, then its solubility will give some indication as to the most likely chromatographic system to employ. If the sample is completely soluble in organic solvents then a liquid-solid system can be employed.

Normally, silica gel would be the choice of adsorbent unless it was already known or was subsequently shown that the solutes were unstable on acidic media. If one or more of the components in the mixture are labile on silica gel, then alumina might be employed or possibly an appropriate bonded phase. If the substances were subsequently shown to be of similar polarity but of different molecular size, then a nonpolar bonded phase would be appropriate. If the solutes decompose owing to the catalytic activity of either silica gel or alumina, then a polar bonded phase having a propionitrile or other polar organic moiety could be used. In the author's experience, however, if adsorption chromatography is to be used, silica gel should be the first stationary phase to be tried. If the sample is insoluble in organic solvents but soluble in water giving a solution that is not neutral or is only soluble in dilute acid or alkali, then ion-exchange chromatography is a likely technique. If the sample is soluble in water and gives a neutral solution, then either ion-exchange chromatography or liquid-liquid systems could be appropriate or possibly reversed-phase chromatography using an appropriate bonded phase. If the sample contains either carbohydrates or peptides, both types of mixtures could be separaed by ion-exchange chromatography.

5 THE CHOICE OF PHASE SYSTEM FOR LIQUID-LIQUID CHROMATOGRAPHY

In liquid-liquid chromatography it is not possible to choose the stationary phase or mobile phase individually since the phases do not only have to exhibit the necessary selectivity for the components of the sample concerned, but the two liquids have also to be immiscible. There is obviously a wide range of solvent mixtures that can provide different polarities and at the same time be immiscible. Unfortunately, in liquid-solid systems, there is no rational method that has so far been devised that allows a suitable phase system to be predicted that would separate a particular mixture of solutes. A particular advantage of liquid chromatography lies in the fact that when a suitable immiscible phase system has been devised, either phase can be used as the stationary phase providing the support has been suitably prepared. This given liquid-liquid systems an added flexibility that liquid-solid systems do not possess.

Liquid-liquid systems are most frequently used where the separations required need to be based on molecular size or where strongly polar materials that are largely hydrophylic are to be separated. In the future, the present major uses of liquid-liquid chromatography are likely to be replaced by liquid-solid chromatography using suitable bonded phases. Although liquid-liquid systems potentially can separate a far wider range of solute types than those mentioned, most of them will lend themselves better to separations by liquid-solid systems and liquid-liquid systems will be resorted to where liquid-solid systems fail to produce a satisfactory separation. The types of separation where liquid-liquid systems are most likely to be employed as an alternative are those given above, and the choice of phase systems that will be discussed will be those pertinent to such samples. An exception will be where liquid-liquid systems are employed because the components of the mixture are labile and the catalytic activity of adsorbents preclude the use of liquid-solid systems. For labile materials the phase selectivity required may cover the complete range of partition characteristics and thus the possible range of phase systems that must be examined will be far more extensive. This will make the selection of the correct phase system more tedious and time consuming, but the basic procedure given below for the more specific types of solute mixtures will still need to be carried out.

To separate substances on a basis of molecular size, the ternary system octane-acetone-water or octane-methanol-water is a likely phase system to examine, the hydrocarbon layer being made the stationary phase using a suitable treated support. An arbitrary mixture of the three solvents is made up such that a two-phase system is produced and well equilibrated by shaking in a separating funnel. Five milliliters of each phase is then taken and placed in a 20-ml stoppered tube; about 1 mg of the mixture to be separated is then added and the mixture well shaken. A sample of each phase is then spotted onto a thin-layer plate and warmed to evaporate the solvent; the spots are then examined in the usual manner by charring or spraying. For a practical phase system the distribution ratio between the two phases should be between 3 and 6 and thus for separations based on London's dispersion forces the spot from the octane layer or upper phase should be 3-6 times as intense as that of the lower or aqueous layer. If such a distribution is not achieved then the acetone or methanol content of the mixture should be adjusted and the experiment repeated until the required distribution is obtained. However, having chosen the phase system the distribution coefficient is only an average value for the solutes and some may be strongly held in the stationary phase and not be eluted in sufficient time to be detectable. A short 6-in. column should, therefore, be packed with the support containing the stationary phase and mobile phase passed through it, and the sample then injected onto the column. No trouble need be taken in packing this column since this is a purely qualitative test. When no further material is eluted from the short column, the mobile pahse is replaced by a flow of stationary phase. This will remove any strongly held materials, and if no materials are eluted by the stationary phase then the solvent system can be used for the analysis. If, however, some material has been retained by the stationary phase, then the distribution experiment must be repeated choosing a solvent that provides a lower distribution coefficient for the mixture with respect to the stationary phase. In this way the liquid-liquid phase system can be selected and further subtle adjustments to the phase system can be made depending on the results obtained using the normal column.

For the separation of strongly polar materials the quarternary mixture of solvents heptane-ethyl acetate-methanol-water is a very useful and flexible phase system to employ. By carefully changing the ethyl acetate and/or methanol content of the mixture, the relative polarity of the two phases can be delicately adjusted. The distribution of the components of the sample between the two phases can be determined in the same manner as that described above using a thin-layer plate as a spotting test.

There are many other immiscible phase systems described in the literature that include solvents such as butanol, propanol, chloroform, and so on, but the possible solvent combinations that are capable of being employed for liquid-liquid systems are so numerous that it is impossible to discuss their individual characteristics. The two solvent mixtures given are extremely versatile and by carefully adjusting their composition, they can provide satisfactory distribution systems for a wide range of solute mixtures. The selection of the correct phase system for liquid-liquid separations can be a far more tedious procedure than for liquid-solid systems where the process of gradient elution using a suitable programing device can greatly facilitate the choice of the correct mobile phase or program for a specific separation.

When using proprietory packings for liquid-liquid chromatography, for example the bonded phase packings, then only the mobile phase solvent program has to be selected, and this can be determined in a similar manner to that described for liquid-solid systems. Some proprietory bonded phase packings for liquid-liquid chromatography can only tolerate certain solvents, which are given in the manufacturers literature on the product. Under these circumstances only the acceptable solvents can be used in the gradient elution system. It should be pointed out that such packing materials may not act strictly as liquid-liquid systems but are more likely to have the characteristics of both liquid-liquid and liquid-solid systems. It should also be noted that such proprietory packings for liquid-liquid chromatography can also exhibit some residual catalytic activity, and in their use with labile materials this should be borne in mind.

6 THE MOBILE PHASE

Ideally, the choice of the mobile phase in a chromatographic separation should be conditioned solely by the separation involved and provide the required selectivity and column efficiency to achieve the analysis in the minimum time. Unfortunately, however, the detector and the column itself impose limitations on the choice of the mobile phase. Generally, the wire transport detector provides the minimum restrictions on the mobile phase that can be employed in that the solvents used must only be reasonably volatile. The refractometer detector does not restrict the choice of the mobile phase under isocratic methods of development, but it cannot be used effectively for gradient elution. Furthermore, the refractometer detector is the least sensitive of all the practical detectors. The most sensitive general detector, the UV detector, cannot be used with solvents having significant UV chromophores; even with solvents having small adsorption in the UV range, gradient elution often results in significant baseline steps, resulting from both changes in extinction coefficient in the solvent and changes in refractive index. For adsorption chromatography it is necessary to have a series of solvents that cover the polarity range from heptane to water so that any particular solvent can be selected for isocratic operation or the whole series employed for incremental gradient elution to separate unknown mixtures or mixtures containing solutes having a wide range of polarities. For the rapid selection of the best mobile phase system it is best to examine the use of a wide number of solvents and not attempt the separation by the use of two or three solvents of extreme polarity by the method of mixing. To understand the reason for this, the processes that occur during gradient elution using three solvents must be considered in detail. However, if it is necessary to employ the UV detector, then it may be necessary to employ mixtures of solvents of diverse polarities to try to simulate the effect of a solvent of intermediate polarity, but this will be a poor but unavoidable compromise.

The introduction of a more polar solvent into the mobile phase passing through a liquid-solid chromatograph column produces two distinct effects in the subsequent development process. At the start, the initial mobile phase is displaced from the adsorbent by the more polar solvent together with those solutes that are less strongly held on the surface than the new solvent. These solutes are eluted together as a group with the front of the polar solvent as it breaks through the column. This part of the dvelopment is called the diplacement effect. Subsequent to the point of breakthrough of the polar solvent front, the polarity of the mobile phase gradually increases until pure polar solvent passes from the column. During this period the polarity of the mobile phase continually increases, whereas the polarity of the adsorbent, being virtually saturated with polar solvent, changes little. While this is occurring, further solutes are eluted; this is the second part of the development and is called the elution effect. When pure polar solute has been eluted for some time, the remaining solutes held on the adsorbent will not be eluted, because their molecular forces with respect to adsorbent are greater than those with respect to the polar solvent. Such solutes will remain in the column for a long period of time or until a solute that is more polar still is introduced into the mobile phase.

The development processes resulting from gradient elution by the method of mixing using three solvents is depicted diagrammatically in Fig. 4.13. It should be stressed that this diagram treats the process of gradient elution in an approximate manner; it is generally correct, but there are secondary effects that are not discussed or represented in the diagram since they would obscure the basic principles and not help in the understanding of the processes concerned. The diagram represents the effect of a linear gradient using three solvents, a nonpolar solvent P, a semipolar solvent Q, and a strongly polar solvent R. The vertical axis

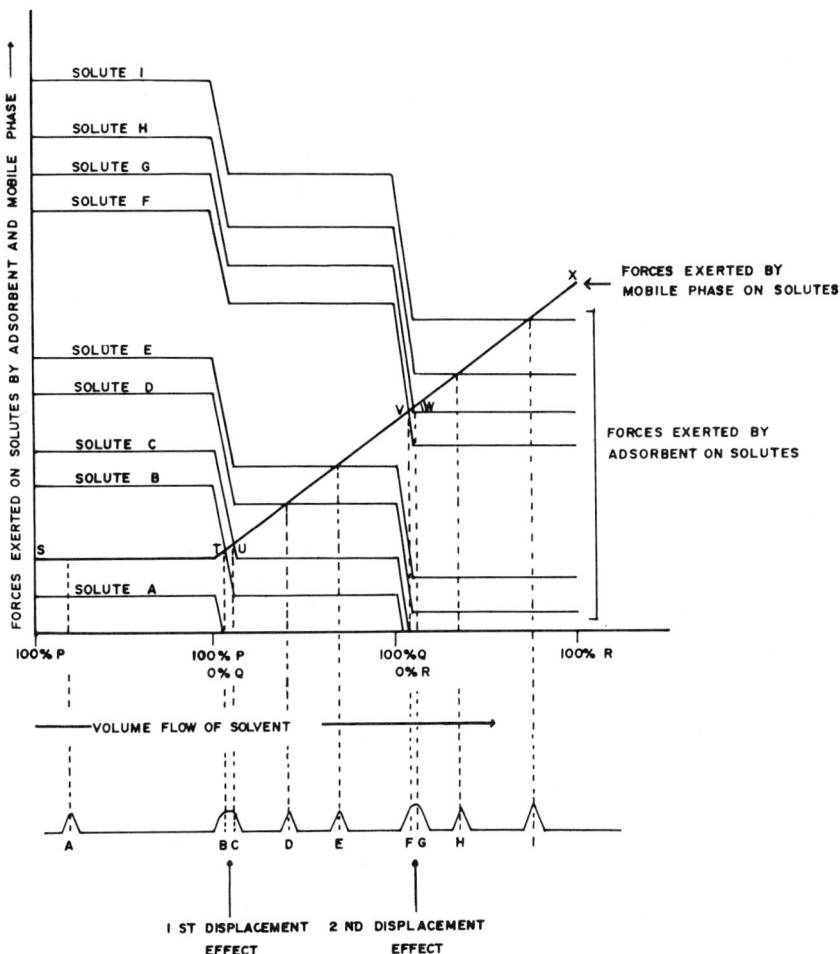

Fig. 4.13. Diagram describing the "displacement effect" resulting from gradient elution using solvents of widely different polarities.

represents the effective molecular forces on the solutes in the mobile and stationary phases and are in arbitrary units.

Consider the curves representing the forces exerted by the adsorbent on the solutes A to I. While a given solvent flows through the column the retaining forces remain constant. However, as soon as a more polar solute is introduced into the mobile phase, it is selectively adsorbed and partially deactivates the stationary phase so the forces acting on the solutes are all suddenly reduced to a new lower level. Exactly the same situation occurs when the third solvent, more polar still, is introduced into the mobile phase, producing another step in the curves. These steps representing the different levels of molecular forces that hold each solute onto the adsorbent during the changes of solvent are shown in Fig. 4.13. Now the forces acting on the solutes in the mobile phase are depicted by the heavy line, and it is seen that after an isocratic period of development (*St*) a linear gradient is used with respect to solvents Q and R (*tuvwx*). Below the curves the chromatogram resulting from the gradient elution is shown; the retention volumes of each solute will be explained by discussing the various development processes that occur.

Isocratic Period: Elution Development

After injection of the sample it is seen that solute A experiences greater forces on it by the mobile phase than the stationary phase and thus is eluted immediately at or close to the dead volume of the column. During the rest of the isocratic period between S and T no further solutes are eluted because all the rest have significantly greater forces acting on them by the adsorbent than by the mobile phase.

Gradient Process: The First Displacement Effect

On the introduction of the semipolar solvent Q into the mobile phase the nonpolar solvent P is immediately displaced from the adsorbent, resulting in its partial deactivation, and the forces between all the solutes and the adsorbent fall to a lower level. Veryl little of the solvent is required to saturate the adsorbent, and thus this will occur during the passage of a relatively small volume of mobile phase between t and u. Scott and Lawrence [18] determined the adsorbtivity of isopropanol on silica gel as a function of the concentration of the alcohol in heptane. The curve they obtained is shown in Fig. 4.14. It is seen that the silica gel is virtually saturated with isopropyl alcohol when its concentration in the heptane is still only between 0.25 and 0.5% v/v. Returning to Fig. 4.13, it is thus seen that the polarity, that is, the force between solute and mobile phase, changes little during the small volume of mobile phase eluted between t and u. However, owing to the rapid deactivation of the adsorbent during this volume flow, the molecular forces on solutes B and C exerted by the absorbent suddenly fall below the level of the forces between the solutes and the mobile phase. Thus between t and u both solutes B and C are eluted simultaneously or very close together and are unresolved. They have in effect been displaced from the adsorbent by the solvent Q together with the original solvent P. For this reason this process can be called the displacement effect of gradient elution.

Gradient Process: The First Elution Effect

During the interval between u and v the polarity of the mobile phase increases

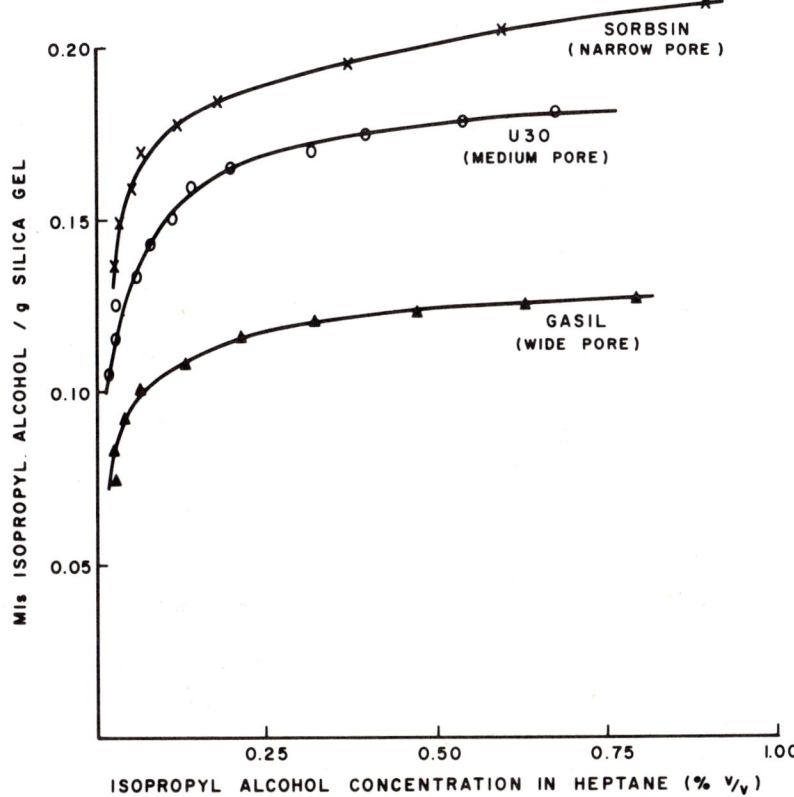

Fig. 4.14. Uptake of isopropanol from heptane by silica gel against the equilibrium concentration of isopropanol in heptane.

approximately linearly from a concentration of about 2% v/v of solvent Q in solvent P to 100% v/v solvent Q. During this interval the forces exerted by the mobile phase on solutes D and E exceed those exerted by the adsorbent and thus solutes D and E are eluted from the column. Since the relative forces between the solutes and the two phases change gradually between u and v, as opposed to the displacement effect between t and u, and furthermore, since the change takes place during the passage of a much greater volume of solvent through the column, solvents D and E are eluted discretely and well resolved. The elution of D and E results almost solely from the change in polarity of the mobile phase and can thus be called the elution effect.

Gradient Process: The Second Displacement Effect

The second displacement effect is exactly analogous to the first; on the introduction of the polar solvent R, the semi polar solvent Q is immediately replaced on the adsorbent by R, which causes its further deactivation, and the forces between the remaining solutes and the adsorbent fall to a new lower level. This process occurs during the passage of a relatively small volume of mobile phase (v and w), and between v and w the forces between solutes F and G and the mobile phase exceed that between these solutes and the adsorbent. Thus solutes F and G are displaced and eluted from the column close together and poorly resolved. This is the second displacement effect resulting from the introduction of the polar solvent R.

Gradient Process: The Second Elution Effect

The second elution effect commences after the second deactivation step is completed at w and extends to point x where 100 v/v pure solvent R is flowing through the column. Because the polarity of the mobile phase and thus the relative forces between the solutes H and I and the two phases change slowly during the stage wx, the solutes H and I are eluted discretely and well resolved. Since this step results again largely from a change in polarity in the mobile phase only, it can be called the second elution effect.

To separate an unknown mixture of solutes that can cover any part of the polarity range it follows that the polarity of the the mobile phase must be changed by use of a suitable gradient system from a nonpolar solute such as hexane to the most polar solvent, water. From the above discussion it can be also seen that if this is attempted using a very limited range of solvents (e.g., 3 or 4) then the displacement effect at each solvent change can result in poor resolution and in practice very often no separation whatsoever. Thus, to obtain a satisfactory gradient program that covers the complete polarity range and which would be applicable to any mixture of solute types, a large number of solvents must be employed. Increasing the number of solvents employed and thus reducing the deactivation steps of the adsorbent will minimize the displacement effect. This will result in a gradual change in mobile phase polarity and adsorbent deactivation and allow solutes having small polarity differences to be effectively separated. Gradient elution using a large number of solvents in this way has been given the term incremental gradient elution; a suitable solvent series for use in such a system has been determined by Scott and Kucera [2].

7 CHROMATOGRAPHIC PROPERTIES OF THE SOLVENT SERIES FOR USE WITH INCREMENTAL GRADIENT ELUTION

The forces holding a solute on the surface of an adsorbent are reflected in the value of its partition coefficient, which in turn is a function of the excess free energy of the solute adsorbed on the solid. Owing to competition between solute and solvent for the active surface, if the excess free energy of the solvent is

7 CHROMATOGRAPHIC PROPERTIES OF THE SOLVENT SERIES

significantly greater than that of the solute with respect to the adsorbent, the solute will no longer be adsorbed and will be eluted from the column. At a constant temperature, the retention volume of a solute will change linearly with the exponent of the excess free energy of adsorption. Thus, if the solvent series is to behave analogously to the Kelvin scale in temperature programed gas chromatography, then the excess free energy of adsorption must increase linearly along the solvent series. The evaluation of such an ideal series of solvents based on the experimental measurement of excess free energies of adsorption would require an inordinately large amount of experimental work. It was, therefore, necessary to compromise between what was theoretically desirable to what is practically possible.

The excess free energy of adsorption of a solvent, ΔG_n, will be a function of K_n, the partition coefficient of solvent n between solvent $n-1$ and the adsorbent for a given series of solvents. For a column of constant phase ratio and constant temperature,

$$-\Delta G_n = RT \ln (a k'_n)$$

where a is the phase ratio of the column, k'_n the capacity ratio of solvent n, and $k'_n = k_n/a$, and T is the temperature.

If the column is operated at constant flow rate, then

$$k'_n = \frac{V_n}{V_0} - 1 = \frac{t_n}{t_0} - 1$$

where V_n and V_0 are the retention volume of solvent n and the dead volume, respectively, and t_n and t_0 are the retention time of solute n and the dead time, respectively. It follows then that

$$-\Delta G_n = RT \ln \left[a \left(\frac{t_n}{t_0} - 1 \right) \right].$$

Thus if a solvent series is chosen such that $(t_n/t_0 - 1)$ for any solvent n measured in solvent $n-1$ is constant, then ΔG_n measured in solvent $n-1$ will also be constant.

Hence for the solvent series

$$\left(\frac{t_n}{t_0} - 1 \right)_{(n-1)} = k'_n \text{ measured in solvent } (n-1) \approx \text{constant.} \quad (4.1)$$

Although this condition provides a rational basis for a choice of solvents, it does little to indicate a useful and practical value for $\frac{t_n}{t_0} - 1$.

Preliminary work carried out by Scott and Kucera with a number of solvents indicated that the magnitude of the change in retention volume of a solute from solvent to solvent could from the basis for chosing a rational series of solvents. In order to restrict the number of solvents to a practical limited, their results indicated that the corrected retention volume of any solute A chromatographed in solvent n should be between two and three times that when chromatographed in solvent $(n + 1)$. Thus for a given column operated at constant temperature and flow rate,

$$(\frac{t_n}{t_0} - 1)_A \Big/ (\frac{t_{(n+1)}}{t_0} - 1)_A \approx 2.5 \qquad (4.2)$$

that is,

$$_A k'_n / _A k'_{(n+1)} \approx 2.5.$$

In practice this means that the corrected retention volume of any solute is about halved each time the solvent is changed. Owing to the limited choice of solvents available resulting from the practical constraints placed on them by column and detector considerations, the conditions imposed on the series by (4.2) cannot be obtained precisely. However, as will be seen later, the conditions can be met sufficiently closely to provide an extremely effective solvent series.

The conditions given in both (4.1) and (4.2) involve some function of ΔG, and as a result suffer the limitations associated with bulk property measurements. The distribution coefficient K, which is a direct function of ΔG, is a measure of the net forces holding the solute, or in this case the deactivating solvent, on the adsorbent. These forces can be polar or nonpolar in nature and neither K nor ΔG differentiates between the two. Depending on the chemical nature of the solutes, two substances having the same partition coefficient in a given liquid-solid system may be held on the adsorbent, one by predominately nonpolar forces and the other by predominately polar forces. It follows that if the solvent series is to elute substances in order of increasing polarity, then the dispersion force effect on each solute must decrease progressively along the solvent series. Now the dispersion force effect of each solvent can be considered as approximately proportional to its molecular size, which in turn can be approximated to its molecular weight; thus a third condition of the solvent series will be

$$M_n < M_{(n-1)} \text{ and } M_n - M_{(n-1)} \approx \text{constant.}$$

where M_n and $M_{(n-1)}$ are the effective molecular wights of solvents n and $n-1$;

7 CHROMATOGRAPHIC PROPERTIES OF THE SOLVENT SERIES

respectively.

Hence the three conditions that had to be met for an effective solvent series were

1. $(\frac{t_n}{t_0} - 1)_{(n-1)} = k'_n \approx$ constant, measured in solvent $n-1$
2. $(\frac{t_n}{t_0} - 1)_A \Big/ (\frac{t_{(n+1)}}{t_0} - 1)_A \approx 2.5$
3. $M_n - M_{(n-1)} \approx$ constant.

Scott and Kucera examined a large number of solvents and selected 12 for use with incremental gradient elution. The series of solvents together with their chromatographic characteristics and compositions are shown in Tables 4.6 and 4.7.

It is seen from Table 4.6 that the average value for k'_n chromatographed in solvent $n-1$ was 0.32, but individual values varied from a minimum of 0.14 for carbon tetrachloride in heptane to 0.51 for acetone chromatographed in methyl acetate. This range of values for k_n had to be tolerated if solvents that conformed to a criteria previously discussed were to be used. It is also seen from Table 4.6 that the average value for $_Ak'_n/_Ak'_{(n+1)}$ was 2.33 but individual values range from a maximum of 3.34 to a minimum of 1.56 (where $_Ak'_n$ is the value of k' for solute A chromatographed in solvent n and $_Ak'_{(n+1)}$ is the value of k' for solute A chromatographed in solute $n+1$). It should be noted that the ratio for water to methanol was obtained from the ratio of the k' values for acetic acid in methanol and water, respectively. Consider a solute that was eluted at a k' of unity with water as the mobile phase; then its potential k' value, Z, if chromatographed using heptane as the mobile phase can be calculated from the following equation:

$$Z = (\frac{_Ak'_1}{_Ak'_2}) \times (\frac{_Bk'_2}{_Bk'_3}) \times (\frac{_Ck'_3}{_Ck'_4}) \times \cdots .$$

Taking the values for $_Ak'_n/_Ak'_{(n+1)}$ from Table 4.6 it can be seen that

$Z = 2.34 \times 3.22 \times 3.34 \times 2.44 \times 2.06 \times 2.06 \times 1.72 \times 1.56 \times$
    $1.96 \times 2.48 \times 2.48$

   $= 8428$

Thus the solvent series can cover a k' range based on heptane as the mobile phase of about 10^4. If the actual k' range employed during the elution procedure is Y, then

Table 4.6 The Basic Solvents used for Incremental Gradient Elution Values for $k'_{(n+1)}$ and $k'_{(n+2)}$ Determined in Solvent n

Solute \ Solvent	n-Heptane 1	Carbon Tetrachloride 2	Chloroform 3	Ethylene Dichloride 4	2-Nitropropane 5	Nitromethane 6	Propyl Acetate 7	Methyl Acetate 8	Acetone 9	Ethanol 10	Methanol 11	Water 12
1 n-Heptane	0											
2 Carbon Tetrachloride	0.144	0										
3 Chloroform	0.651	0.286	0									
4 Ethylene Dichloride	1.800	0.750	0.233	0								
5 2-Nitropropane			0.450	0.300	0							
6 Nitromethane			2.15	0.612	0.148	0						
7 Propyl Acetate				2.812	1.003	0.485	0					
8 Methyl Acetate					1.162	0.565	0.232	0				
9 Acetone						1.082	0.638	0.370	0			
10 Ethanol							1.483	0.863	0.512	0		
11 Methanol								1.161	0.817	0.402	0	
12 Water									1.463	0.882	0.355	0
Average Value of $k'_{(n)} k'_{(n+1)}$	2.34	3.22	3.34	2.44	2.06	2.06	1.72	1.56	1.96	2.48	2.48	

Table 4.7 Physical Properties of Solvent-Solvent Mixtures Used for Incremental Gradient Elution

	n-Heptane	Carbon Tetrachloride	Chloroform	Ethylene Dichloride	2-Nitropropane	Nitromethane	Propyl Acetate	Methyl Acetate	Acetone	Ethanol	Methanol	Water	Theoretical Molecular Weight of Solvent-Solvent Mixture	Actual Molecular Weight of Solvent-Solvent Mixture	Actual Density of Solvent-Solvent Mixture
Basic Solvents Used for Series															
Molecular wt. of Solvent	100	154	119.5	99	89	61	88	74	58	40	32	18			
Density of Solvent	0.684	1.589	1.484	1.252	0.988	1.132	0.883	0.934	0.788	0.816	0.791	1.000			
Solvent No.	1	2	3	4	5	6	7	8	9	10	11	12			
1	100												100	100	0.684
2		100											154	154	1.589
3		57.8	42.2										140.4	140.4	1.545
4		36.1	26.1	17.6									126.8	126.8	1.435
5		19.9	14.5	20.7	44.9								113.2	113.2	1.214
6		14.4	10.5	14.4	32.4	27.8							99.6	99.6	1.206
7						16.3	61.7						86	86	1.047
8								100					72.4	74	0.928
9									100				58.8	58	0.788
10										100			45.2	46	0.816
11											100		31.6	32	0.791
12												100	18	18	1.000

Composition (% v/v) of Actual Solvent/Solvent Mixtures used for Incremental Gradient Elution

237

$$Y = 2.34 + 3.22 + 3.34 + 2.44 + 2.06 + 2.06 + 1.72 + 1.56 + 1.96 + 2.48 + 2.48$$

$$= 25.66.$$

Thus a mixture of solutes that contained substances that were eluted by heptane at one extreme to water at the other would have the total analysis time effectively reduced by a factor of $\frac{8428}{25.6} = 328$ compared with the theoretical analysis time if separated on heptane alone.

In the series, the solvent carbon tetrachloride is allowed to raise the mean molecular weight of the eluant to 154, and for the subsequent 10 solvents the molecular weight of the eluant was arranged to fall linearly to a molecular weight of 18 for water in accordance with condition 3. The increment of decrease in molecular weight from solvent to solvent was 13.6 and to achieve this, mixtures were made up using solvents previously employed in the series to achieve the correct mean molecular weight. The basic solvents and their properties together with the composition of the actual solvents used are shown in Table 4.7. It is seen that only solvents 3-7 needed to be mixtures, the molecular weight of the remaining solvents being sufficiently close to the required value to be acceptable. Where solvent mixtures were employed to obtain the desired mean molecular weight, the new more polar solvent introduced in the mixture was always maintained in excess of 25%. This ensured that the polarity effect of the new solvent was maintained at a sufficiently high level to achieve the necessary deactivation of the adsorbent and the elution of the appropriate solutes.

Expressing a series of solvents in order of their deactivating capacity toward silica gel, that is, expressed as their k' values relative to heptane (the k' values being related to a standard column where a is constant), provides a rational series of solvents based on the retentive characteristics of the silica gel. Such a series of solvents would allow a logical choice or lead to an experimental procedure that would permit the rapid identification of the best solute to use for a particular separation. A test separation of an unknown mixture using an intermediate solvent n would either provide the required separation, result in a chromatogram in which all solutes were eluted close to or at the dead volume or alternatively a chromatogram where the solutes were strongly retained or not eluted at all. In the former case, the solvent number would be reduced to one of less polarity to provide greater retention and selectivity, and in the latter case the solvent number would be increased to one of greater polarity to reduce retention times and elute the whole of the sample from the column. A series of solvents applicable to liquid-solid chromatography in order of their deactivating capacity toward silica gel is shown in Table 4.8 together with those physical properties pertinent to liquid chromatography. It should be noted that, with few exceptions, solvents of intermediate polarity between butyl chloride and tetrahydrofuran all absorb strongly in the UV region normally used for detection purposes. It is obvious that this range of solvent polarity is not available for use with the UV detecting system and indicates that a complete polarity scan of an unknown sample using incremental gradient elution is not possible. When using the UV detector large polarity steps must be taken over various portions of the polarity scale, resulting in significant displacement effects and loss of resolution. One way of partly solving this problem with the UV detector is to use very low concentrations of polar solvents in nonpolar solvents such as heptane or chloroform, as in the moderated mobile phase systems described by Maggs and Young [19]. Maggs and Young attempted to simulate intermediate polarities by using mobile phases consisting of fractions of a percent of isopropanol in heptane in an attempt to extend the displacement steps resulting from the deactivation of the silica gel and thus achieve an elution type development. As explained already, this is a poor compromise with regard to the elution procedure and, furthermore, the required level of moderator is extremely difficult to control in practice.

Other gradient elution systems have been suggested [20, 21] that offer alternative solvents and methods, but if the choice of system is restricted by the use of the UV detector or other detectors that respond to the changing properties of the solvent, then the elution procedure must be a compromise between what is necessary for the effective separation of the sample and what is possible with the detector employed.

When using the UV detector most separations of mixtures of substances of *similar* polarity can be achieved by isocratic development using a carefully selected solvent mixture containing largely heptane and chloroform together with 1-5% v/v of one or more of the solvents, ethyl acetate, tetrahydrofuran, propanol, or methanol. However, it is often time consuming to determine the correct solvent composition to obtain the separations required. A method based on such a system of solvents for scanning an unknown sample has been described by Rabel [22].

The less sensitive UV settings (nonlinear or ×64) are used initially in order to eliminate baseline drift due to the changing solvent systems. More sensitive settings can be used for quantitative work, because the solvent mixtures finally selected will probably not cause such steep changes in the gradient or baseline. The series of solvents and solvent mixtures selected are shown in the following list:

1. Heptane.
2. 5% Chloroform in heptane.
3. 2% Ethyl acetate in heptane.
4. 1% 2-Propanol in heptane.

Table 4.8 The Physical and Chromatographic Properties of Some Common Solvents

Solvent	Relative k' SiO$_2$ + Heptane	UV Cut off (nm)	RI	Density (20°C)	η (cP 20°C)	Dielect Const. (20°C)	Solvent Strength ϵ^o (SiO$_2$)	Dipole Moment (25°C)	Boiling Point (°C)	Mol. Wt.	Solubil Param.	UV Cut off (nm) 250 300 350
n-Pentane	0.00	205	1.358	0.6214	0.23	1.844	0.00	0.00	35.4	72.15	7.1	
n-Hexane	0.00	195	1.375	0.6548	0.32	1.880	0.00	0.09	68.7	86.18	7.3	
n-Hylum	0.00	197	1.388	0.6795	.42	1.924	0.01	0.00	98.4	100.21	7.4	
n-Octane	0.00	210	1.397	0.6985	0.55	1.948	0.01	0.00	125.7	114.23	7.0	
Cyclohexane	0.05	200	1.427	0.7739	1.00	2.023	0.03	0.00	80.7	84.16	8.2	
Carbon tetrachloride	0.14	265	1.466	1.5844	0.97	2.238	0.11	0.00	76.7	153.82	8.6	
Carbon disulphide	0.25	380	1.626	1.2555	0.37	2.641	0.20	0.00	46.3	76.14	10.0	
n-Heptyl chloride	0.50	220	1.427	0.8758	—	—	—	1.95	159.0	134.65	—	
n-Butyl chloride	0.05	220	1.402	0.8809	0.47	7.39	—	1.90	78.5	92.57	—	
Chloroform	0.66	245	1.443	1.4799	0.57	4.806	0.26	1.15	61.1	119.38	9.1	
Chlorobenzene	1.13	300	1.525	1.1009	0.80	5.708	0.23	1.54	131.7	112.56	9.6	
Benzene	1.13	280	1.501	0.8737	0.65	2.284	0.25	0.00	80.1	78.12	9.2	
Methyl iodide	1.13	350	1.531	2.2649	0.52	7.00	0.35	1.48	42.4	141.94	9.9	
Toluene	1.20	285	1.496	0.8623	0.59	2.379	0.23	0.31	110.6	92.14	8.9	
Dichloromethane	1.30	232	1.424	1.3168	0.44	9.08	0.32	1.14	39.8	84.93	9.6	
Techlorethylene	1.30	280	1.438	1.3292	0.90	3.42	—	0.80	87.2	133.41	—	
n-Propyl chloride	1.33	225	1.389	0.8850	0.35	7.70	0.23	1.97	46.6	78.54	8.3	
n-Butyl bromide	1.38	350	1.440	1.2686	—	—	—	2.08	101.3	137.03	—	
1,2-Dichloropropane	1.38	230	1.439	1.1590	—	—	0.25	2	96	112.99	—	
Ethyl Bromide	1.44	350	1.424	1.4505	0.41	9.39	0.29	1.90	38.4	108.97	8.8	
Isopropyl Bromide	1.51	350	1.425	1.3060	0.54	9.46	—	2.04	59.4	123.00	—	
1,2-Dichlorethylene	1.52	230	1.445	1.2463	0.79	9.2	0.38	1.76	83.8	96.94	9.7	
1,2-Dichlorethane	1.80	225	1.445	1.2458	0.89	10.65	0.38	1.86	83.5	98.96	9.7	
Phenetole	3.12	300	1.507	0.9605	1.36	4.22	0.32	1.36	170.0	122.17	—	
Anisole	3.85	300	1.517	0.9893	1.13	4.33	0.38	1.245	153.8	108.14	9.7	

Solvent											
2-Nitropropane	5.47	380	1.394	0.9829	0.77	25.52	0.41	3.73	120.3	89.10	—
Nitroethane	8.90	380	1.392	1.0446	0.68	28.06	—	3.60	114.1	75.07	—
Nitromethane	13.56	380	1.394	1.1313	0.67	35.87	0.49	3.56	101.2	61.04	11.0
n-Butyl ether	43.6	210	1.399	0.7641	0.74	3.083	—	1.18	142.2	130.23	—
n-Propyl ether	51.8	200	1.381	0.7419	0.44	3.39	0.30	1.23	89.6	102.18	—
Methybutyrate	56.9	250	1.388	0.898	—	5.60	—	1.7	102.0	102.13	—
n-Butyl acetate	62.0	254	1.396	0.8764	0.77	5.01	0.33	1.84	126.1	116.16	8.7
n-Propyl acetate	64.6	260	1.384	0.8830	0.56	6.00	—	1.78	101.6	102.13	—
Methyl propionate	67.6	260	1.377	0.915	—	5.50	—	1.7	79.7	88.10	—
Ethyl acetate	87.3	260	1.370	0.8946	0.47	6.02	0.38	1.88	77.1	88.10	8.6
Ether	90	215	1.353	0.7076	0.23	4.34	0.38	1.15	34.5	74.12	7.4
Methyl acetate	94	260	1.362	0.928	0.37	6.68	0.46	1.61	56.3	74.04	9.2
Methyl ethyl ketone	150	330	1.318	0.7997	0.43	18.51	0.39	2.76	79.6	72.11	9.3
Acetone	156	330	1.359	0.7844	0.32	20.70	0.47	2.69	56.5	58.08	9.4
Tetrahydrofuran	160	225	1.408	0.8842	0.55	7.58	0.35	1.75	66.0	72.11	9.1
n-Propanol	168	205	1.380	0.7998	2.3	20.3	0.63	3.09	97.2	60.10	10.2
Isopropanol	193	205	1.380	0.7813	2.3	19.9	0.63	1.66	82.3	60.10	11.4
Ethanol	377	205	1.361	0.7850	1.20	24.6	0.68	1.66	78.3	46.07	11.2
Methanol	546	205	1.329	0.7866	0.60	33.6	0.73	2.87	64.7	32.04	12.9
Water	1146	180	1.333	0.9971	1.00	80.3	—	1.86	100.0	18.02	21.0
Aceticacis	8430	210	1.329	1.0437	1.26	6.15	—	1.68	118	60.05	12.4

5. 5% 2-Propanol in heptane.
6. 10% 2-Propanol in heptane.

The above solutes are used sequentially and changed manually by turning off the pump and switching the tubing to the next solvent flask. Some care is exercised during this switchover so that air bubbles are not introduced into the system. Once the preliminary screening is done, the final form of the gradient is easily adjusted: Where peaks are crowding together, the slope of the gradient should be reduced, either by selection of different solvents or by lengthening the solvent changing time or by increasing the mixing chamber volume. Where too great a peak separation occurs, the reverse procedure is required. An example of the application of this procedure to the separation of a variety of mixtures is included in the authors original work.

Results from thin-layer separation can also help in the selection of a suitable solvent system in liquid chromatography. If the separation has been achieved by thin-layer chromatography then a similar solvent is likely to be effective for a silica gel column. However, since the development of a thin-layer plate is brought about by a form of gradient elution, the polarity of the solvent must be reduced slightly for use with the silica gel column. The solvents used for development in columns packed with nonpolar bonded phases are usually mixtures of methanol and water or acetonitrile and water. The relatively high viscosity of H_2O and MeOH is one of the reasons for the poor efficiencies from bonded phase columns. The range of composition appears to be between 5 and 50% v/v of water, and the correct solvent mixture can be assessed by experiment using progressively higher concentrations of water until the desired separation is effected. Because the viscosity of liquids decreases with increasing temperature, and the diffusion coefficient is increasing, the efficiency of these columns can be improved by operating them at higher temperatures, for example, around 60°C. This is discussed in Chapter II where mass transfer in the mobile phase is shown to be a function of

$$\frac{\nu^{1/6}}{D_m^{2/3}} \sqrt{u} \ .$$

REFERENCES

1. R. E. Majors, *Anal. Chem.*, **44**, 1722 (1972).
2. R. P. W. Scott and P. Kucera, *Anal. Chem.*, **45**, 749 (1973).
3. R. P. W. Scott and P. Kucera, *J. Chromatog. Sci.*, **12**, 473 (1974).
4. Halasz in *Modern Practice of Liquid Chromatography*, J. J. Kirkland, Ed., Wiley, New York, 1971, p. 329.
5. J. C. Giddings, *Anal. Chem.*, **34**, 1338 (1964).
6. *Perry's Chemical Engineers Handbook*, R. M. Perry, C. M. Chilton, and S. D. Kirkpatrick, Ed., McGraw Hill, New York, 1963, pp. 14-21.

7. G. J. Kennedy and J. H. Knox, *J. Chromatog. Sci.*, **10**, 549 (1972).
7a. B. G. Linsen, *Physical and Chemical Aspects of Adsorbents and Catalysts*, Academic, New York, 1900.
8. C. Horvath and S. Lipsky, *Anal. Chem.*, **41**, (10), 1722 (1969).
9. R. K. Iler and S. Pinksney, *Ind. Eng. Chem.*, **39**, 1379 (1947).
10. C. Rossi, S. Munari, C. Cengarie, and G. F. Tealdo, *Anal. Chem.*, **35**, 1253 (1963).
11. I. Halasz and I. Sebastian, paper presented at the Fifth International Symposium on Advances in Chromatography, Las Vagas, Nevada, January 1969.
12. E. Grushka, Ed., *Bonded Stationary Phases in Chromatography*, Ann Arbor Science, Ann. Arbor, 1974.
13. R. P. W. Scott and P. Kucera, *J. Chromatog. Sci.*, **13**, 337 (1975).
14. F. M. Rabel, *Anal. Chem.*, **45**, 957 (1973).
15. J. F. K. Huber et al., *Anal Chem.*, **44**, 111 (1972).
16. A. J. P. Martin and R. L. M. Synge, *Biochem. J.*, **35**, 1358 (1941).
17. J. J. Kirkland and C. H. Dilks, Jr., *Anal. Chem.*, **45**, 1778 (1973).
18. R. P. W. Scott and J. G. Lawrence, *J. Chromatog. Sci.*, **8**, 619 (1970).
19. R. J. Maggs and T. E. Young, *Gas. Chromatography 1968*, C. L. A. Harbourne, Ed., Institute of Petroleum, London, 1969, p. 217.
20. L. R. Snyder and D. L. Saunders, *Advances in Chromatography*, A. Zlatkis, Ed., Preston Technical Publications, 1969, p. 289.
21. L. R. Snyder, *Principals of Adsorption Chromatography*, Marcel Dekker, New York, 1968, p. 28.
22. F. M. Rabel, *Amer. Lab.*, **6**, 33 (1974).

Chapter V

CHROMATOGRAPHIC PROCEDURES

1 Column Packing 246

  Dry Packing 246
  Wet Packing 247
  Slurry Packing Procedure 247
    Preparation of Balance-Density Slurry 248
    Packing the Column 299
    Column Conditioning 249

2 Column Length and Linear Mobile Phase Velocity 250

  Preparation of Sample 256
  Injection of Sample 257
  Chromatogram Data 259

3 Measurement of Column Efficiency, Retention Ratios, and Peak Areas 259

  Measurement of Flow Rate 259
  Measurement of Dead Volume 260
  Measurement of Mobile Phase Velocity 260
  Measurement of Column Efficiency 261
  Measurement of Retention Ratios 263
  Measurement of Peak Areas 263

4 Practical Notes on the Operation of a Liquid Chromatograph 265

  Spurious Peaks, Their Source and Elimination 265
    Contaminated Syringe 265
    Contaminated Septum 265
    Air Dissolved in Sample 265
    Elution of Sample Solvent 265
    Detector Displacement Effects 265
    Random Spides on Recorder Trace 265
  Bubbles in Detector Cell 266
  Baseline Instability of the Recorder Trace 267

5 Qualitative and Quantitative Analysis 268

  Qualitative Analysis 268
  Quantitative Analysis 268
    Quantitative Measurements Using Peak Heights 269
    Quantitative Measurements Using Peak Area 269
    Procedure for Quantitative Analysis 270

The technique of liquid chromatographic analysis has not been developed to nearly the same extent as that of gas chromatography. The modern gas chromatograph is a very sophisticated, well-engineered instrument that can be operated successfully by a relatively inexperienced technician. The liquid chromatograph, however, is at a stage where the gas chromatograph was several years ago. For its effective use specialist knowledge is required and the degree of success that is obtained for the instrument will vary largely depending on the experience of the operator. The purpose of this section is to give some guidelines on how to operate a liquid chromatograph and, in particular, to help the novice who has little experience.

1 COLUMN PACKING

Contrary to popular opinion there is no magic associated with packing a column, but the procedure does need a little experimental skill, patience, and some experience. The packing procedure can only affect band dispersion and it follows that column efficiency, the HETP/u curve or, in particular, the reduced plate height curve are the pertinent methods of evaluation that can be made to indicate the quality of the packing. There are basically three methods of packing a column: the dry method, the wet method, and the slurry packing technique. Generally, adsorbents with particle diameters in excess of 20 μ can be packed by the dry method whereas for adsorbents having particle diameters of 20 μ or less, the slurry packing method of Majors [1] should be employed. The wet packing procedure is usually employed for liquid-liquid columns. If the dry packing procedure is used for liquid-liquid columns, then great care must be taken to ensure that the stationary phase is not lost by evaporation during the packing process.

Dry Packing

The end of the column should be closed by a small compressed wad of glass or Teflon wool about 2 mm in length contained by the column terminal union or, preferably, as an alternative to a Teflon wool plug, by a stainless steel frit. The pore size of the frit should be no greater than one-third of the mean particle diameter of the packing. The column is clamped in a vertical position and sufficient packing should be added to the column to fill about 2-3 cm of column length. The column should then be tapped laterally with gentle but firm strokes for 2-3 min, accompanied by occasional vertical taps on a hard surface. The column is then filled with an equivalent amount of packing and the process repeated until the column is filled. In the packing process the maximum packing density should be aimed for, which in simple terms means that as much packing as possible must be put into the column. The mobile phase supply (preequilibrated with stationary phase if a liquid-liquid column has been prepared) is then connected to the column and the mobile phase passed through it at about the maximum flow rate that will be required during the analysis. Flow rates of 2.0-5.0 ml/min are usually adequate for 4-5-mm-diameter columns and proportionately less for smaller diameter columns. The flow of the mobile phase should be continued for about 15 min, during which time the column is tapped. After the 15-min period of tapping or when the last of the air has been removed from the column, the mobile phase supply is turned off and disconnected. On examining the top of the column it may be found that the packing has fallen slightly and thus more packing must be added by the wet packing method described below to fill the column. Some workers [2] do not dry pack their columns by tapping but by ramming with a Teflon-tipped ramrod that just fits inside the column. The column is again packed in 2-3-cm lengths at a time, but the packing is gently compressed by the ramrod at each filling. This packing process is difficult since the required pressure must be great enough to compress the packing but not so great as to fracture the support or adsorbent particles and thus produce fines. This latter procedure is best tried by the more experienced liquid chromatographer, but the highest column efficiencies reported for dry packing have been obtained by this method. It should be pointed out that the dry packing method should never be used for ion-exchange resin packings because the resin swells when in contact with the mobile phase, and thus the wet packing method must be used to ensure that the swelling process is complete before preparing the column.

Wet Packing

The wet packing procedure is more tedious than the dry packing method but is recommended for liquid-liquid columns where the stationary phase is volatile and dry packing the column could result in loss of stationary phase by evaporation. The column should be filled with mobile phase that has been equilibrated with the stationary phase and sufficient packing added to fill about 3 cm of the column. The packing is allowed to fall by sedimentation and the mobile phase (equilibrated with the stationary phase) passed through the column at a flow rate slightly in excess of that to be used during development. The flow is continued for about 5 min with lateral tapping and the procedure is repeated, packing 3 cm of the column at a time until the packing is complete. Such a column is equilibrated during the packing procedure and is ready for immediate use. However, it is important to use a precolumn in liquid-liquid chromatography to ensure complete equilibrium between the two phases and prevent the stationary phase from being stripped from the column.

The Slurry Packing Procedure

At present, the only effective method of packing stationary phases having particle diameters below 20 μ is by the slurry packing procedure. Because of the importance of this packing technique a detailed method will be given (for a 50-cm by 4.5-mm I.D. column) based on that described by Majors [1]. However, other workers, for example, Kirkland [3, 4] have also demonstrated the use of this technique for packing bonded phases and porous silica microspheres. For columns of other dimensions the method can be modified by taking quantities proportional to the ratio of the column volume to that of the standard column described.

Equipment and Materials

Pump	Haskel air operated Model #17082 or similar
Reservoir	Stainless steel type 316, 50 cm × 15 mm I.D.
Column	Stainless steel type 304, 50 cm × 4.5 mm I.D. precision bore
Column outlet frit	2 μ stainless steel end frit
Pressure gauge	Gauge reading to 10,000 lb/in.^{2}
Compression fittings and tubing	As required to assemble the components
Ultrasonic bath	Any model of convenient bath size to accommodate a 250-ml tall-form beaker
Silica gel	Appropriate 10-μ-diameter silica gel heated at 200°C for 4 hr and cooled down to room temperature in a vacuum dessicator
Balanced density solvent	228 ml tetrabromoethane 270 ml tetrachloroethylene 2.5 ml methanol (All spectral quality)
Distilled water	
Ethanol	200 proof
Acetone	Spectral grade or distilled in glass. Each run through a 60 × 8 cm I.D. column packed with activated silica gel (150-mesh chromatographic grade). One column for each solvent.
Ethyl acetate	
Ethylene dichloride	
n-Heptane	

Preparation of Balanced-Density Slurry

The procedure is described with the experimental detail used for packing silica gel as an example. Thirty-five grams of the 10-μ silica gel is slurried with 350 ml of balanced density solvent and kept as a stock slurry. At least 4 hr should elapse between making up a new stock solution and using it for packing a column. Any variation in the density of the silica gel will require adjustment of the solvent composition, which is best made by a trial and error procedure. Having slurried a batch of the silica gel, the direction of migration of the silica is observed when the slurry is left undisturbed for 30 min. If the particles rise, more tetrachloroethylene is added and if they fall, more tetrabromoethane is added.

One hundred milliliters of homogeneously dispersed slurry is taken from the stock and placed in a 250-ml tall-form glass beaker and degassed by vibrating it in the ultrasonic bath for 15 min. The slurry is allowed to cool down to room temperature.

Packing the Column

1. Fit the 2-μm end frit to the end of the column via a 1/4-1/16-in. reducer and close the outlet end of the reducer with a 1/16-in. plug.
2. Fill the empty 50 cm × 4.5 mm I.D. column with the slurrying solvent (note this is the solvent only) taking care to ensure that there are no trapped air bubbles.
3. Connect the reservoir to the top of the column and carefully introduce 100 ml of the balanced density slurry, again taking care to avoid trapping or introducing air bubbles.
4. Complete the filling of the reservoir by adding distilled water.
5. Fill the pump liquid-end and connecting tubing with distilled water and connect to the reservoir.
6. Pressurize the system to 7000 lb/in.^{2} (i.e., approximately 170 lb/in^2 on the air piston).
7. Remove the 1/16-in. plug from the end of the column (caution: avoid solvent spray.)
8. Maintain the pressure until 100 ml of water has been collected (note that the water breaks through after all the solvent has been forced through the column, and it is at this point that water collection is commenced).
9. When 100 ml of water has been collected, shut off the pump at the air pressure source and allow the pressure to dissipate through the column. The column is then removed for conditioning with solvent.

A 50 cm × 4.5 mm I.D. column satisfactorily packed by this procedure should contain approximately 8 g of silica gel.

Column Conditioning

The packing in the column is conditioned by passing through it, in succession, approximately 7 column void volumes (about 40 ml) of the following solvents at a flow rate of 2 ml/min. Larger volumes will be required if pump and tubing dead volumes are not kept minimal:

1. Acetone.
2. Water.

3. Ethyl alcohol.
4. Acetone.
5. Ethyl acetate.
6. Ethylene dichloride.
7. *n*-Heptane.

The column is then ready for use.

An alternative method of slurry packing is to employ a solvent of high viscosity to keep the particles suspended, as opposed to a solvent having a density similar to that of the silica gel. This procedure was investigated by Anhauer and Halasz [5], who employed solvents having viscosities lying between 40 and 60 cP. For packing purposes, Halasz used pressures of about 7000 psi and completed the packing procedure in 15-30 min. The use of high-viscosity solvents for slurry packing without doubt can produce excellent columns, but generally higher pressures are required for packing purposes. If packing pressures of 12,000 psi are available then a methanol solution containing about 40% v/v of glycerol employed with 10-μ silica gel will provide in excess of 10,000 theoretical plates from a 25 cm × 4.5 mm column for solutes eluted at k' values between 3 and 5.

It would appear that the role played by viscosity in the packing procedure is somewhat complex. In the first instance the viscosity has to be sufficiently high to ensure that the movement of the support particles relative to the packing solvent must be small, and that significant segration of particles does not occur during the period of packing. However, if the viscosity is too high then the packing procedure is slow and thus results generally in poor column efficiencies. It is possible that higher pressures employed with solvents of greater viscosity would produce columns of significantly greater efficiency. However, the optimum conditions for viscosity slurry packing must await the results of a systematic study of the relationship between packing pressure, solvent viscosity, and particle diameter on the efficiency obtained from the packed column before such conditions can be given. Slurry packing using viscous solvents is generally recommended for workers inexperienced in slurry packing techniques, because high column efficiencies are more easy to obtain and columns packed in this manner are more reproducible. Subsequent to packing, the columns are conditioned by the same procedure as that used for columns packed by the balanced-density method.

2 COLUMN LENGTH AND LINEAR MOBILE PHASE VELOCITY

Both the linear mobile phase volocity and the column length control the efficiency from a given column and thus its potential resolution.

In order to increase the number of theoretical plates from the column, it is

2 COLUMN LENGTH AND LINEAR MOBILE PHASE VELOCITY

possible either to increase the column length or to reduce the liquid mobile phase velocity; owing to the form of the HETP curve for liquid chromatographic columns, halving the flow rate nearly doubles the column efficiency. The linear relationship between column efficiency and the increase of the mobile phase velocity is an approximation. It will be true for solutes eluted at a k' value exceeding 3 to 4 since the major factor controlling the band dispersion will be the resistance to mass transfer in the stationary phase, which is linearly related to the mobile phase velocity. For solutes eluted at k' values less than 3, the convective mixing term and the resistance to mass transfer in the mobile phase are the controlling factors and, as has been shown in Chapter II, these are not linearly related to the mobile phase velocity. Reduction in mobile phase velocity will reduce these dispersion terms and thus increase the column efficiency for solutes eluted at a k' of less than 3 but not linearly. However, since most mixtures separated in liquid chromatographic columns extend from a k' of zero to at least 10 or 20, reduction in flow rate is as valid a way of improving the resolution as increasing the column length.

In gas chromatography there is a minimum in the HETP curve, and thus reduction of the mobile phase velocity is limited to a value of about 2-3 cm/sec. Below this mobile phase velocity the value of the HETP will increase again, owing to the longitudinal diffusion effect, and the efficiency again falls. In liquid chromatography the minimum mobile phase velocity is much lower and, in fact, is in practice hardly usable, so with this provision the liquid mobile phase velocity can be progressively reduced to obtain theoretically whatever efficiency one requires. Why, therefore, it is necessary to double the column length to obtain twice the efficiency? It would be simpler to reduce the flow rate to ½, and the alternative would result in the same retention time and resolution.

This question can be quantitatively examined using the plate theory. Consider a column of N plates with the plate volume for the most retarded solute being $(v_m + Kv_s)$. Let the most retarded solute be injected under conditions such that the adsorptive capacity of the stationary phase is just not exceeded. Now from the plate theory the characteristics of the solutes eluted from this column (1), are as follows:

$$\text{Retention volume of solute} = n(v_m + Kv_s) = Vr_1$$
$$\text{Width of solute peaks} = 2\sqrt{n}(v_m + Kv_s) = 2\sigma_1$$
$$\text{Maximum volume of charge} = 0.1\sqrt{n}(v_m) = Vs_1.$$

Now consider a column of twice the length but packed under identical conditions so that only n changes by a factor of two but the plate volume remains the same. Thus for column 2,

$$\text{Retention volume} = 2n(v_m + Kv_s) = Vr_2$$

Peak width of solvent peak = $2\sqrt{2n}(v_m + Kv_s) = 2\sigma_2$

Maximum volume of charge = $0.1\sqrt{2n}(v_m)$ = Vs_2.

Finally, consider the original column operated at ½ the flow rate. Assuming the HETP curve is approximately linear then for column 3 its efficiency will be doubled and its plate capacity halved. Thus

Retention volume of solute = $\dfrac{2n(v_m + Kv_s)}{2}$ = Vr_3

Peak width of solute band = $\dfrac{2\sqrt{2n}(v_m + Kv_s)}{2}$ = $2\sigma_3$

Maximum volume of charge = $\dfrac{0.1\sqrt{2n}(v_m)}{2}$ = Vs_3.

Now the resolution obtained from a column will be proportional to $V_R/2\sigma$. Furthermore, the peak height of a solute band can be taken as proportional to $1/2\sigma$. The comparison of the resolving power, concentration of solute at peak maximum, charge volume, and retention volume for columns 2 and 3 with respect to column 1 is shown in Table 5.1. Considering the results shown in Table 5.1, it is clearly seen that the same improvement in resolving power is achieved for both columns 2 and 3, namely, 1.414 and, because the flow rate has been halved for column 3, both columns 2 and 3 have achieved this at the expense of doubling the analysis time. Column 2 will produce a solute band having a reduced peak height but can now tolerate a proportionally larger volume of charge. Conversely, column 3 will produce solute bands having an increased peak height but will only tolerate a proportionally reduced volume of charge. Thus the total mass of solute that can be placed on all three columns will remain the same but as a result of the increase in efficiency both columns 2 and 3 will provide the same increase in resolution.

What are the advantages, therefore, of using longer columns? The longer columns can be used to advantage where a solute contained in the mobile phase has a solubility so small that, as there is a maximum sample volume that can be used, it limits the quantity that can be placed on the column without impairing resolution and consequently limits the column loading. The column length can also be increased where the required flow rate to give the necessary resolution on a shorter column is so small (less than 0.05 ml/min) that effective control of the flow rate becomes practically difficult or where the value of the linear mobile phase velocity is reduced to the point where the approximation that h varies linearly with u no longer applies.

One final limitation of reducing the flow rate to provide higher efficiency occurs when using extremely high-efficiency columns where the peak dispersion

Table 5.1 The Comparative Properties of Columns of Different Lengths and Operated at Different Flow Rates

Column	Relative Resolution	Relative Peak Concentrations	Relative Maximum Charge Volume	Relative Retention Volumes
2	$\dfrac{\dfrac{2n(v_m+Kv_s)}{2\sqrt{2n}(v_m+Kv_s)}}{\dfrac{n(v_m+Kv_s)}{2\sqrt{n}(v_m+Kv_s)}}$ $=\sqrt{2}=1.414.$	$\dfrac{2\sqrt{n}(v_m+Kv_s)}{2\sqrt{2n}(v_m+Kv_s)}$ $=\dfrac{1}{\sqrt{2}}=0.707$	$\dfrac{0.1\sqrt{2n}(v_m+Kv_s)}{0.1\sqrt{n}(v_m+Kv_s)}$ $=\sqrt{2}=1.414.$	$\dfrac{2n(v_m+Kv_s)}{n(v_m+Kv_s)}$ $=2$
3	$\dfrac{\dfrac{2n(v_m+Kv_s)/2}{2\sqrt{2n}(v_m+Kv_s)/2}}{\dfrac{n(v_m+Kv_s)}{2\sqrt{n}(v_m+Kv_s)}}$ $=\sqrt{2}=1.414.$	$\dfrac{2\sqrt{n}(v_m+Kv_s)}{2\sqrt{2n}(v_m+Kv_s)/2}$ $=\sqrt{2}=1.414.$	$\dfrac{0.1\sqrt{2n}(v_m)/2}{0.1\sqrt{n}(v_m+Kv_s)}$ $=\dfrac{1}{\sqrt{2}}=0.707$	$\dfrac{2n(v_m+Kv_s)/2}{n(v_m+Kv_s)}$ $=1$

is very small. If the efficiency is increased further by reducing the flow rate, the extra column dispersion effects become the predominant dispersion effects and, thus, further reduction of the flow rate produces little improvement in efficiency. This reason for using longer columns is probably the most important.

It is of interest to consider the extra column dispersion effects that can be tolerated from the 25-cm column, 4.6 cm I.D. packed with 10-μ silica gel and producing 10,000 theoretical plates. The dead volume V_0 of such a column will be about 3 ml and thus a solute eluted at a k' of 2 will have a retention volume of 9 ml. Now

$$\sigma^2 = \sigma_C^2 + \sigma_E^2$$

where σ is the standard deviation of the eluted peak, σ_C the standard deviation of the peaks resulting from column dispersion, and σ_E the standard deviation resulting from extra column effects. Now if it is assumed that a 5% increase in σ_C can be tolerated, then

$$(1.05\sigma_C)^2 = \sigma_C^2 + \sigma_E^2$$

or

$$1.1\sigma_C^2 - \sigma_C^2 = \sigma_E^2;$$

thus

$$0.1\sigma_C^2 = \sigma_E^2.$$

Now form the plate theory,

$$\sigma_C = \sqrt{n}(v_m + KA_s)$$
$$V_R = n(v_m + KA_s)$$

where n is the column efficiency, v_m is the volume of mobile phase per plate, A_s is the surface area of the adsorbent per plate, and K is the distribution coefficient of the solute. Thus

$$\sigma_C = \frac{V_R}{\sqrt{n}}$$

and

$$\sigma_E^2 = 0.1 \frac{V_R^2}{\sqrt{n}}.$$

2 COLUMN LENGTH AND LINEAR MOBILE PHASE VELOCITY

Thus for the column considered,

$$\sigma_E^2 = 0.1 \left(\frac{9}{\sqrt{10000}}\right)^2 = 0.00081$$

and

$$\sigma_E = 0.028 \text{ ml} = 28 \text{ } \mu\text{l}.$$

Thus the standard deviation resulting from extra column effects is only 28 µl, and it follows that the column detector connections and detector volumes must all be kept to a minimum and the injection system designed to provide the minimum band dispersion or injection. It is advisable for columns having HETP values of 0.035 mm or less that all connection tubes between column and detector be 0.010 in. I.D., the detector volume should be no more than 8 µl and preferably 3 µl, and furthermore an injection system employed that does not provide significant band dispersion on injection of the sample. If such precautions are not taken, efficiencies of 10,000 theoretical plates, although available from the column, may in fact never be realized. The first indication that extra column band spreading effects are becoming significant is when the measured efficiencies of the solutes eluted from the column *increase* with the k' value of the solute between k' values of 0 and 3. Whenever the efficiencies of peaks eluted at k' values between zero and one are less than the efficiency of peaks eluted later in the chromatogram, the extra column dispersion effects are significant and steps must be taken to reduce them. In practice when using microparticulate columns it is suggested that column lengths of 25 or 50 cm be employed. These lengths will generally provide sufficient efficiency for most separations. If adequate resolution is not obtained, greater efficiency can be achieved by reduction of the flow rate or the selectivity may be improved by adjustment of the mobile phase polarity.

An interesting corollary of the alternative use of low mobile phase velocities to longer columns to provide high efficiency lies in the fact that at low flow rates the column inlet pressure is low and, therefore, high pressures in liquid chromatography may not be necessary. Considering present-day columns, to some extent this is true and many separations obtained on a 50-cm column at a flow rate of 1 ml/min with an inlet pressure of 1500 psi could equally well be achieved on a 10-cm column operated at about 0.2 ml/min at about 60 psi. Such an argument tacitly assumes that extra column band spreading effects can be kept sufficiently low to realize the column efficiency at the resulting low plate heights. In practice this may indeed not be possible, and the extent to which high efficiencies can be obtained by reducing linear mobile phase velocities may be very limited. Furthermore, as column technology develops, particle diameters are likely to be reduced further and, together with other column modifications that may be introduced, even shorter length columns are likely to require high pressures to provide even smaller flow rates. Thus high pressures are likely to stay with the technique of liquid chromatography but, as stated before, this should not be considered a disadvantage since the pump engineers appear quite capable of producing the pressures that are likely to be needed.

Preparation of Sample

In liquid chromatography it can be normally assumed that the sample will take the form of an involatile liquid or solid. If the sample is supplied in solution with a solvent more polar than the mobile phase, the solvent should be removed by evaporation if isocratic development is being employed; otherwise, repetitive analysis of the sample may result in deactivation of the adsorbent. If an evaporation procedure is carried out, care must be taken to use as low a temperature as possible to minimize thermal decomposition. It is also advantageous to carry the evaporation out under nitrogen. If the sample is relatively insoluble in the mobile phase, it should be borne in mind that the sample volume can be reasonably large as discussed in the Chapter II. If the sample has so low a solubility in the mobile phase that it does not permit adequate sample to be placed on the column in a reasonable sample volume, then its solubility may have to be enhanced by using a solvent mixture other than the mobile phase. However, as stated above, it must be remembered that if the sample is dissolved in a solvent of a more polar nature than the mobile phase then, if isocratic development is employed, repetitive sampling can result in a continuous deactivation of an adsorbent (e.g., silica gel) and subsequent changes in solute retention.

On the first injection the chromatogram produced is unlikely to have the correct distribution of peak heights; some peaks may be off-scale or some too small for accurate measurement. A chromatogram having peaks the correct size can be obtained by adjustment of the sensitivity or the charge size. It is, therefore, of prime importance that the original sample be made up accurately to a known concentration so that the mass of material placed on the column is precisely known. If high-efficiency columns are being used, to separate complex mixtures, a time period of a hour or more may be needed before all the components are eluted from the column; to economize on time it is therefore essential that the second chromatogram have the correct peak sizes for analysis. Thus, the correct sample size for the second chromatogram must be exactly determined. If the mobile phase flow rate is changed to produce shorter analysis times or greater resolution, it should be remembered that when using concentration-sensitive detectors doubling the flow rate will reduce peak heights by as much as 40% and reducing the flow rate by one-half will increase the peak heights by as much as 40%. The effect of flow rate on peak height should be carefully noted since a popular fallacious belief is that reducing the flow rate decreases the peak height. It is fairly obvious, however, that if you decrease the flow rate the efficiency increases and the peak dispersion decreases. Thus, if the peak width in milliliters is reduced for the same sample size, the peak height or concentration must increase to maintain the same peak mass. It should also be pointed out that all present-day detectors are concentration-sensitive devices.

In liquid-liquid chromatography the sample can sometimes be placed on the column dissolved in a small quantity of stationary phase. However, this procedure is not recommended for routine analyses because many such samples if placed on the column will result in a significant change of the stationary phase content of the column, which may eventually bleed from the column and result in detector instability.

The amount of sample injected onto the column must be sufficient to allow the detector to provide an adequate signal for each solute peak; thus the quantity of the sample injected must also be considered in the light of the sensitivity of the detector employed. For the preliminary assessment of the correct sample size, detector sensitivity should be set at a medium range. If the wire transport detector is being used in conjunction with columns 50 cm long and 4.6 mm I.D., then about 1-2 mg of sample should be used. If the refractometer detector is being employed the sample mass may have to be increased to 2-3 mg. For a UV detector the charge size may be between 0.1 and 0.2 mg. These charges must be proportionally changed for columns of different dimensions; otherwise, overloading may occur. These figures can only be approximate; they assume that there are between five and ten components present in the mixture and none are present at low concentrations. The necessary charge will also depend on the nature of the substances to be separated and the column characteristics; however, the sample sizes suggested above (sample mass as opposed to sample volume) will usually provide a chromatogram from which a more accurate assessment of the most satisfactory sample size can be made.

Injection of the Sample

The simplest and most reliable method of injecting a sample onto a chromatographic column is by the stop flow method. However, this method will not provide an accurate value for the column dead time or dead volume. Furthermore, if the dead volume is not known then values taken for retention ratios may not be reliable. A design of a suitable injection head is given in the section on chromatographic apparatus; p. 126. The sample should be taken up into a suitable microliter syringe of appropriate volume and there should be a range of such syringes with volumes from 1 to 100 μl in capacity available. It is essential that all syringes used for injection be kept scrupulously clean. After completing the analysis of a sample, any syringe employed must be meticulously cleaned with solvent and dried. The cleaning procedure is particularly difficult with a 1-μl syringe and, therefore, they must be given special attention. If syringes are left contaminated in any way by previous samples these invariably produce spurious peaks on the chromatogram that change their level as the analyses are repeated. If the injection head is fitted with a guide tube, which is advised for columns having diameters greater than 3 or 4 mm, a mark should be made on the hypodermic needle so that on insertion of the needle to the mark the tip is allowed to project just beyond the guide tube. The column flow is stopped either by a suitable valve or by switching off the mobile phase pump. If the mobile phase system contains pulse dampening devices, a few seconds should be allowed to elapse for the pressure at the inlet of the column to fall to atmospheric pressure before removing the cap. The injection head cap is then removed and the sample injected about 3-5 mm below the surface of the packing. The sample should be injected slowly so as not to disturb the packing. Unless the sample is injected well into the packing, band diffusion into the head of the column is likely, which will produce diffuse peaks and can seriously impair the efficiency of the column. The needle should also be withdrawn from the injection head slowly so as not to disturb the mobile phase in the column and thus produce band spreading. It is also important to ensure that when the cap is replaced no air is entrained in the injection head. Since the diffusion of substances in liquids is very slow, there is no need to hurry this procedure and when completed the flow of mobile phase is restarted.

If the septum-type injector system is employed, the same procedure is used but the flow of mobile phase is not arrested and, in this case, the sample should be discharged into the column as rapidly as possible. It is also useful to wait a few seconds after having discharged the sample before withdrawing the needle from the injection head. This will allow the mobile phase to carry the sample away from the point of injection and thus when the needle is removed no sample is taken into the guide tube owing to the piston effect of the needle withdrawal. The effect of injection methods on band dispersion has been elegantly studied by Kirkland [6], who tabulated the various effects of different methods of injection on plate height. Kirkland concluded that injection onto a wire screen situated on the top of the column gave somewhat more repeatable results with efficiency close to that obtained from on column injection.

The third method of injection involves the use of a sample valve. To use this method of injection it is necessary to have an excess of sample to wash the valve prior to injection. The procedure is as follows: A sample of pure mobile phase is placed in a tube and drawn through the sample chamber of the valve by means of the hypodermic syringe. This process is then repeated with a tube containing the sample to be separated. A sufficient quantity of the sample must be drawn through the valve to completely flush the system. This quantity is approximately 10 sample volumes of the solution. The valve is then actuated to place the sample chamber in line with the mobile phase supply. This method of sampling is particularly useful for routine analyses but the valve itself must be operated with care, and precautions taken to ensure that no solid particles of support or adsorbent enters the valve as the sealing surfaces can easily be scratched, which will result in the mobile phase leaking past the seals of the valve.

Chromatogram Data

Immediately before injection and marking the injection point, the pertinent information with respect to the sample and the chromatographic conditions are written directly on the recorder chart. This is necessary not only for reporting purposes; the information will also be valuable for assessing the best operating conditions for future samples of a similar nature. It is an unfortunate fact that a significant portion of the separations described in the literature can never be repeated because there is insufficient information provided to simulate the original operating conditions.

The minimum data that should be recorded directly on the chromatogram is as follows:

Sample identification.
Column dimensions.
Column temperature (if other than ambient).
Stationary phase (partical diameter, support, loading, etc.).
Mobile phase or gradient conditions.
Mobile phase flow rate.
Composition of sample solution, % v/w or % w/v.
Sample volume.
Detector sensitivity and the type of detector employed.
Chart speed.

3 MEASUREMENT OF COLUMN EFFICIENCY, RETENTION RATIOS, AND PEAK AREAS

Measurement of Flow Rate

The usual method of measuring the mobile phase flow rate is by means of a small graduated cylinder and stop watch. Care should be taken with this procedure, however, because small flow rates of volatile solvents can provide values as low as 10% of the true flow rate due to evaporation losses. The tube from the detector should be led to the bottom of the graduated cylinder and the top of the cylinder should be closed (but not sealed) to prevent evaporation. A method of measuring flow rates that does not include this evaporation loss is by the bubble flow method described in the section on apparatus. p. 185.

Measurement of Dead Volume

The measurement of dead volume can be one of the most difficult measurements to make in liquid chromatography. In gas chromatography it is fairly easy to select a completely unretained substance such as permanent gas and measure the dead time (that is, the time of the elution of an unretained solute) and thus calculate the dead volume from the mobile phase flow rate. In liquid chromatography, however, it is difficult to select a solute that is completely unretained. The nearest approach to this for liquid-solid chromatography is to operate the column with a very polar mobile phase such as alcohol and use benzene as the solute, assuming a UV detector is being employed, because benzene will be completely unretained on silica gel deactivated with alcohol. The time between injection and peak maximum for benzene will be the dead time, and the product of this time and the flow rate will give the dead volume. This experiment must be carried out using on-line injection employing the septum method or the sample valve method. If a refractometer detector is being used with heptane as the mobile phase, pentane can be used as a marker for the dead volume since pentane will be detected owing to the difference in its refractive index from that of heptane. If the stop flow technique is being used, then actual volume measurement must be employed because on starting the flow there is a significant time required for the pressure to build up in the front of the column and for the normal flow rate to be obtained.

Having determined the dead volume of the column, for example, by using the mobile phase, alcohol, and solute benzene, the column can be reconditioned for any solute and the dead time calculated for the new solvent from the flow rate. If such a system of benzene and alcohol cannot be employed, the best compromise is to measure the retention time of the least retarded solute chromatographed and take this as the column dead time. It should be emphasized, however, that this approach can lead to serious errors, particularly where nonpolar solvents are being used in liquid-solid chromatography.

In liquid-liquid chromatography the last method is probably the only effective one that can be employed for measuring the dead volume of the column; it may be an inaccurate procedure, but other alternatives are not possible since the mobile phase cannot normally be changed.

The dead volume of a column includes all volumes of mobile phase from the injection point to the detector and include that of the connecting tubes. However, the dead volume remains constant for all conditions of operation until the column is either changed or repacked.

Measurement of Mobile Phase Velocity

The liquid mobile phase velocity is calculated from the dead time and is taken as the ratio of the column length to the dead time. This value includes the time taken to pass through any connecting tubes and will, therefore, be slightly less than the true value. However, the error will be insignificant and constant and thus linear velocities measured in this way can be used for plotting HETP curves.

Measurement of Column Efficiency

The efficiency of every column should be measured after packing to determine how well the column has been packed and to be able to compare it with other columns that have been in prior use. The number of theoretical plates given by a column will vary with the mobile phase flow rate and with the retention volumn of the solute on which the efficiency measurement is made. It is therefore necessary to standardize on the mobile phase flow rate unless a complete HEPT curve is determined for each column, which can be very time consuming. It is also necessary to select a solute retention volume for efficiency measurements if the comparison of efficiency between one column and another is to be significant. The choice of flow rate is arbitrary and for 1/16 or 1/8 mm I.D. columns can range between 0.2 and 1 ml/mm, but whatever the flow rate selected within this range, this same flow rate must be used for all measurements of efficiency for columns of the same diameter. It is not possible to compare the efficiency of different diameter columns at the same flow rate, but it is possible to compare efficiencies from different diameter columns if the measurements are made at the same linear velocity of the mobile phase. In liquid chromatography the efficiency given by a column will generally decrease with the retention volume of the solute used for measurement, the highest efficiency being obtained for an unretained peak and the lowest for the last peak eluted. As discussed earlier, if it is found that the column efficiency increases with the k' of the solute, then extra column band dispersion effects are now the predominant factors affecting column efficiency. An average value for the column efficiency will be obtained for peaks eluted at k' values lying between 3 and 5, that is, at a retention time lying between the time for 4 to 6 dead times. Solutes eluted at such k' values will usually be symmetrical, and for columns having values of H greater than 0.025 mm will not be significantly affected by extra column band spreading processes. The choice of solute will depend on the phase system and the detector employed, and so specific solutes cannot be recommended. It should be emphasized, however, that any solute used for efficiency measurements should be very pure since traces of impurities can broaden the peak and give apparent low efficiencies. Solutes having cis-trans isomers should be particularly avoided when measuring efficiency in liquid-solid chromatography as the two isomers can often be partially separated with such systems.

The necessary measurements that have to be made on the chromatogram are shown in Fig. 5.1. The retention distance y can be measured directly with a good quality steel rule to the nearest 0.02 cm. It should again be emphasized that the

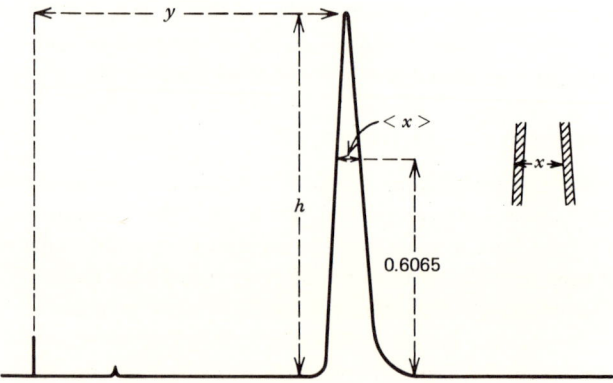

Fig. 5.1. The measurement of efficiency.

sample must be placed on the column using an on-line injection procedure. The stop-flow method of injection should not be employed when measuring efficiency, since the flow rate will not be constant for the whole of the chromatogram owing to the time lag for the pump to achieve a constant inlet pressure to the column. Efficiencies measured by the stop flow injection method will be significantly greater than the true value as the value for y (Fig. 5.1) will be extended during the pressure build-up period. If columns of low permeability are used, however, it may be necessary to tolerate this error. To calculate the column efficiency, the baseline is projected beneath the peak and the peak height measured. A mark is made in the center of the peak at 0.607 of the peak height and it has been shown in Chapter II that this corresponds to the position of the points of inflection of the error function curve. The peak width is measured at this point and this is best achieved by means of a comparator (magnifying graticule), if available. The comparator should have a 1-cm scale graduated in 0.02 cm and the distance between one edge of the recorder trace on one side of the peak to the other edge of the trace on the other side of the peak should be measured as shown in Fig. 5.1. This eliminates errors resulting from the finite width of the recorder trace. If a comparator is available, the chart speed of the recorder should be adjusted such that the peak width x lies between 0.5 and 1 cm; otherwise the chart speed should be adjusted to provide a peak width of 2-3 cm and the peak width can then be measured directly by means of a good quality steel rule.

As given in Chapter II the column efficiency is calculated from the equation

The alternative method of measuring efficiency using the peak width at the base should not be employed since it entails the geometric construction of the tangents to the peak, which can result in significant error.

The value of h, the variance per unit of the column length l, can be calculated from the equation

$$h = \frac{1}{n}.$$

It should be remembered that the value of h calculated in this manner includes all extracolumn effects that are contributing to the band width.

Some adsorbents when fully activated by a solvent regeneration procedure exhibit relatively poor efficiencies compared with those expected from the nature and particle size of the adsorbent used. This effect is most pronounced when using nonpolar phases such as heptane and can vary between adsorbents from different suppliers and between different batches of the same adsorbent from the same supplier. Synder [7] reported this effect and showed that the efficiency would be improved by using a mobile phase containing trace quantities of water. According to Synder, traces of water in the mobile phase resulted in improved efficiency and reduced k' values for a given solute.

Measurement of Retention Ratios

The retention ratio of two peaks is the ratio of the respective distances between the dead point and their peak maxima. The dead point can be identified by adding a small quantity of an unretained solute to the sample or as previously discussed. Because pressure equilibrium is usually achieved by the pump in a period less than the dead time of the column, the stop flow method can normally be used when measuring retention ratios if so desired. In some instances it is useful to measure retention ratios relative to a standard solute, in which case the standard must be added to the sample with the unretained solute. All retention ratios are thus measured relative to the retention distance between the dead point and the peak maxima of the standard solute. Retention ratios can only be compared for identification purposes from one chromatogram to another where the same phase system or where exactly identical gradient elution conditions are employed for both separations. Retention values are the most useful measurements for identification purposes in liquid chromatography since the technique has not been yet developed sufficiently for functions such as retention indexes to be usefully employed.

Measurement of Peak Areas

There are six general methods for measuring peak areas:

1. The peak height and the peak width at one-half the peak height are measured and the peak area is taken as the product of these values. This method is satisfactory for symmetrical peaks but tends to give errors when the later peaks become asymmetrical.

2. This method is similar to (1) but takes the product of the peak height and the peak width at 0.607 of the peak height as porportional to the peak area. Because the peak width is measured at the points of inflection of the curve, this method will still provide an accurate measurement of peak area when the peak becomes asymmetrical providing the form of the curve remains approximately Gaussian. If the peaks become significantly assymetrical, for example where their concentration is such that the absorption isotherm is no longer linear, then errors will be involved.

3. Tangents are constructed to the points of inflection of the Gaussian curve and the baseline produced beneath the peak. The area is taken as proportional to the product of the height and one-half base width of the triangle so formed. This method is not recommended. It is the least accurate since it involves geometric construction and gives serious errors for asymmetric peaks.

4. The area of the peak can be measured by means of the planimeter after projecting the baseline beneath the peak. This method is very tedious and has to be repeated a number of times before accurate results are acquired; it can obviously cope with all shapes of peaks.

5. This method involves projecting the baseline beneath the peak and cutting the peak out of the chart paper and weighing it. This is probably one of the most accurate methods of measuring peak areas and can cope with peaks of any shape. It is, however, a little tedious and unfortunately destroys the chromatographic record. The chromatogram should be photocopied before using this method, but the chart paper itself must be used for measurement since the inhomogeneity of the photocopy paper can produce errors if the copies are used for the measurement.

6. Automatic integration can be employed using the standard integrators available for measurement of peak areas in gas chromatography. However, the present rate of output from most liquid chromatographs does not, at present, warrent the incorporation of expensive electronic integrators for the measurement of peak areas except for carefully controlled routine analyses. Where electronic integrators are used they will, without doubt, provide the most accurate measurement of peak areas.

For chromatograms having good peak shapes methods (1) or (2) are recommended. However, if asymmetrical peaks are involved then method (5) should be employed.

4 SOME PRACTICAL NOTES ON THE OPERATION OF A LIQUID CHROMATOGRAPH

Spurious Peaks, Their Source and Elimination

Contaminated Syringe

Contaminated syringes usually provide spurious peaks and these are most frequently met when using 1 μl syringes. Careful cleaning before use completely eliminates this problem.

Contaminated Septum

This source of spurious peaks occurs after prolonged use of a septum and results from the contamination of the pierced hole in the septum by previous samples. Replacement of the septum eliminates this difficulty.

Air Dissolved in Sample

Dissolved air in the sample can give peaks at the dead time of the chromatogram when using the refractometer or the UV detector. It results from the change in refractive index of the solvent when air is eluted. The UV detector is sensitive to refractive index changes in the mobile phase as well as changes in extinction coefficient. The sample should be degassed or made up in degassed solvent.

Elution of Sample Solvent

This can occur when the sample is dissolved in a solvent other than the mobile phase and can result in peaks occurring anywhere in the chromatogram depending on the polarity of the solvent employed. The peaks can be negative or positive when using the refractometer or UV detector but will not occur with the wire transport detector. The only solution to this problem is to try other solvents for dissolving the sample that are more similar to the mobile phase.

Detector Displacement Effects

When using a mobile phase that contains a small percentage of a polar solvent contained in a nonpolar solvent, for example ethyl acetate or isopropanol in heptane, a layer of the more polar solvent is often adsorbed on the surface of the quartz or glass detector cell. If a solute is eluted that is more polar than the polar solvent, the latter is displaced from the cell walls and is replaced by the solute. This displacement results in a spurious peak or more often a distorted peak shape. Changing the mobile phase solvent mixture is the only way to eliminate this effect.

Random Spikes on the Recorder Trace

This effect is usually a result of small bubbles being eluted; it can be eliminated, if caused by trapped air in the packing, by pumping the column with a high flow of mobile phase. If the spikes are still not eliminated then the solvent should be degassed.

Bubbles Held in the Detector Cell

Bubbles held in the detector cell are indicated by a permanently off-scale peak. They occur when the column is initially being brought to equilibrium with the mobile phase or as a result of using solvents that have not been degassed. Alcohol-water mixture are particularly prone to produce bubbles in the column if the solvents have not been degassed. Solutes can be easily and quickly degassed by shaking them in a flask under slightly reduced pressure. There are various ways of removing bubbles from the detector and the number of methods that can be used are as follows:

1. Adjust the pump to a high flow rate and restrict the outlet from the detector slightly. This will apply a back pressure to the detector cell reducing the volume of the bubbles, which may then be dissolved and/or forced out of the cell by the high flow of mobile phase. Care should be taken to ensure that the back pressure applied to the detector does not exceed the specification level for the cell; otherwise the cell gaskets will be disturbed and the cell will leak mobile phase into the cell housing.

2. If the bubble persists, loosen the union connecting the column to the detector and with a hypodermic syringe force a more polar solvent back through the detector cell and out through the loosened union. Care must be taken to ensure that the air emerges from the union because it can collect in the union and when the column is reconnected pass back into the cell.

There are several ways of reducing the possibility of bubbles collecting in the detector cell. The eluent from the column should be made to enter the bottom of the detector cell and pass out through the top connection. This allows the bubbles to rise freely out of the cell. The exit tube from the cell should be made vertical and bend in a U and the top over the container that is used to collect the fractions or mobile phase. This system applies a slight back pressure to the cell due to the head of solvent above it and reduces the extent of expansion of the bubble when leaving the column and allows it to pass freely through the cell. The alternative procedure of connecting the column to the top of the cell and out through the bottom applies a reduced pressure to the column outlet causing bubbles to expand and block the cell. Furthermore, the reduced pressure can also cause air to be sucked into the detector cell if the cell gaskets are not fitted as tightly as they should be. Poorly fitted cell gaskets are a very common source of bubbles in the detector cell and when reassembling a detector cell it is imperative to ensure that cell and gaskets are assembled tightly.

Baseline Instability of the Recorder Trace

Baseline instability can generally take two forms, long-term drift in the baseline that can be either up or down the chart and short-term noise or "grass" on the baseline, the period of which can vary from a few seconds to milliseconds. Baseline instability normally becomes a problem when working at the higher sensitivities of the instrument. There are two main sources of long-term drift. The column has not reached equilibrium with the mobile phase, which results in the composition of the column eluent slowly changing. This usually occurs when the mobile phase has been changed and more time is required for complete equilibrium to take place. The other common source of long-term drift results from ambient temperature changes or using a mobile phase that is not at the same temperature as the column, detector, and its surroundings. Thermal drift can be eliminated by thermostating the detector cell and, if necessary, the column as well with a proportional controller. The use of an on/off temperature-controlled thermostat will result in steps in the baseline occurring as the controller switches on and off.

Short-term noise usually arises from pulsations due to insufficient damping of the pump, and it is usually possible to identify the time period of the short-term noise with the piston movements of the pump. Short-term noise from pumps can be almost completely eliminated from UV detectors by thermostating the cell and in some cases the connections between the column and detector cell. In the case of refractometer detectors, thermostating the cell will reduce the short-term pump noise but in some instances may not completely eliminate it. Short-term noise, usually appearing as a "fuzz" on the baseline, can also be due to the recorder sensitivity being too high or conversely the recorder may have insufficient damping.

In the case of the wire detector, noise and instability can arise from several factors. The entrainer and associated tubing should be regularly cleaned and the alignment of the wire through the tubes and over the pulleys should be regularly checked for correct adjustment. Unfortunately, the wire provided is not always of consistent quality and occasionally a spool of wire is supplied that produces a high noise level on the recorder. A poor spool of wire can usually be identified by the appearance of the used wire on the winding spool. The wire should all be of the same even color, but a poor spool of wire will show uneven or mottled color on the winding spool. Such a spool of wire should be rejected. When leaving the wire stationary for any length of time the furnace temperatures should be turned to their standby position. The coating block adjustment should also be regularly checked and the coating block and coating needle regularly cleaned. The detector should be operated in a dust-free environment devoid of solvent vapor and employed with clean solvents free of high-boiling impurities.

5 QUALITATIVE AND QUANTITATIVE ANALYSIS

Qualitative Analysis

At the present stage in the development of liquid chromatography the only effective means of identifying substances is by use of retention ratios. The more sophisticated forms of identification used in gas chromatography, for example, the Kovats [8] retention indexes, cannot be usefully employed because a suitable series of reference solutes have not been determined to meet the wide range of phase systems available. Even under those circumstances where retention ratios are used, great care must be taken to maintain the same phase system and, as stated before, where gradient elution methods of development are employed, the elution program must be precisely repeatable. To identify or confirm the presence of a particular solute it is necessary to add a standard solute to the mixture and measure retention ratios of the unknown to the standard solute. Another solution is then prepared containing the standard solute together with the expected solute and the mixture chromatographed and retention ratios calculated. If the retention ratios of the unknown and suspected substance differ by less than 5%, then this can be considered as evidence supporting the identification of the unknown. However, the sample should then be run under slightly different phase conditions and the two measurements of retention ratios repeated. If the retention ratios again match, then the two solutes are probably identical. However, it must be stressed that in liquid chromatography the certainty of identifying specific solutes by retention ratios is far less certain than in gas chromatography. Repeated separation using still further phase systems can help to increase the reliability of identification, but for certain and unambiguous identification, the solute needs to be collected and examined by suitable spectroscopic techniques.

Quantiative Analysis

To carry out quantitative analysis in liquid chromatography the detector must be linear, the meaning of which has already been discussed in the section dealing with detectors. No liquid chromatography detector presently used provides the same or similar response to all compounds. It follows that the normalization procedure common in gas chromatography, where the percentage composition of a component is obtained by expressing the area of the peak to the total area of all the peaks, cannot be employed in liquid chromatography. Generally, calibration of the detector for all components to be quantitatively determined must be carried out. Where the moving wire detector, fitted with the methane conversion system, is employed, it is possible to assume the response factor from the carbon content of the solutes eluted, but the validity of this assumption has not be verified over a wide range of solute types, so it should be used with caution.

There are two methods for obtaining quantitative measurements from a chromatogram, one using peak heights and the other using peak areas. The relative merits of these methods are the subject of some controversy.

Quantitative Measurements Using Peak Heights

Most detectors are concentration sensitive devices and, therefore, for quantitative accuracy the concentration at the peak maximum must reproducibly relate to the total mass of solute if peak heights are to be the basis of measurement. Now the peak height is inversely related to the peak width, and thus the peak widths must also be reproducibly constant relative to other peaks in the chromatogram. If follows that the k' of the solute must remain constant, which entails carefully controlled stationary phase activity and mobile phase polarity together with good temperature control. The method of injection must be highly reproducible since poor injection can broaden early peaks in the chromatogram relative to peaks that are eluted later in the chromatogram. However, owing to the shape of the HETP curve the pump control need not be excessively precise since the relative peak heights will not change significantly with small changes in flow rate. Thus, peak heights would give accurate quantitative measurements providing the phase system was well controlled with respect to stationary phase activity, mobile phase polarity, and column temperature and a reproducible injection technique was employed by adequately skilled personnel. However, the pump used could be relatively inexpensive since a highly precise flow rate need not be maintained.

Quantitative Measurements Using Peak Area

If the peak concentration (peak height) is integrated with respect to the volume flow of mobile phase through the peak then the result provides a value relative to the absolute mass of solute contained in the peak. Furthermore, because the peak height is always inversely related to the peak width in volume flow of mobile phase, the integral is entirely independent of changes in bandwidth resulting from different phase selectivity, temperature, or even changes resulting from poor reproducibility of injection. Thus it is no longer necessary either carefully to control the stationary phase activity, mobile phase polarity, or column temperature or even to employ a reproducible injection system other than to ensure that the required resolution is maintained. However, the concentration/volume integral in practice involves the measurement of peak area on the chart, and the retention axis is not directly related to volume flow of mobile phase but to time. It follows that, for accurate quantitative analysis, volume flow of mobile phase must be precisely related to elapsed time and thus a precisely controlled mobile phase flow rate must be ensured. Therefore, if peak areas are to be used for quantitative analysis, a relatively expensive pump must be used to provide precisely controlled constant flow rates, but carefully controlled chromatographic conditions, reproducible injection procedures are no longer essential. Thus, both peak height and peak area measurements can provide accurate quantitative analysis and the choice will depend on the apparatus available and the skill of the personnel involved.

Procedure for Quantitative Analysis

The procedures outlines below are given for peak area measurements but apply equally to peak height determinations. In the equation given, appropriately designated terms for A (peak area) can be replaced by corresponding terms of H (peak height) and the resulting equations will be applicable for quantitative analysis using peak height measurements.

To determine the percentage level of a particular solute, its area must be compared with that of a standard that has been added to the sample in a known concentration. The standard must be extremely pure and must be chosen so that it is eluted at a point in the chromatogram where it does not overlap the peaks of other substances present in the sample.

Consider the determination of a substance A in a particular sample using a standard B. Let a two component mixture be made up having $X_1^A\%$ of compound A and $X_1^B\%$ of compound B and let the sample be injected on the column providing two peaks or area Y_1^A and Y_1^B, respectively. If the detector response factors for substances A and B are a and b, respectively, then

$$\frac{X_1^A}{X_1^B} = \frac{aY_1^A}{bY_1^B} = \frac{\psi Y_1^A}{Y_1^B} \quad \text{where } \psi = \frac{a}{b}$$

and ψ is the relative response factor of substance A to substance B. Hence

$$\psi = \frac{X_1^A}{X_1^B} \frac{Y_1^B}{Y_1^A}.$$

Thus the proportionality factor ψ for the solute A and B can be calculated.

Now consider the unknown sample containing substance A at a percentage composition X_2^A and add to the sample solute B at a percentage concentration X_2^B. Let the chromatogram for this mixture provide peak areas for A and B of Y_2^A and Y_2^B, respectively. Then the consentration of A in the sample will be given by

$$X_2^A = \psi \frac{Y_2^A}{Y_2^B} X_2^B.$$

The same principle can be used for estimating any number of solutes in the mixture providing calibration samples are prepared and chromatographed to determine the relative response factors ψ for the standard and each of the solutes to be measured.

For simple mixtures having a limited number of solutes that are completely resolved on the chromatogram, the absolute response factor for each solute can be determined and each peak area corrected for its response factor and a normalization procedure carried out. For example, if a mixture contains solutes A, B, C, D, ... that are completely resolved by the column and their respective absolute response factors are $a, b, c, d,...$ then the percentage of any solute (e.g., C) in the mixture will be given by

$$\frac{Cc}{Aa+Bb+Cc+Dd+\cdots}.$$

Unfortunately, the condition for complete resolution of all components in the mixture does not often arise in practice so that the latter method for quantitative analysis cannot be frequently used.

REFERENCES

1. R. E. Majors, *Anal. Chem.*, **44**, 1722 (1972).
2. J. F. K. Huber et al., *Anal Chem.*, **44**, 111 (1972).
3. J. J. Kirkland, *J. Chromatog. Sci.*, **9**, 206 (1971).
4. J. J. Kirkland, *J. Chromatog. Sci.*, **10**, 593 (1972).
5. J. Anhauer and I. Halasz, *J. Chromatog. Sci.*, **12**, 139 (1974).
6. J. J. Kirkland, *Gas Chromatography 1972*, S. G. Perry and E. R. Adlard, Eds., Applied Science Publishers for the Instiute of Petroleum, London, 1972, p. 39.
7. L. R. Snyder, *Principles of Adsorption Chromatography*, Marcel Dekker, New York, 1968, pp. 88 and 103.
8. E. Kovats, *Helv. Chim. Acta*, **41**, 1915 (1958).

Chapter VI

THE COMBINATION OF LIQUID CHROMOTOGRAPHY WITH SPECTROSCOPIC TECHNIQUES

1 The On-Line Liquid Chromatography UV Spectrometer System (LC/UV) 275
2 The On-Line Liquid Chromotograph-Mass Spectrometer System (L/MS) 228
3 The LC/MS Using Direct Inlet Sampling 280
4 LC/MS Using the Wire Transport Method of Sample Introduction 283
   Design of the Interface 283
   Operation 286
   Performance of Chromotograph-Mass Spectrometer Combination 287
5 Future of LC/MS 293
6 The Combination of LC with Other Spectroscopic Techniques 294

Modern liquid chromotography can now produce excellent separations of complex mixtures, but the chemical nature of the components separated are often completely unknown. It is obvious for such mixtures that elucidation of the structure of the individual soltues requires the support of one or more spectroscopic techniques. In fact, the technique of liquid chromatography is in exactly the same situation that gas chromatography was a decade ago; having achieved the seaparation, the indentity of the components remained to be determined. The problem was solved in the field of gas chromatography by linking the gas chromatograph directly to the mass spectrometer and subsequently to an IR spectrometer. Alternatively the eluates from the gas chromatographic column were condensed into a trap and the material examined using the normal spectroscopic sampling techniques. Up to about 1973 the collection and subsequent concentration of the eluate from a liquid chromatograph was the only known method for obtaining spectroscopic data for an eluted component.

Some spectroscopic techniques make more demands on the chromatography systems then others from the point of view of sample size. The size of sample necessary to provide adequate data also depends to some extent on the nature of the sample itself, but the normal average size required for satisfactory measurements to be made (for each of the more common spectroscopic techniques) is given in Table 6.1.

Table 6.1 Sample Size for Different Spectroscopic Techniques

Spectroscopic Technique	Sample Size
60 Hz NMR	20 mg (micro cell 5 mg)
100 Hz NMR (continuous wave)	5 mg
100 Hz NMR (time averaging)	150 µg
100 Hz NMR (fourier transform)	40 µg
Low-resolution mass spectrometer (solid probe injection)	0.1 µg
High-resolution mass spectrometer (solid probe injection)	1 µg
IR spectrometry (normal scan)	200 µg
IR spectrometry (Fourier transform)	10 µg
UV spectrometry	1 µg

It is seen from Table 6.1 that nuclear magnetic resonance spectroscopy, by itself, the most informative spectroscopic technique for determining molecular structure, requires the largest sample size. Unless Fourier transform facilities are available, the sample has to be separated by semi preparative scale chromatography. Furthermore, the sample volume necessary is relative large even when using micro cells, which, considering the time required to scan the sample and the need for special solutes, makes the NMR spectrometer an unlikely candidate for practical on-line use with a liquid chromatograph.

Similar limitations apply to the use of the IR spectrometer in line with a liquid chromatograph. Most solvents used as mobile phases in liquid chromatography exhibit strong absorption in IR regions of the spectrum, which, combined with the relatively large sample volume necessary, also renders the IR spectrometer unsuitable for on-line operation with a liquid chromatograph.

The two remaining spectroscopic techniques, UV spectroscopy and mass spectrometry, have the minimum requirements with respect to sample size, have adequate sensitivity, and lend themselves particularly to on-line operation with the liquid chromatograph. Owing to the high sensivity of the UV spectrometer, its small sample volume, and its established use as a liquid chromatograph detector, it was naturally the first spectroscopic technique to be directly associated with the liquid chromatograph. The association of the mass spectrometer with the liquid chromatograph, however, presented some problems with respect to the method of sample introduction. In spite of these interfacing problems, two suitable sample systems have been devised and shown to be effective and a third has been proposed. In keeping with the chronological order of their development the combination of the UV spectrometer and the liquid chromatograph will be described first.

1 THE ON-LINE LIQUID CHROMATOGRAPH UV SPECTROMETER SYSTEM (LC/UV)

Unlike the LC/MS system the LC/UV has to be operated using the interrupted elution technique, because the scan time for the UV spectrometer may be several minutes. When the peak is sensed at its maximum, the flow of mobile phase is arrested and the sample scanned; at the completion of the scan the flow of mobile phase is continued until the next peak is eluted and then the procedure is repeated. The mobile phase can be stopped and the scan initiated automatically using an appropriate peak-sensing device, but so far this feature has not been incorporated in the presently available commercial instruments. Stopping the mobile phase flow in a liquid chromatographic column does not impair the resolution of peaks still remaining in the column. Under no-flow conditions the only band dispersing process that can occur is that due to longitudinal diffusion, and because the solute diffusivity in the mobile phase is between 10^{-5} and 10^{-6} cm^2/sec, diffusion effects are not significant even if the mobile phase flow rate is arrested for several hours.

The spectrometer used can be an ordinary unmodified UV spectrometer fitted with suitable low-volume flow-through sample cells. To ensure stability the cells must be thermostated and to prevent extra column band broadening effects, the cells should have a total capacity of less than 10 μl. The column can be connected directly to the sample cells of the spectrometer, which can then be used during development as the detector by operating the spectrometer at some fixed wavelength appropriate to the analysis. When the recorder registers a peak maximum the mobile phase can be stopped and the sample scanned.

A good commercial example of an effective LC/UV system is the Variscan manufactured by Varian Aerograph, which has also been designed to function as a multiwavelength detector. A diagram of the layout of the optics is shown in Fig. 6.1.

Either a tungsten-halogen or deuterium lamp can be selected by push-button. The light passes through a Czerny-Turner monochromator with selectable slits, one-half of a dual chopper, the sample compartment, the second half of the dual chopper, and is sensed by a single photomultiplier tube. The optical paths through the sample and reference cells are symmetrical and contain equal numbers of transmitting and reflecting elements.

The two halves of the chopper are synchronized so light passes alternately through the sample and reference cells. It is important to note that 100% of the light passes through a given cell when the light transmission of the cell is being measured. This is a distinct advantage in HPLC, as contrasted to beam-splitter arrangements where only 50% of the available light passes through each cell

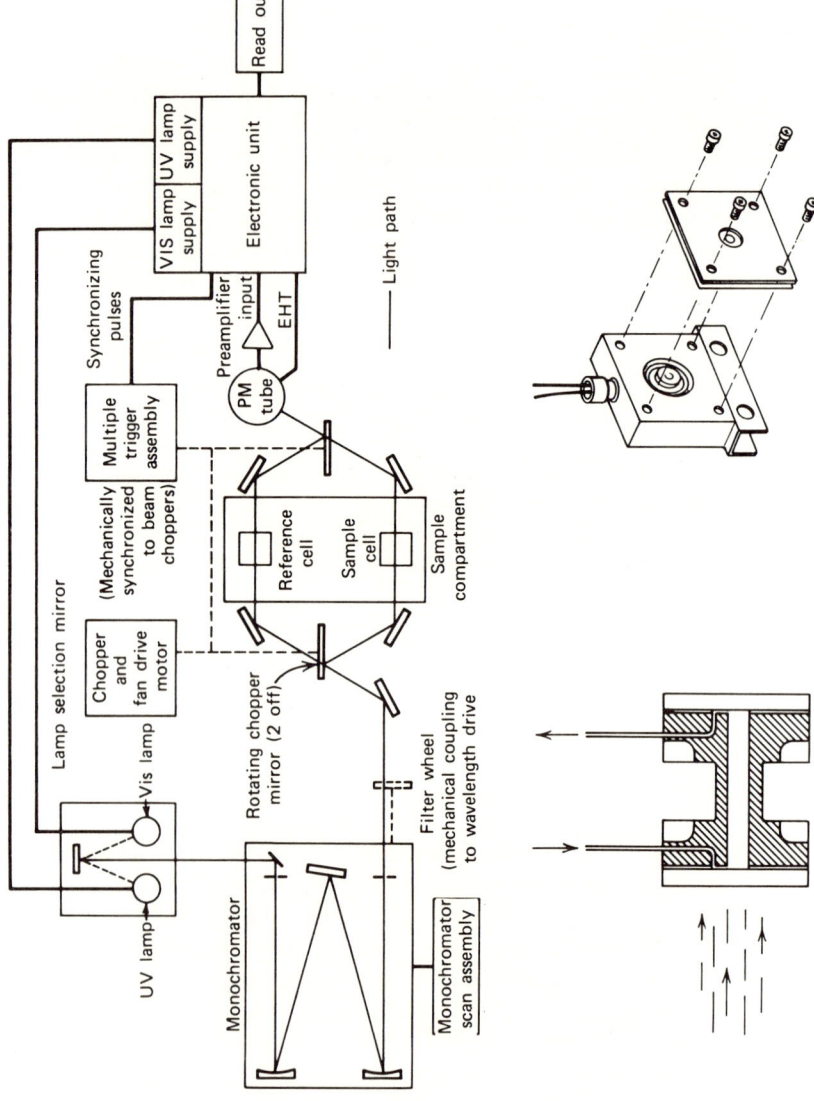

Fig. 6.1. The optical system of the Variscan multiwavelength detector.

1 CHROMOTOGRAPH UV SPECTROMETER SYSTEM (LC/UV)

during measurement.

Stray light from the monochromator is rejected by a seven-segment filter that is automatically actuated by the wavelength drive. The sample compartment is isolated from the rest of the optical system by matched pairs of fused silica windows.

The system has four ranges, linear in absorbance with sensitivities of 0.1, 0.5, 1.0, and 2.0 absorbance full scale on a 100-mV recorder. A 10-mV recorder provides scale expansion of a factor of 10. The system has a continuous adjustable sensitivity range that is proportional to concentration as well as linear in transmittance. The cell assembly is shown in the lower left diagram in Fig. 6.1. The cell volume is 8 μl with a path length of about 1 cm and diameter 1 mm.

The flow cell assembly consists of two water-jacketed flow cells, a baseplate that bolts into the sample compartment of the spectrophotometer, and a mounting rail. The flow cells are a matched pair, identical and interchangeable. One cell contains the sample solution and the other the reference solution. It is easy to adjust the cells both laterally and vertically in order to achieve optimum alignment with the beam. The cell inlets and outlets, each 0.002 in. I.D., enter the cells through fittings at the top of the water jacket. Cell gasket pressure is easily adjusted by means of a threaded insert in the center of the water jacket cover. The lower right-hand diagram in Fig. 6.1 is a cutaway of the stainless steel flow cell body. Liquid enters the cell through the left channel, flushes the window, traverses the 1-mm-diameter × 1-cm-long bore, flushes the right-hand window, and exits through the right channel.

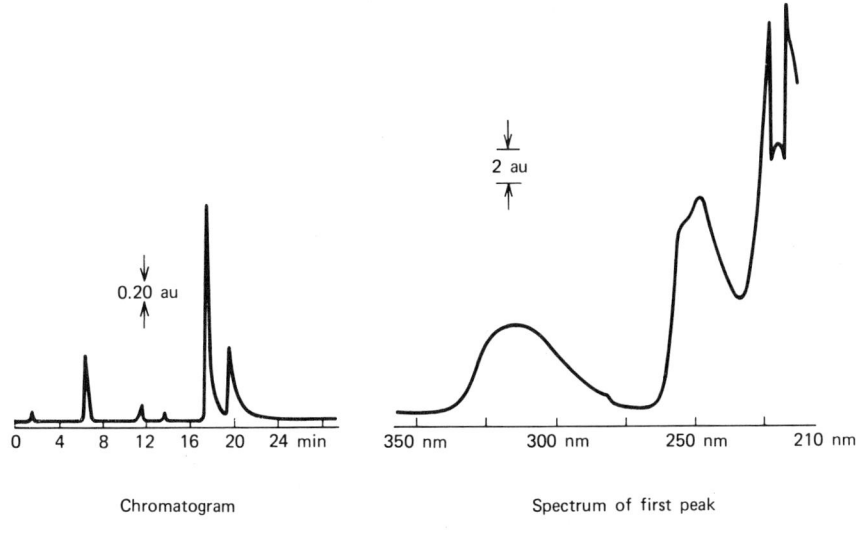

Fig. 6.2. Plant flavonones.

An example of the use of the instrument to separate a mixture of plant flavonones is shown in Fig. 6.2. The MicroPack column was 50 cm long × 2.4 mm in diameter and packed with silica gel. Gradient elution was employed for development commencing with hexane and terminating with at 10% v/v solution of isopropanol in dichloromethane. Flavonones have an absorption maximum at about 250 nm, but the detector was operated at a wavelength of 310 nm to exclude solvent effects. This procedure, however, also excluded some of the impurities that did not absorb at 310 nm. The spectrum shown was obtained by the stop flow technique for the first eluted peak in the chromatogram.

In Fig. 6.3 the spectrum for chlorophyll is shown obtained by the same technique, but in this instance the scan was taken from 220 to 700 nm providing both UV and visible spectra. The chromatographic conditions employed are included in Fig. 3.6.

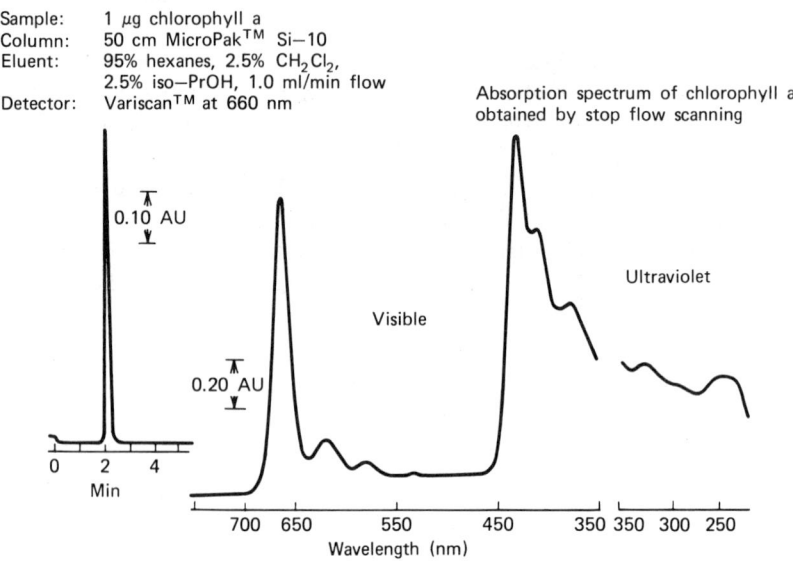

Fig. 6.3. Chlorophyll a analysis.

It is clear that by using the stop flow technique with on-line LC/UV, good spectra can be obtained from microgram amounts of solute. The only disadvantage of the system is that inherent in UV spectroscopy: UV spectra provide very limited information for the elucidation of molecular structure.

2 THE ON-LINE LIQUID CHROMATOGRAPH-MASS SPECTROMETER SYSTEM (LC/MS)

The first chromatographic apparatus to be successfully linked to a mass spectrometer was the gas chromatograph. The problems associated with interface between the two instruments for gas chromatography were, however, not serious since the mobile phase is a gas and thus, provided the pumping system of the mass spectrometer was adequate, a portion of the eluent could be fed directly into the mass spectrometer without impairing its performance. The interface between the liquid chromatograph and the mass spectrometer, however, of necessity must be more involved since the eluent from the column is liquid and the quantity of liquid that can be passed into the mass spectrometer without affecting its performance is severely limited.

There are basically two proven approaches for directly linking a mass spectrometer to the liquid chromatograph. One, a method developed by McLafferty et al. [1-3], continuously vaporizes a portion of the column eluent together with the contained solute into the mass spectrometer and utilizes the solvent vapor as the chemical ionization agent to produce chemical ionization (CI) spectra. This system, without doubt, should provide the highest sensitivity but its use with gradient elution development may be limited. Because the solvent composition will be continually changing throughout the development of the chromatogram, the chemical ionization agent will also change and tend to affect the chemical ionization processes and thus confuse interpretation. The system also depends on the substances chromatographed having some volatility at normal temperatures since there is no means of heating the solutes to vaporize them. Thus, the only vapor that will be formed will result from the partial pressure of solute at the mass spectrometer temperature. Finally, for structural elucidation of eluted solutes, CI spectra provide very limited information since the spectra consists largely of an $M^+ + 1$ ion and one or two fragments only. It should be emphasized, however, that this procedure is probably the most sensitive and would be ideal for identifying the presence of suspected substances whose structure was already known.

The other method of interfacing a liquid chromatograph and mass spectrometer was developed by Scott et al. [5-7] and utilizes the wire transport system to carry a sample into the ion source of a quadrupole mass spectrometer. The eluent from the column passes over a moving wire and the coated wire then enters a differential pump interface and into the mass spectrometer ion source. The sample is volatilized from the wire into the electron beam by the heat generated by a current passing through the wire. The wire then passes out of the source via a second and similar interface to a winding spool. This system, even when fully developed, will be significantly less sensitive then the CI mode already described; however, the spectra produced are electron impact spectra (EI) and are far more useful for structure determination. Furthermore, as the solvent is evaporated in the interface the solvent plays no part in the ionization procedure, and thus gradient elution development with any type of solute is possible with the only provision that the solutes are reasonably volatile. The two LC/MS systems will now be considered in detail.

3 THE LC/MS USING DIRECT INLET SAMPLING

The chemical ionization LC/MS system was developed by McLafferty's group, Baldwin and McLafferty [1], Arpino, Baldwin, and McLafferty [2] and Arpino, Dawkins, and McLafferty [3] and a diagram of the interface they employed is shown in Fig. 6.4. The interface was connected to a Hitachi RMH2 modified for

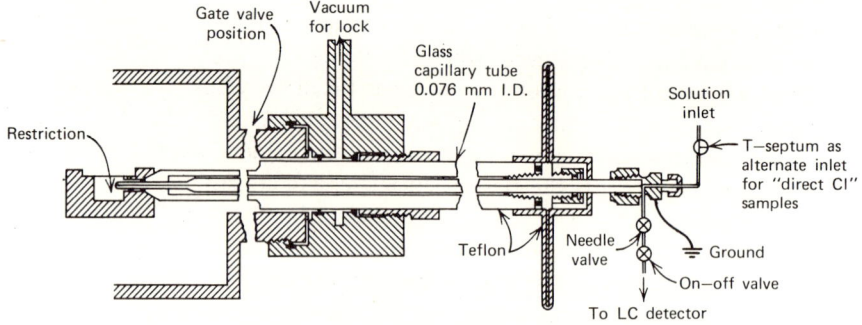

Fig. 6.4. Inlet probe for continuous introduction of solutions; replaces the direct introduction sample probe of the RMH-2, utilizing the same vacuum lock system.

chemical ionization [4] and a 2500-liter/sec diffusion pump was added to the ion source region. The solution is introduced into the ion source through a glass capillary 0.076 mm I.D. that passes through the center of a Teflon rod. This rod is inserted through the vacuum lock made for the direct solids introduction probe to provide a vacuum-tight seal to the ion source. Excess glass at the ion source end of the capillary tube can be convenieintly removed with HF, and a suitable flow contriction if formed by drawing out the tip in a small flame. Delay time in the capillary is about 6 sec at a flow rate of 0.01 ml/min; this rate can be readily maintained for times typical of those required for LC runs. Flows of several times this rate could be achieved, but the possibility of high-voltage breakdown in the source would be increased. Current flow through the capillary using pure methanol was < 1 μA at an ion source potential of 9600 V. For solutions of much higher conductivity, McLafferty considered the dangerous conductance through the capillary could be avoided by using a quadrupole or other spectrometer with a low ion source potential. The solvent serves as the ionizing reagent, although other reagents could be added concurrently to alter the type of ionization.

Compounds that have insufficient vapor pressure for direct chemical ionization would not be detectable by this technique. The ion source block was modified by increasing the exit slit from 0.05 X 4 mm to 0.5 X 8 mm. This lowers the ion source pressure and yields less high molecular weight solvent clusters; it also provides a more intense ion beam, thus increasing the overall sensitivity of the system. The acutal ion source pressure was not known; at the tip of the inlet capillary it probably corresponds to the vapor pressure of the solvent at the source temperature of 200°C and decreases through the ion source and ion exit slit down to about 10^{-4} torr in the ion source housing. There was some effect of capillary tip position relative to the path of the high-energy electrons (500 eV) on the CI spectral data.

The interface and mass spectrometer was used in conjunction with a Waters Associates ALC202 liquid chromatograph fitted with two M6000 pumps and provided with a gradient elution control unit. The sample of column eluent to the interface was taken from a T junction positioned between the column exit and the UV detector. A fine metering needle valve (1SGD, Nupro) controls the split ratio of liquid entering the MS. Flow rates between 0.5 and 1.5 ml/min were employed, which permitted flows of 10 to 12 μl/min to enter the interface. A fine-mesh filter at the interface entrance eliminated most capillary plugging problems except for unusually high solute concentration. McLafferty states that the mass spectrum depends on the nature of the solvent-solute couple, but with all the examples studied so far, a mass spectrum indicative of the solute could be obtained. In general, polar solvents such as tetrahydrofuran, acetonitrile, methanol, and water produce simple mass spectra where (M + H) is the abundant, if not the only, detected ion. Saturated hydrocarbon solvents such as n-pentane give abundant hydride abstraction products, $(M - H)^+$.

Chloroform, a solvent often used in adsorption LC, is attractive for LC/MS because useful fragmentations are often induced in the solute molecule. As an example, the LC analysis of three steroids on a C_{18}- Corasil II, 2-ft column monitored by the UV detector, and independnetly recorded and processed by the MS computer system are shown in Fig. 6.5. A mixture of 5-α-3-androstanone, estrone methyl ether, and androstanolone dissolved in tetrahydrofuran was injected in 10 μl quantities onto the column; this mixture was equivalent to about 0.2-0.25 mg of each solute. A linear gradient from 40% CH_3CN-60% H_2O to pure acetonitrile at 1 ml/min was employed over a time period of 10 min. The mass sepctrometer was run at resolution 1500, source temperature 200°C, ionizing electron energy 500 eV, and emission current 0.7 mA. Repetitive cyclic scan speed was 10 sec/decade from mass 600 to mass 120, with a flyback time of 2 sec.

Referring to Fig. 6.5 it is seen that androstanone gave no appreciable peak from the UV detector but a significant peak for this compound is to be seen on the total ion current chromatogram. Conversely the large peak detected by the UV monitor (probably on injeciton) was undetected by the MS. In the example given, the overall sensitivity of the MS system is not high but the sensitivity with respect to ion unit signal is a function of both the solvent or ionizing agent and the nature of the solute. For example, McLafferty claims that 1 μg of cholesterol injected onto the column using acetonitrile as the mobile phase was easily

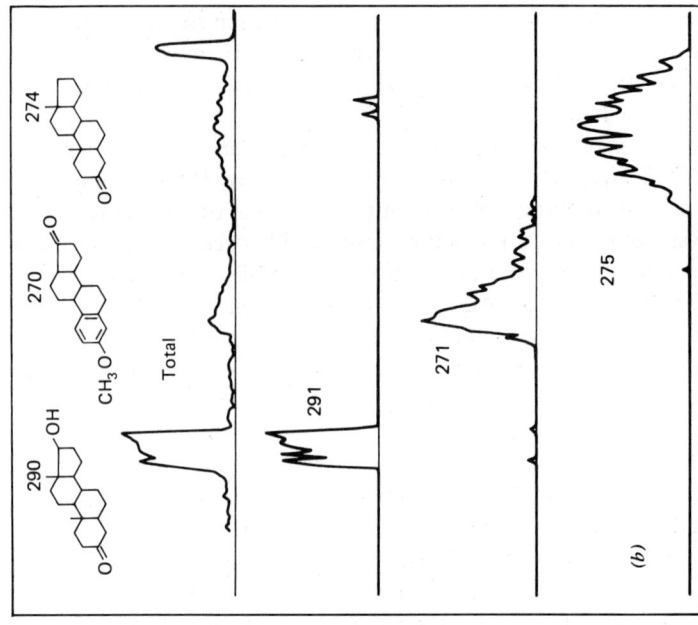
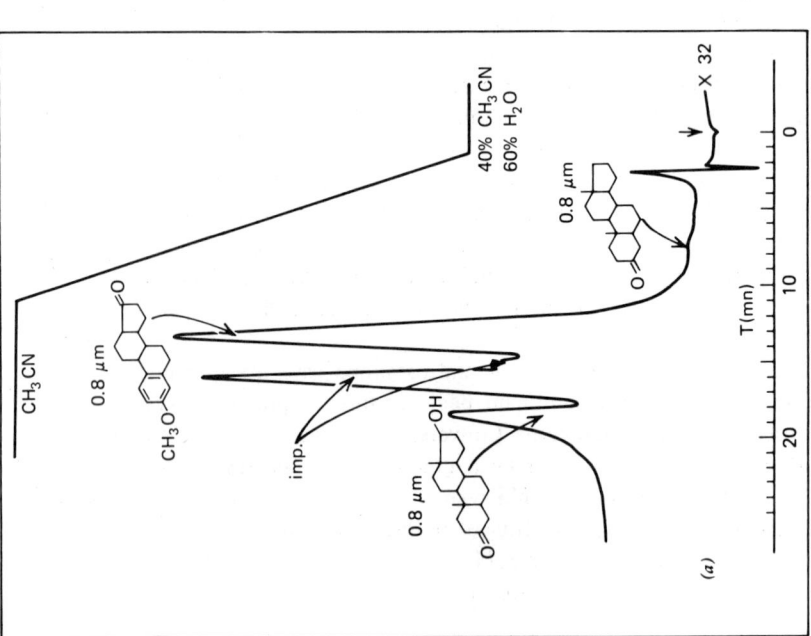

Fig. 6.5. (*a*) Steroid analysis: LC chromatogram recorded from the UV detector response. (*b*) Steroid analysis: Total MS ion current chromatogram and mass chromatograms of the quasimolecular ions retrieved from the computer using the same sample and column conditions.

detected as was 0.2 μg of tertiary-butyl anthraquinone monitoring solely on peak mass 265 (single ion detection). It can be concluded that the chemical ionization mode of LC/MS operating as described by McLafferty has been established as a viable system. In its present form it does not appear to realize the full potential for sensitivity that would be expected. Further gradient elution development can confuse the spectra obtained and the soltues examined must have a singificant vapor pressure at the source temperature to provide satisfactory spectra.

4 LC/MS USING THE WIRE TRANSPORT METHOD OF SAMPLE INTRODUCTION

This approach to coupling a liquid chromatograph to a mass spectrometer was developed by Scott, Scott, Munroe, and Hess [5, 6] and utilizes the wire transport system similar to that in the wire transport detector. In principle the eluent passes over a moving wire, which would carry a sample of the eluent into an evaporation chamber where the solvent would be removed leaving the solute deposited on the wire. The wire would then pass through a suitable interface directly into the ionization chamber of the mass spectrometer where it would be heated and the vapor ionized to provide the necessary spectra. Such a system would provide electron impact spectra or chemical ionization spectra as required, but would be completely independent of the solvent used for development in the liquid chromatograph and thus miantain the versatility and capabilities of the separation technique.

Design of the Interface

In the standard deflection mass spectrometer the ion optics are critically controlled by the potential gradients between the electrodes in the ion source. Such a situation would, therefore, be very sensitive to an earthed wire passing in close proximity to these electrodes. For this reason the mass spectrometer used was a quadrupole type and that chosen was the Finnigan Quadrupole Mass Spectrometer where the ion optics of the source are very simple and insensitive to the proximity of earthed conductors.

A diagram of the interface is shown in Fig. 6.6. These are made in pairs, one for the entry of the wire into the ion source and the other for its exit to the winding spool. Each interface consists of two small chambers separated and terminated by ruby jewels carrying a 0.01-in. central aperture. These jewels are standard watch jewels used for bearings in small time pieces. They are 0.1 in. in diameter and 0.014 in. thick and are set into the chambers by means of a high-temperature epoxy resin cement. The hard ruby surfaces of the jewels completely eliminate abrasive effects of the wire. Each chamber is connected to an evacuating port. The outside chambers directly in contact with the atomsphere

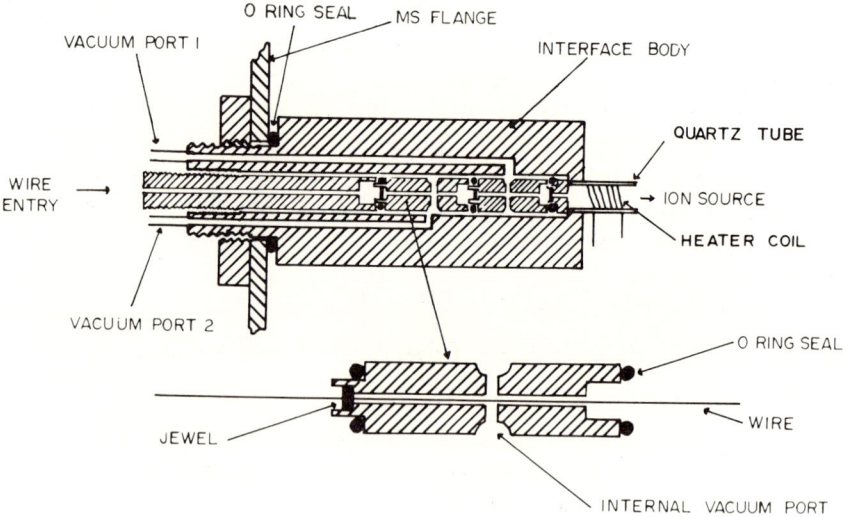

Fig. 6.6. The LC/MS interface. Courtesy of the CIBA Foundation.

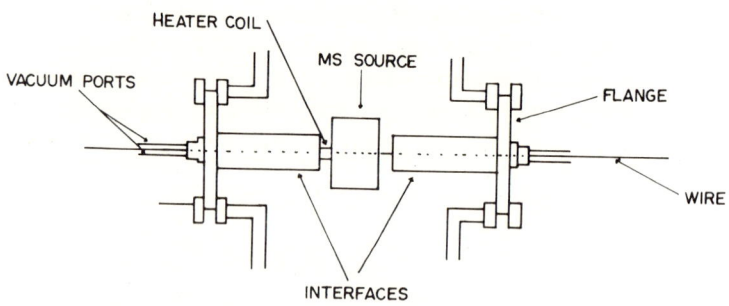

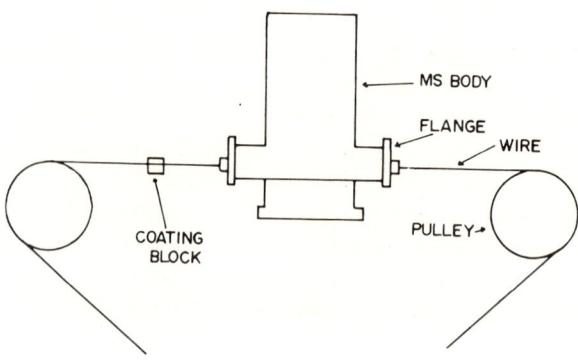

Fig. 6.7. Diagram of interfaces located in MS source.

are connected in parallel to a 300-liter/min rotary vacuum pump. The second chambers between the first chambers and the mass spectrometer source are also connected in parallel and to a 2-in. oil diffusion pump backed by a rotary pump with a 160-liter/min capacity. This system of differential pumps reduces the gas pressure in stages. In the first chamber it is reduced to about 0.2 mm of mercury and in the second to about 2×10^{-3} mm of mercury. A diagram of the interfaces situated in the mass spectrometer head is shown at the top of Fig. 6.7 and it is seen that they are arranged in a reentrant design so that the wire, on leaving the interface, is within 1-2 cm of the electron beam that will produce the ions.

The wire train employed was the standard unit from the Philips Chromatography moving wire detector and a diagram of the complete arrangement is shown at the bottom of Fig. 6.7. The wire from the train system passes around an insulated pulley, through the column eluent, and directly into the first interface and thence into the mass spectrometer. The wire leaves the mass spectrometer through the second interface around another pulley and into the capstan winding system of the wire train mechanism. The insulated pulleys were constructed using roller bearings and particular care was taken in their alignment to minimize friction. A photograph of the first version of the apparatus in operation is shown in Fig. 6.8 in which the interfaces were not reentrant in design. They are basically of the same form as the interfaces described but were prototype models and protrude external to the mass spectrometer. Because of

Fig. 6.8. The LC/MS apparatus utilizing the Philips moving wire transport system.

the insensitivity of the ion source to slight changes in potential in its neighborhood, the wire could be heated to vaporize the solute by passing a current through it. The wire transport detector wire supplied by Philips Chromatography for the wire transport detector is made of stainless steel, 0.005 in. diameter. Passing a current of 250 mA through this wire situated in air, at atomspheric pressure, results in little heating, because of the cooling effect of the surrounding air. However, when the wire carrying this current is situated in the high vacuum of the mass spectrometer, heat can only be lost by radiation and its temperature rises to about 250°C. Thus, by passing a current through the wire during its passage through the mass spectrometer the temperature of the wire increases in the mass spectrometer where volatilization is required. The voltage is applied between the left pulley just before the coating block and to the pulley situated subsequent to the exit interface and can be adjusted to provide heating current up to a maximum of 400 mA.

Operation

The mass spectrometer is operated under normal conditions except that the filament current employed is about 3 mA. The instrument can be used with the standard data handling system supplied for the Finnigan Mass Spectrometer and a typical set of operating conditions is given in Table 6.2. The chromatographic

Table 6.2 Mass Spectrometer Operating Conditions

Filament current	3 ma
Electon energy	60 V
Source pressure	3×10^{-6} mm of mercury
Mass range	60-199, 200-399, 400-650
Integration time	3, 5, 8
Sample/AMU	1, 1, 1
Threshold	1
Attenuation	5
Mass range setting	High
Mass run time	400 min
Delay between scans	3 sec

column can be situated such that the terminal 1/16-1/4-in. Swage lock column coupling rested about ½ mm above the moving wire. Under these conditions the wire was always immersed in the column eluent despite eluent leaving the column in drops. In the example given the column employed was 50 cm long, 4.6, mm I.D. and packed with 10-μ Partisil silica gel, a slurry packing procedure being used. The column was connected to a Milton Roy minipump, which in turn was fed by the apparatus used for incremental gradient elution. A stop flow method of injection was used and samples of approximately 2-6 mg were employed for each chromatogram depending on the complexity of the mixture. Subsequent to a separation the column was regenerated by a solvent sequence in the usual manner.

In the determination of sensitivity, 0.01% solutions of diazepam in a 10% tetrahydrofuran chloroform mixture were dropped onto the wire in a continuous stream and the total ion current monitored. The sensitivity was taken as that concentration of diazepam that would provide a constant ion current signal equivalent to twice the noise level. The sensitivity measured in this way for diazepam was 4×10^{-6} g/ml. The wire carried 10 μl/min of eluent into the mass spectrometer at the maximum wire speed 15 cm/sec. Thus, assuming a complete spectrum was obtained in 1 sec, this would correspond to a total mass of approximately 7×10^{-10} g of solute per spectrum. The jewel system interface resulted in the pressure in the ion source being about 3×10^{-6} mm of mercury.

Performance of Chromatograph-Mass Spectrometer Combination

An example of a separation of a synthetic mixture using the LC/MS system is shown in Fig. 6.9. The mixture contained normal triacontane, cholesteryl laurate, and cholesterol and phenobarbital. The total sample was about 2 mg and placed on the column as a 0.6% solution in a mixed solvent heptane, acetone, and methanol. Beneath the total ion current chromatograms in Fig. 6.9 are shown the reconstructed chromatograms based on peak masses of 386 and 204, significant ions of cholesterol and phenobarbital, respectively. It is seen that good spectra were produced and these are obtained on the original prototpye instrument having a sensitivity of 2×10^{-5} g/ml. In Fig. 6.10 the chromatogram obtained for a fermentation extract in methanol-chloroform is shown, and this sample contains solutes that cover a wide range of polarity. The mixture was first chromatographed using the incremental gradient elution technique [7] and the optimum conditions previously described used for the development of the chromatogram. A charge of 2 mg was injected onto the column and the detector used was the wire transport detector. The chromatogram produced is shown at the top of Fig. 6.10. During the chromatogram development fractions were collected for each peak and numbered as shown in the diagram. These were concentrated in a current of nitrogen to about 0.1 ml and each fraction examined using the LC/MS system as a probe injection device. A drop of each sample was spotted onto the wire for a few seconds and the total ion current traces of these probe samples are shown on the right-hand side of Fig. 6.10. All the fractions were placed on the moving wire over a period of 8½ min. Thus, the spectra for an individual probe sample was obtained in approximately 25 sec. The sample was then chromatographed on the LC/MS system using a 6-mg charge and the same operating conditions. The chromatogram obtained from the

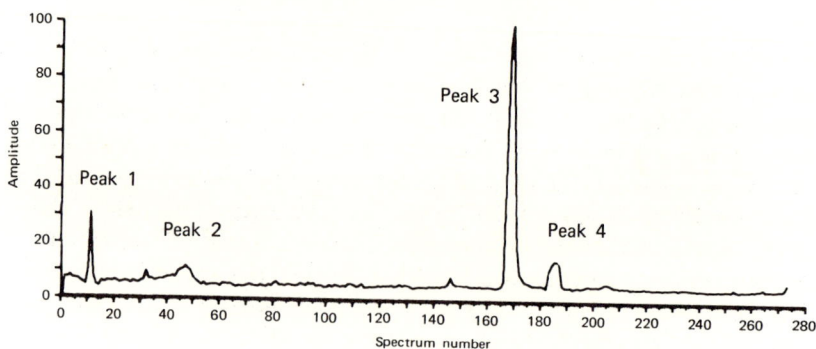

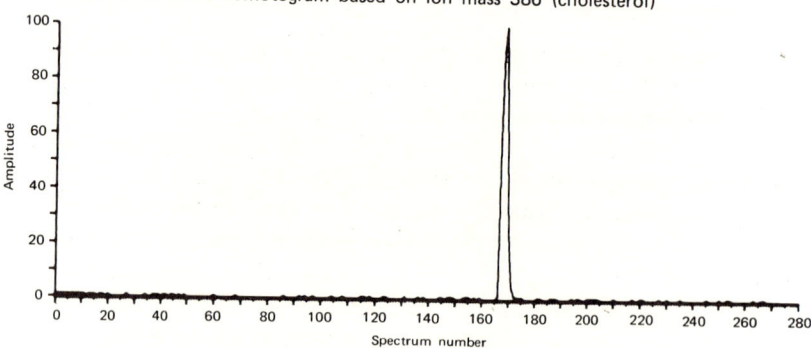

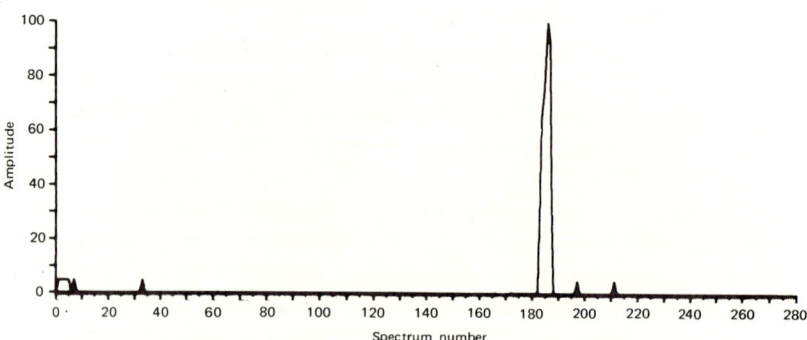

Fig. 6.9. LC/MS analysis of a synthetic mixture.

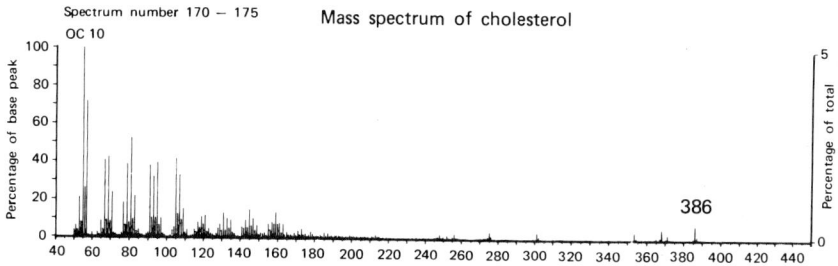

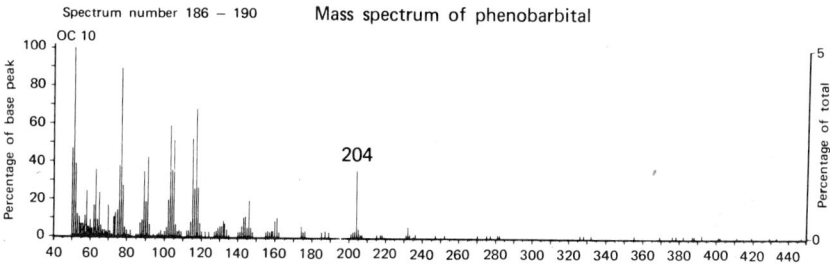

Fig. 6.9. (continued)

total ion current monitor is shown as a hard copy on the left-hand side of Fig. 6.10. It is seen that due to the nature of the hard-copy presentation of the chromatogram it is somewhat compressed in size relative to the original using the wire transport detector. However, the same pattern of peaks are obtained although owing to the larger charge size employed, the resolution is not as good.

A spectrum was obtained from the probe injection sample number 7 and is shown as the lower spectrum in Fig. 6.11. It is seen that it has a significant ion mass of 327; thus a reconstructed chromatogram from the ion current to mass 327 was obtained for the probe samples. This is seen as the lower chromatogram on the right-hand side of Fig. 6.10. Peak 7 is clearly selected and using the same mass 327 a reconstructed chromatogram was obtained from the original ion current chromatogram from the LC/MS run. This is shown at the bottom of the right-hand side of Fig. 6.10, and again it is seen that the peak is clearly and unambiguously selected from the mixture. A spectrum was then taken from the respective peak in the LC/MS chromatogram and this is shown at the top of Fig. 6.11 and can be compared with the spectra from fraction 7. It is seen that basically the two spectrums are identical although the spectrum from fraction 7 obviously contains traces of a contaminating material.

The authors stated that the quadrupole mass spectrometer that was utilized for

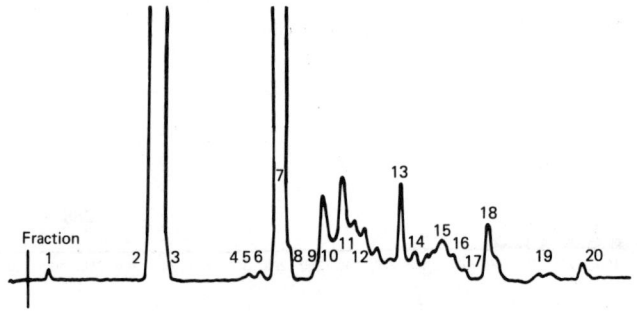

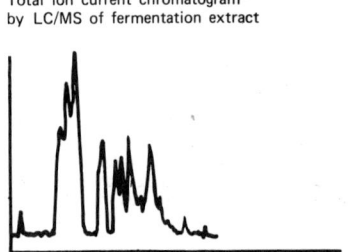

Total ion current chromatogram by LC/MS of fermentation extract

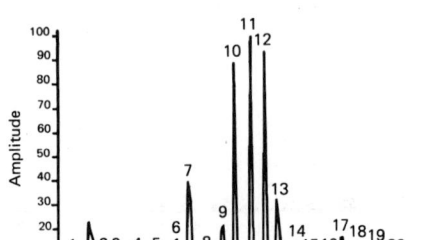

Total ion current trace of fractions sampled directly onto wire

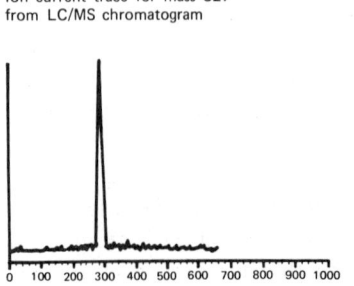

Ion current trace for mass 327 from LC/MS chromatogram

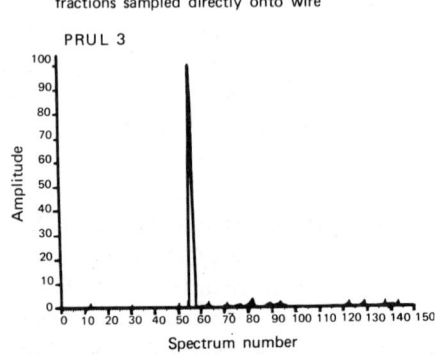

Total current trace for mass 327 from fractions sampled directly onto wire

Fig. 6.10. Chromatogram of a fermentation extract by incremental gradient elution (IGE).

this work was an old version and had very poor resolving power. As a result of this low resolution, the authors' claimed sensitivity of the instrument could not be increased to its normal level. In an attempt to determine the sensitivity

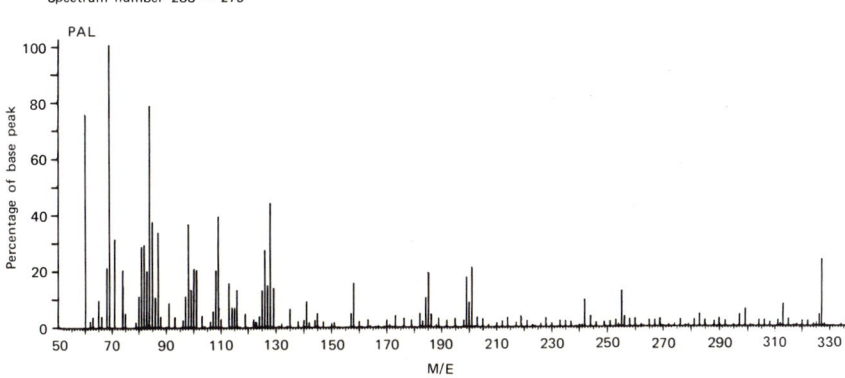

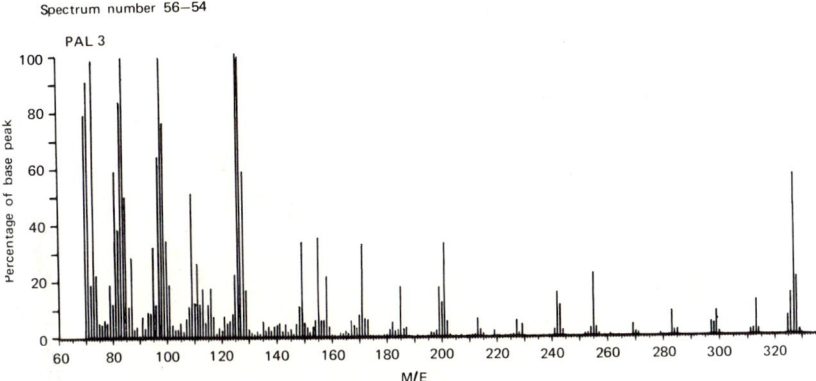

Fig. 6.11. Spectrum 285-262 from LC/MS chromatogram.

of the system, when a modern instrument with specified resolution was used, the sensitivity of the older instrument was increased at the expense of resolution until a single peak became a doublet (owing to the increased widths) when processed by the computer. The sensitivity with the apparatus operating under these conditions was estimated at about 5×10^{-7} gm/ml. Under the same conditions a sample of vitamin A mother liquor was chromatographed again using incremental gradient elution development. The total ion current chromatogram obtained for the sample is shown in Fig. 6.12; the charge placed on the column was about 1 mg.

It is clear that two-stage differentially pumped interfaces will permit a wire to pass through the ion source of a quadrupole mass spectrometer while maintaining source pressure at about 3×10^{-6} mm of mercury.

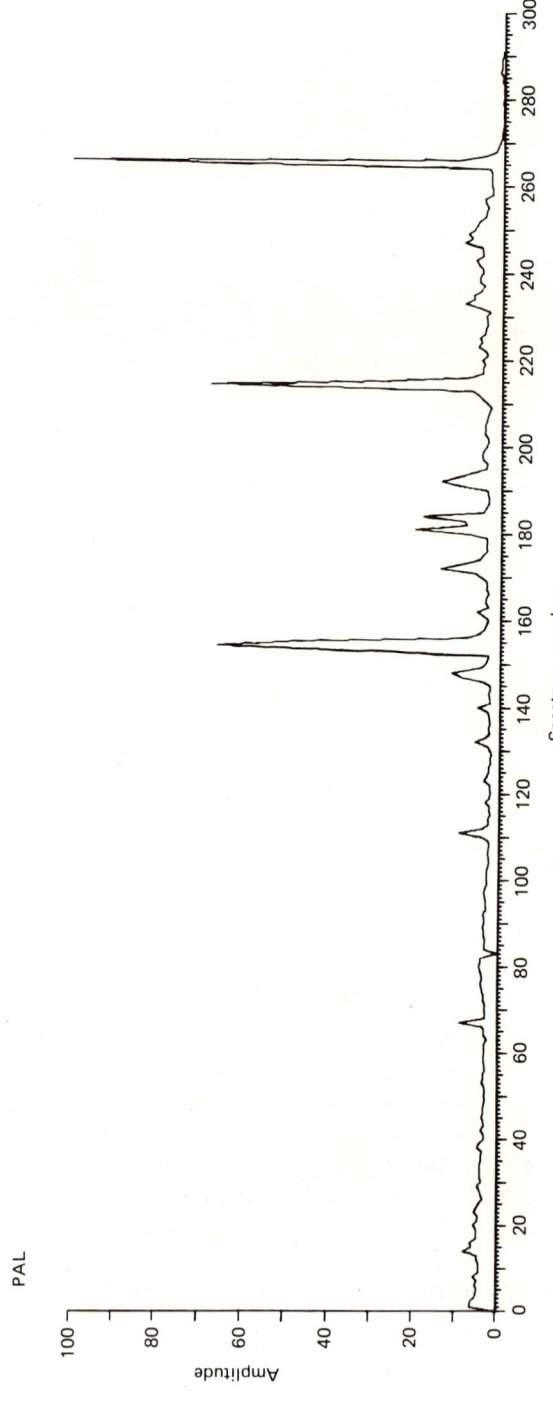

Fig. 6.12. Total ion current chromatogram of vitamin A acetate mother liquor.

The alternative use of the apparatus to provide a probe sampling system together with on-line LC/MS monitoring of column eluents is another interesting application of the system and could be of value as an automated probe sampling system for general mass spectrometry work. The system provides electron impact spectra that as far as has been investigated are the same as those obtained from normal probe injection. It appears that the apparatus in its present form is practical for providing LC/MS monitoring but requires significant charges to be placed on the column where trace materials are being identified, and this may cause a reduction in column performance by way of lower resolution. The total mass of material required to provide spectra from the present instrument assuming 1-sec sampling time was about 500 pg. A more effective spectrometer together with an improved method of volatilizing the sample could provide a sensitivity to about 5×10^{-7} g/ml, which would be satisfactory for most LC separations. Further sensitivity enhancement might be obtained if required by modification of the interfaces to reduce dead volumes. An alternative method of vaporizing the sample might include the use of heater coils, the use of infrared radiation focused on the wire in the ion source, or alternatively by the use of a laser.

5 FUTURE OF LC/MS

Mass spectrometry is an ideal spectroscopic technique to combine with liquid chromatography since it can not only provide a means of identification, but it can at the same time be highly sensitive.

Two practical systems have been shown to be effective but probably still need some development before they can be produced commercially. However, it has obviously been established that LC/MS will be a common technique in the future.

There is, however, another possible method of introducing samples of eluted solutes into a mass spectrometer which was suggested by Horning et al. [8]; the diagram of their apparatus is shown in Fig. 6.13. The method, at the time of writing this book, has only been confirmed using direct sampling but would obviously be used for the direct and continuous injection of a sample of a liquid chromatographic column eluent. The eluent would be vaporized in a heated gas stream, which would then pass over a radioactive source whereby positive ions are formed by a complex series of ion molecular reactions. The ions then pass through a micropore aperture directly into the ion source of a quadrupole mass spectrometer providing the direct spectra. The authors claim on a direct sampling procedure a sensitivity of 5-10 pg per spectrum. It would appear that the solute being examined must have some significant vapor pressure at the ionizing chamber temperatures, and the effect of solvent change during gradient elution is not clear. However, this is obviously an interesting alternative method

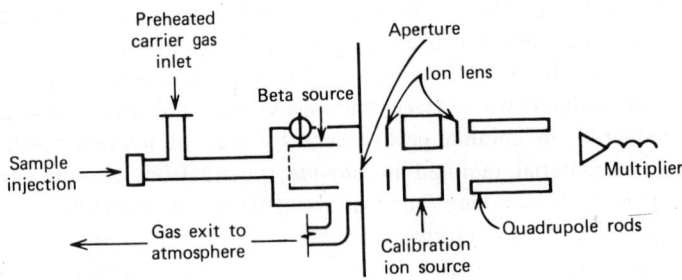

Fig. 6.13. Schematic diagram of the apparatus. The reaction volume is about 1 cm in diameter and 1 cm in length.

for LC/MS operation and will, without doubt, be followed with great interest by those workers interested in utilizing LC/MS as an analytical tool.

6 THE COMBINATION OF LC WITH OTHER SPECTROSCOPIC TECHNIQUES

The development of an LC/IR system faces many practical difficulties. The instrument itself is relatively insensitive and if flow-through cells are employed they would place impossible restrictions on the solvents used in the chromatographic development. The sampling of a column eluent by some form of transport system to bring the solute to the infrared spectrometer is theoretically possible, but the sample must then be either volatilized and the IR spectrometer sensitivity to vapor precludes this possibility or the solute must be washed from the transport and the sample examined in solution in the normal way. The last alternative is fraught with practical problems and it must be concluded that a practical LC/IR apparatus is an unlikely system to be developed in the near future. For the present, off-line production of IR spectra appears the best if not the only practical solution.

For similar reasons it seems unlikely that a practical LC/NMR system will be developed for sometime to come. Although the development of a flow-through system for a spinning sample tube is feasible, the restraints the spectroscopic technique itself places on the solvents that can be employed for the chromatographic development would seriously impair the efficiency of the chromatograph. The use of deuterated solvents is theoretically possible but would be inordinately expensive in practice. These disadvantages, combine with the technique's insensitivity, render it necessary to wait until significant improvements in the NMR apparatus have been developed before a LC/NMR system can become a practical reality.

The LC/Laser Raman system does, however, appear a slightly more feasible proposition. Employing a laser of sufficient power can reduce the problem of background fluorescence and under such circumstances the laser Raman spectrometer has a reasonable sensitivity, samples of only a few micrograms being necessary. However, for this sensitivity the samples have to be neat; if solutions are employed then the sample has to be at relatively high concentration in the solvent, and this precludes the use of flow-through cells. In Raman spectroscopy the spectra are not obtained by absorption; they are emission spectra resulting from molecular excitation by the incident light. It follows that Raman spectra could be obtained from the sample coated on a surface, for example, a ribbon-type transport system. If the eluent from the column was evaporated from the surface of a moving ribbon and the ribbon passed through the sample chamber of the Raman spectrometer, it might be possible to obtain spectra by the stop flow technique. If pulsed Fourier transform Raman spectroscopy were developed in the near future it would enhance the sensitivity of such a system and/or reduce the scan time necessary to provide the spectra.

REFERENCES

1. M. A. Baldwn and F. W. McLafferty, *Org. Mass Spectry.*, **7**, 1111 (1973).
2. P. Arpino, M. A. Baldwin, and F. W. McLafferty, *Bio. Mass Spectry.*, **1**, 80 (1974).
3. P. Arpino, B. G. Dawkins, and F. W. McLafferty, *J. Chromatog. Sci.*, **12**, 574 (1974).
4. F. H. Field, *Ion Molecular Reaction* J. L. Franklin, Ed., Plennum Press, New York, 1972, Chapter 6.
5. R. P. W. Scott, C. G. Scott, M. Munroe, and J. Hess, Jr., *The Poisoned Patient: The Role of the Laboratory*, Elsevier, New York, 1974, p. 155.
6. R. P. W. Scott, C. G. Scott, M. Munroe, and J. Hess, Jr., *J. Chromatogr. Sci.*, **99**, 395 (1974).
7. R. P. W. Scott and P. Kucera, *J. Chromatogr. Sci.*, **11**, 83 (1973).
8. E. Horning et al., *J. Chromatogr.*, **99**, 13 (1974).

Chapter VII

PREPARATIVE LIQUID CHROMATOGRAPHY

1 **Factors that Affect the Charge That Can be Placed on a Given Liquid Chromatographic Column** 298

   Column Radius r 299
   Distribution Coefficient k 300
   Packing Density d 300
   Surface Area per Gram of Adsorbent A_s 300
   Column Length (l) and Particle Diameter dp 300

2 **Column Overloading** 301

   Overload due to Sample Feed Volume 301
   Overload due to Sample Mass 306

3 **Apparatus for Preparative Liquid Chromatography** 309

   Solvent Reservoirs 309
   Injection Methods 310
   Columns 310
   Detectors 312
   Fraction Collection 313
   Hazards 313
   Commercial Equipment 313

4 **Stationary Phases for Preparative Liquid Chromatography** 313

5 **Packing Preparative Columns** 316

6 **Preparative Chromatographic Procedure** 317

   The Mobile Phase 317
   Sample Injection 317
   Recycling Development 318
   Gross Overload with Peak Cutting 319

7 **The Future of Preparative Liquid Chromatography** 320

The term preparative scale chromatography "means different things to different men." To the biochemists, preparative scale chromatography may mean the separation of a few milligrams of sample for subsequent structure elucidation by appropriate spectroscopic techniques. To the organic chemists, preparative scale chromatography will probably mean the isolation of 5-50 g of an intermediate for subsequent synthetic work. It follows that preparative scale chromatography encompasses a wide range of sample loads from a few milligrams to tens, or perhaps even hundreds of grams of material being separated at one time. High levels of sample load will often require a range of special column to be available, chromatographic apparatus specially designed to take such columns together with appropriate operating conditions to run them. Considering the four attributes that can be obtained from a chromatographic system, namely, resolution, speed, scope, and load, it is fairly obvious that to obtain any two of these attributes in excess, the other two will have to be sacrificed. Thus, in preparative scale chromatography, load must be obtained at the expense of speed and scope together with, in some instances, reduced resolution.

For any chromatographic system there is a limiting charge that can be placed on the column before the resolution in impaired, and it is obviously important to determine those paramenters of the column system that will control the magnitude of the limiting charge. Loss of resolution from column overload can arise from two causes, either excessive sample feed volume or excessive sample mass, the former ususally resulting from poor solubility of the sample in the mobile phase. When considering the maximum mass of charge that can be placed on a column, it is the relative masses of the individual solutes in the mixture that conditions the maximum charge. During development the individual solutes become separated and it is the capacity of the column to carry the individual solutes that limits the initial charge. Thus, the maximum sample load is limited by the level of the major component or components in the mixture together with the properties of the particular column and mobile phase employed.

1 FACTORS THAT AFFECT THE CHARGE THAT CAN BE PLACED ON A LIQUID CHROMATOGRAPHIC COLUMN

For any column operating under given conditions, it has been shown in Chapter II and elsewhere [1] that both the maximum sample feed volume and the maximum solute mass that can be placed on the column without impairing the column efficiency are directly proportional to plate volume and the square root of the efficiency. Thus

$$M = A \sqrt{n}(v_m + Ka_s)$$

where M is the limiting mass of sample, n is the column efficiency, v_m is the volume of mobile phase per plate, a_s is the surface area of adsorbent per plate, K is the distribution coefficient of the solute, and A is a constant. Now from the plate theory $V_r = n(v_m + Ka_s)$ where V_r is the retention volume of the solute. Thus

1 FACTORS THAT AFFECT THE CHARGE

$$M = \frac{AV_R}{\sqrt{n}}.$$

Now if the peak is well retained it can be considered that $v_m \ll Ka_s$ and furthermore,

$$V_r \rightarrow nKa_s \rightarrow VKdA_s \rightarrow \pi r^2 lKdA_s$$

where V is the volume of the column, r is the radius of the column, l is the length of the column, d is the packing density of the adsorbent (gm/ml), and A_s is the surface area of the adsorbent (m²/gm). Hence

$$M = \frac{A\pi r^2 dlKA_s}{\sqrt{n}}.$$

Assuming n is a fixed value $(n)_c$ that just provides the necessary resolution, then

$$(n)_c = (\frac{1}{h})_c$$

where h is the height equivalent to the theoretical plate at a constant linear mobile phase velocity. Furthermore, for a significantly retained peak, according to the HETP equation, $h = \alpha dp^2$ where α is a constant and dp is the particle diameter of the support. It follows that

$$M = A\pi r^2 lKdA_s(\frac{\alpha dp^2}{1})_c^{1/2} \tag{7.1}$$

Equation (7.1) gives the various column parameters that condition the maximum load that can be placed on a column; the effect of these parameters will now be considered individually.

Column Radius r

The loading capacity of a column will increase with the square of the radius and directly with the column cross-sectional area. Increasing the cross-sectional area of the column also increases the quantity of stationary phase per plate and thus the plate volume. Increasing r does not increase the retention time of the solute providing the column flow rate is also proportionally increased to maintain the same linear mobile phase velocity. It follows that by increasing the column radius, more mobile phase is utilized, and solvent economy is sacrificed for the greater loading capacities obtained.

The Distribution Coefficient K

Increasing the distribution coefficient K results in an increase in the retention volume of the solute and thus the column loading capacity. K can only be increased for a given column by changing the nature or composition of the mobile phase and will result in both an increase in retention time and a greater volume of mobile phase being used. Obtaining a higher load capacity in this way will be done at the expense of both speed and solvent economy.

The Packing Density d

The effect of packing density is theoretically important but in practice is not a significant variable since all columns that are reasonably well packed have packing densities close to their maximum.

The Surface Area Per Gram of Adsorbent A_s

In Chapter IV it was shown that the loading capacity of a silica gel increases with the surface area of the adsorbent for those silica gels currently in use for liquid chromatography. However, high surface area adsorbents result in increased retention volumes and, under isocratic conditions of development, a narrow solute polarity range in which to operate; the polarity of the mobile phase is necessarily close to that of the soltues being separated. Thus, employing adsorbents of high surface area to increase the loading capacity of the column requires that both the speed and scope of the chromatographic system be scarificed.

Column Length (l) and Particle Diameter dp

It is seen from (7.1) that l and dp are related, insomuch that the ratio dp^2/l must remain constant to maintain the necessary efficiency to achieve the required resolution. It follows that the effect of both l and dp on the column loading capacity must be considered together, but it should be emphasized that this is true only when loading effects of preparative scale columns are being considered. From (7.1) it is seen that the loading capacity of the column will increase as the square root of the column length, but dp has to be increased appropriately together with l if the effeciency is to be kept to the same minimum requirement for resolution. It follows that the column will become longer but be packed with coarser particles. Using (7.1) the fractional increase in load with column length and particle diameter is shown in Table 7.1 for a 25-cm-long column 4.6 cm I.D. packed initially with 10-μ particle diameter silica gel.

It is seen from Table 7.1 that the loading capacity of the column can be doubled by increasing the column length by a factor of 4 and the particle diameter of the absorbent by a factor of 2. Increasing the loading capacity in this way also increases the solute retention volume and thus sacrifices both speed of separation and solvent economy. However, in the example given, the

Table 7.1 Relationship Between Column Length, Particle Diameter, Relative Permeability, and Relative Loading Capacities for Columns Having a Constant Cross-Sectional Area and the Same Efficiencies

Column Length (cm)	Particle Diameter (μ)	Relative Loading Capacity	Relative Permeability
25 cm	10	1.0	1.0
50 cm	14	1.4	2.0
75 cm	17	1.7	3.0
100 cm	20	2.0	4.0

column permeability is also increased by a factor of 4 as the column length is increased by the same ratio and thus no increase in column inlet pressure is required. When working with large-diameter preparative columns it is highly desirable to maintain high column permeability in order to make less stringent the specifications of wall thickness to provide adequate mechanical strength.

It should be emphasized that the above conditions apply to columns that carry the maximum load but are not overloaded in the sense that the normal bandwidth is not exceeded by more than 5% as described in Chapter II. If the 25-cm-long column is initially overloaded, then the advantages derived from increasing the column length may well be in excess of the relative loading capacities given in Table 7.1.

2 COLUMN OVERLOADING

In practice it is not usually possible to construct a specific column for each preparative scale sample presented for separation. Normally a limited number of columns are available for preparative work and the conditions of separation have to be adjusted for each sample to meet the limitations of the columns that are available for use. The mobile phase is usually chosen to provide a suitably high separation ratio between the solute peaks of interest and then the column is overloaded until the peak of the solute to be isolated is just separated from its closest neighbor. It was pointed out earlier that a column can be overloaded in two ways. When samples that are relatively insoluble in the mobile phase are being separated, the column may be overloaded with respect to sample volume, whereas for soluble samples, column overload will arise from excess of sample mass.

The effect of overload due to sample feed volume will first be quantitatively examined.

Column Overload due to Sample Feed Volume

Consider the situation explicitly depicted in Fig. 7.1. To determine the band

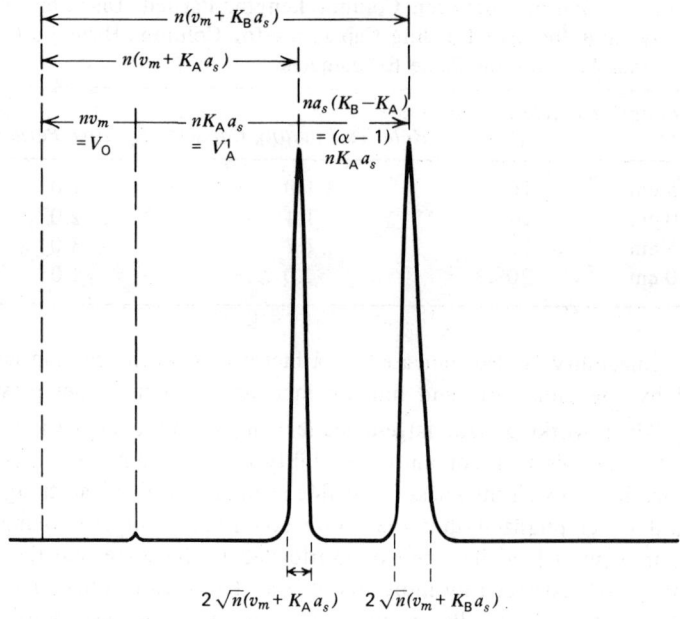

Fig. 7.1. Theoretical diagram for volume overload.

spreading effects resulting from significant sample volumes, the principle of the summation of variances should be employed as demonstrated in Chapter II. However, when the sample volume becomes excessive, the band spreading that results from column overload becomes equivalent, to the first approximation, to the sample volume itself. Thus, referring to Fig. 7.1 the peak separation in milliliters of mobile phase will be equivalent to the volume of sample plus the sum of half the base width of the respective peaks. Bearing in mind that half the peak width at the base is approximately twice the standard deviation of the peak, then from the plate theory, assuming that both peaks have the same efficiency n,

$$(\alpha-1)nK_A a_s = V_L + 2\sqrt{n}\,(v_m + K_A a_s) + 2\sqrt{n}\,(v_m + K_B a_s)$$

where n is the column efficiency, K_A and K_B are the distribution coefficients of solutes A and B, respectively, α is the separation ratio of solute B and solute A, and V_L is the sample overload volume. Rearranging,

$$V_L = (\alpha-1)nK_A a_s - 2\sqrt{n}\,[(v_m + K_A a_s) + (v_m + K_B a_s)].$$

Noting that

$$nK_A a_s = V'_A, \quad nK_B a_s = V'_B, \quad \frac{V'_A}{V_0} = k'_A, \quad \frac{V'_B}{V_0} = k'_B \quad \text{and}$$

$$k'_B = \alpha k'_A$$

where $V_0 = nv_m$ is the column dead volume and

$$V_l = V_0 \left[(\alpha - 1) k'_A - \frac{2}{\sqrt{n}} (2 + k'_A + \alpha k'_A) \right] \quad (7.2)$$

Equation (7.2) allows the calculation of the maximum volume of sample that can be placed on the column to maintain the separation of solutes A and B (B being eluted later than A) in terms of the column dead volume, the column efficiency, the separation ratio of solute B to solute A, and the respective k' values of solutes A and B.

Taking a practical example of a preparative column 4 ft long, 1 in. I.D. having a dead volume of 300 ml and an efficiency of 2000 theoretical plates, the maximum sample volume was calculated using (7.2) for solutes of different α values and k'_A values; the results are shown in Fig. 7.2. It can be seen from Fig. 7.2 that the maximum overload charge varies widely from just over 200 ml for an α value between the solutes of 1.2 and a k'_A value of 2 to well over 2 liters for solutes having an α value of 2.0 and the first solute having a k'_A value of 6. For solute pairs having high separation ratios chromatographed on large columns having dead volumes of 300 ml or more, the volume of sample can be exceedingly high before the resolution of the solutes is impaired.

A practical use of (7.2) is given by the following example. A 25-cm-long 4.6-mm-I.D. column having a dead volume of 3.48 ml was used to separate a mixture of benzene, naphthalene, and anthracene. The chromatographic properties of the solutes chromatographed on this column are given in Table 7.2. The column was packed with Partisil silica gel 10 μ in diameter and was fitted with a sample valve and sample loop. Owing to the band spreading from the sample loop injector it is seen that expected efficiencies are not realized until the k' value of the solute is at a value of 4 or more. The significant increase in efficiency with the k' value of the solute exphazises the importance of maintaining low extra column dispersion effects when employing microparticulate columns of high intrinsic efficiency. The maximum volume of sample that could be employed to separate both benzene from naphthalene and naphthalene from anthracene caluclated from (7.2) are included in Table 7.2.

Sample solutions were made up such that on injecting 10 μl, 1 ml, 2 ml, and 3 ml of solution onto the column, each injected sample contained 176 μg of benzene, 9.0 μg of naphthalene, and 0.3 μg of anthracene; the chromatograms

Table 7.2 The Chromatographic Properties of Different Solutes Chromatographed on a Standard Column

	Benzene		Naphthalene		Anthracene
k'	1.18		2.33		4.31
n	1850		4480		5470
α	<	1.97	><	1.85	>
V_L	3.1 ml		6.1 ml		—

obtained are shown in Fig. 7.3. The top chromatogram is of a 10-μl sample of the mixture representing no overload. The chromatograms obtained are shown on the left-hand side of Fig. 3. It is seen that calculated sample volume of 3 ml just permits the separation of benzene and naphthalene as (7.2) predicts.

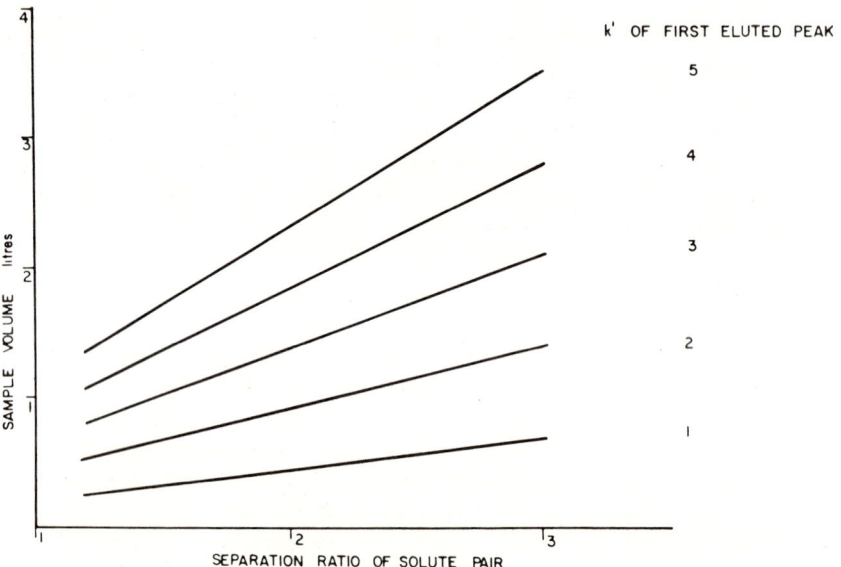

Fig. 7.2. Graph of maximum sample volume against separation ratio of solute pair for solutes eluted at different k' values.

In the same manner 2, 4, and 6 ml of solutions containing a total of 9.0 and 0.3 μg of naphthalene and anthracene, respectively, were injected onto the column and the chromatograms obtained are shown on the right-hand side of Fig. 7.3. It is seen that again the maximum volume of charge that can be used while maintaining resolution between naphthalene and anthracene is accurately predicted by (7.2).

2 COLUMN OVERLOADING 305

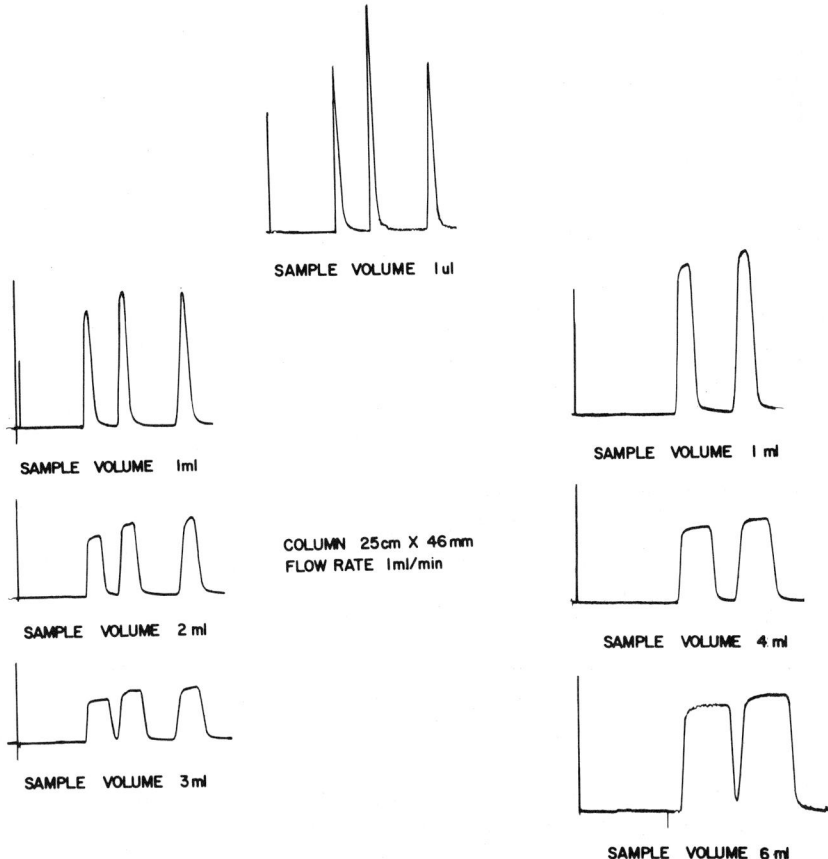

Fig. 7.3. Chromatograms demostrating volume overload.

In Fig. 7.4 the retention of the front and back of the three peaks, measured at 0.6065 of the peak height, and plotted against the volume of charge. It is seen that the peak front for each solute has a constant retention distance irrespective of the volume of the charge. The retentions of the backs of the peaks, however, only remain constant up to a charge volume of 0.5 ml and, subsequent to this, increase linearly with the volume of the charge. The spread of the peaks is also the same for each solute and is not dependent on the nature of the solute or its k' value. It should be noted that at the predicted charge volume of 3 ml, the back of the benzene peak has just reached the front of the

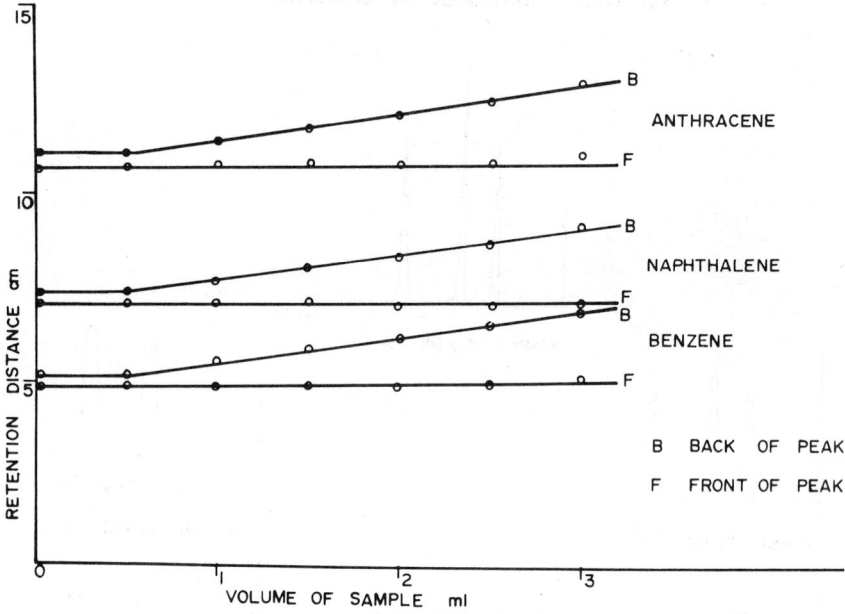

Fig. 7.4. Graph of retention distance of benzene, naphthalene, and anthracene against volume of injected sample.

naphthalene peak. The spread of the peaks toward greater retention times is characteristic of volume overload and, as will be seen later, is in contrast with the effect of mass overload, where the retention times of the peaks are reduced as the mass of charge is increased.

Destefano and Beachell [2] also investigated the effect of volume overload on column resolution and concluded that it is better to overload a column by using a large volume of a dilute solution of the solute than a small volume of a concentrated solution of the solute. The validity of this conclusion, however, depends somewhat on the k' values of the solutes eluted and the separation ratios of the solutes of interest.

Column Overload due to Sample Mass

The effect of excessive sample mass on the development process that takes place in the column is extremely complicated and for this reason it is difficult to treat mass overload quantitatively. A large mass of sample will modify the development process in basically three ways. First, there will be a dispersive effect resulting from the limited capacity of the stationary phase. The sample will spread along the column, carried by the mobile phase, until it contacts sufficient adsorbent to permit it to be held on the stationary phase surface under equilibrium conditions. This will result in band spreading similar in form to sample volume overload and, if it were the sole effect of mass overload, could be treated quantitatively in a like manner. The peaks so formed would be square topped and of similar shape to those shown in Fig. 7.2. However, superimposed on this band spreading process is the second effect of mass overload that results from the deactivation of the adsorbent and the increased effective polarity of the mobile phase. If the charge is massive the sample occupies a significant portion of the column immediately after injection resulting from the first effect described above. The adsorbent thus becomes partially deactivated all the while the overloaded solute is in the column, causing all solutes contained in the mixture to be eluted more rapidly at reduced retention times. This effect is further aggravated by the higher polarity of the mobile phase resulting from the high concentration of the overloaded solute in mobile phase, which will also cause other solutes to be eluted more rapidly. Thus, as a result, the retention times of all the solutes are reduced. Finally, the high concentrations of solute on the adsorbent surface that result from mass overload will, as discussed in the theory of asymmetric peaks, result in nonlinear adsorption isotherms and the eluted peaks will exhibit prounced tails.

Chromatograms demonstrating the effect of mass overload are shown in Fig. 7.5. The column was the same as that used to produce the sample volume overload chromatograms shown in Fig. 7.3 but the sample volume, in this case, was kept at 200 μl for all three chromatograms. The chromatogram on the

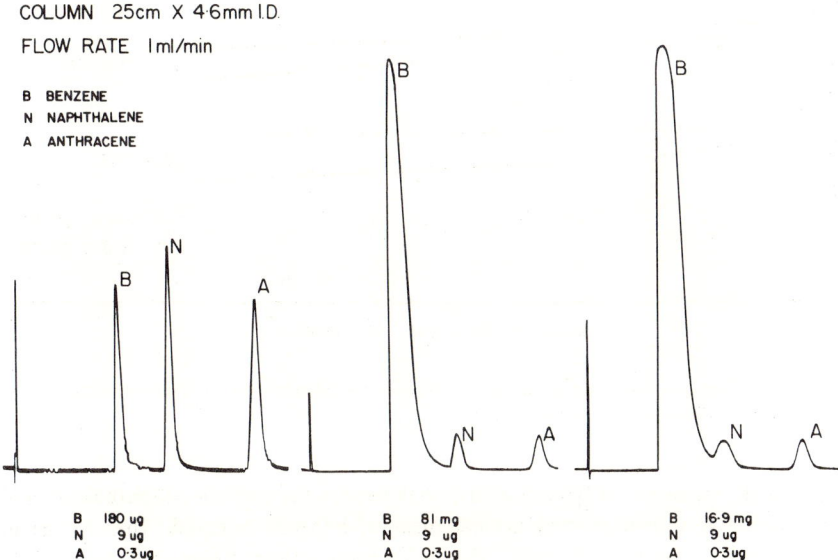

Fig. 7.5. Chromatograms demonstrating mass overload.

308   PREPARATIVE LIQUID CHROMATOGRAPHY

left is that of the reference sample that contained 176 μg of benzene, 9 μg of naphthalene, and 0.3 μg of anthracene. The center and right-hand side chromatograms are for the same sample but with 8 mg and 16 mg of benzene added. The three mass overload effects are clearly demonstrated on both chromatograms. In both chromatograms the benzene peaks have singificantly broadened, as would be expected from the massive charges employed, and they exhibit gross asymmetry together with long tails. It is obvious that the concentrations of benzene in both the mobile and stationary phases have reached those levels where the distribution isotherm is no longer linear. It is also clearly seen that the retention times of all three solutes have been significantly reduced. As discussed previously, this results from both the deactivation of the absorbent by the excessive amount of benzene in the column together with the increased polarity of the solvent since the mobile phase is no longer pure heptane but a solution of benzene in heptane. In Fig. 7.6 the retention distance of the front and rear of

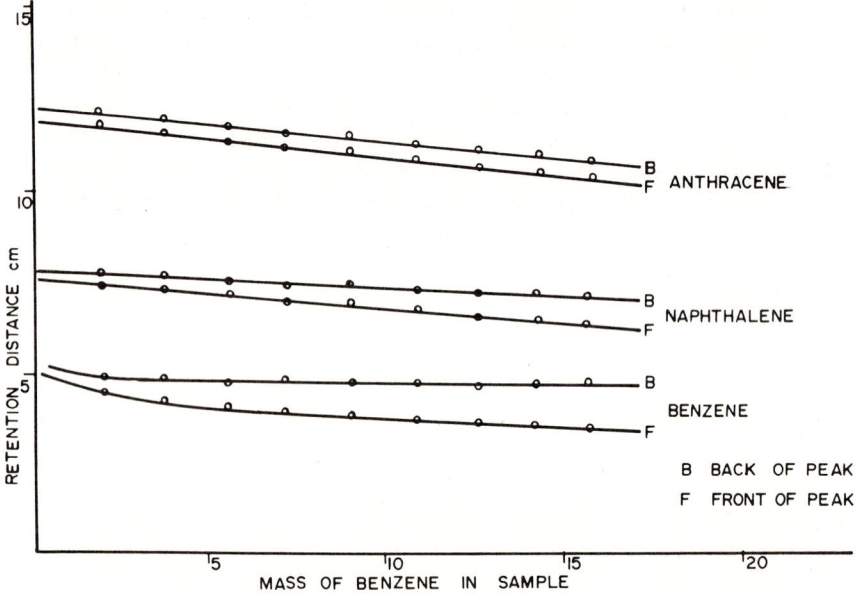

Fig. 7.6. Graph of retention distance of benzene, naphthalene, and anthracene against mass of benzene contained in injected sample.

each peak, measured at 0.6065 of the peak height, for benzene, naphthalene, and anthracene is shown plotted against mass of benzene injected. The change in retention is clearly indicated, the maximum effect being for the solute anthracene and the minimum for the overloaded solute benzene. It should be also noted that there is little change in bandwidth of the most retained solute anthracene with increase in mass of benzene injected. There is, however, a significant increase in the bandwidth of the naphthalene peak and a massive spread in the band width of the benzene peak. The lower chromatogram also shows that the short 25-cm column can cope with a charge of 16 mg and still achieve an effective separation from naphthalene at a level of 0.056% and anthracene at a level of 0.002% of the original mixture. The naphthalene peak, however, will be significantly contaminated with benzene because it is eluted in the tail of the benzene peak.

3 APPARATUS FOR PREPARATIVE LIQUID CHROMATOGRAPHY

Solvent Reservoirs

The basic layout of a preparative liquid chromatograph is shown in Fig. 7.7.

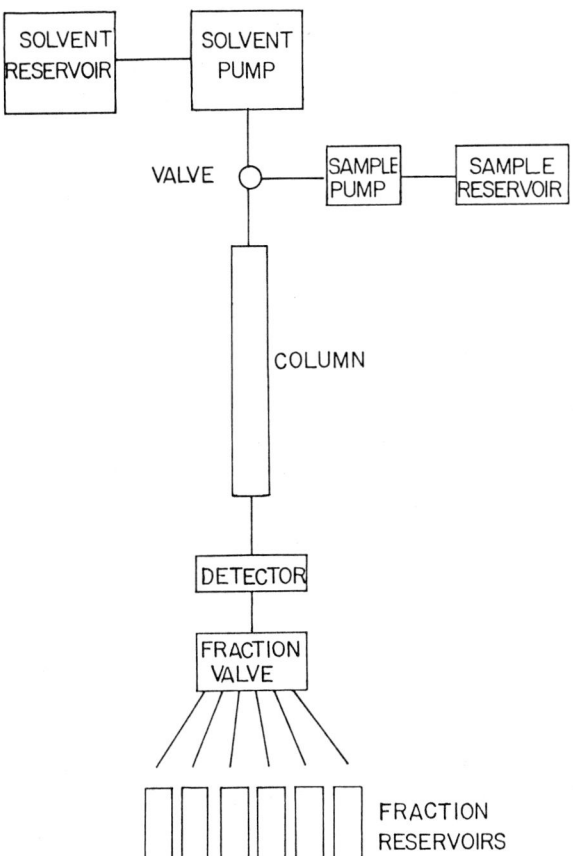

Fig. 7.7. The basic preparative liquid chromatograph.

The solvent reservoir is made of glass or stainless steel; for use with large-diameter columns it should have a capacity of several gallons. The mobile phase passes to an appropriate high-capacity pump, which for use with columns up to 1 in. I.D. should have pumping rates of up to 150 ml/min. If columns of 3 in. I.D. are to be used, then pumping rates of 600 ml/min may be necessary. Owing to the limited mechanical strength of large-diameter columns, maximum pumping pressure of about 3000 psi should be satisfactory. The material from which the pumps are fabricated should be inert (stainless steel) and any glands incorporated in the pump should be of PTFE or similar material.

Injection Methods

There are two forms of injection systems usually employed in preparative liquid chromatography. For injecting large sample volumes an optional second pump can be connected by a T junction to the top of the column. This pump does not need to have the same capacity as the solvent pump and is used solely for sampling purposes but requries a small reservoir of 1 or 2 liters capacity to provide the supply of sample. The required volume of sample is injected onto the column by operating the pump at constant flow rate for a predetermined time. Alternatively, sample valves fitted with appropriate sample loops can be employed for injecting smaller charges. Sample loops of up to 20 or 30 ml are practical, but for sample volumes greater than this the use of a sample pump is recommended. As discussed in Chapter II, open tubes can cause significant band dispersion even with preparative columns, so the sample loop should not be left in line with the column during development. The valve should be turned back to the "flush" position immediately after injection so that the "tail" of the sample left in the loop does not cause dispersion and tailing on the column. The time interval for sampling can be predicted from a knowledge of the column flow rate and volume of sample to be injected.

Columns

The columns should be made of stainless steel and fitted with low-dead-volume unions carrying stainless steel frits of the appropriate pores size to contain the packing. It is important not to exceed the maximum pressure rate for the column tubing. In Table 7.3 the bursting pressure is given for stainless steel tubing of different diameters and wall thickness so that the maximum column pressure can be determined. Column pressures should never exceed 75% of the bursting pressures given in Table 7.3.

The diameter of the column must be chosen to be commensurate with the loads to be used if a choice of columns is available. The loading capacities for a given column are difficult to predict since the sample load will depend on the nature of the mixture to be separated. If the solutes of interest are well separated, loads of 10-20 mg can be placed on a 25-cm-long 4.6-mm-I.D. column

Table 7.3 Stainless Steel Tubing Theoretical Internal Bursting Pressures Based on Barlow's Formula for 75,000 psi Ultimate Bursting Pressure (Welded Tubes)

O.D. (in.)	Wall Thickness (in.)						
	0.025	0.028	0.032	0.035	0.042	0.049	0.058
1/2	7,500	8,400	9,600	10,500	12,600	14,700	17,400
3/4	5,000	5,600	6,400	7,000	8,400	9,800	11,600
1	3,750	4,200	4,800	5,250	6,300	7,350	8,700
1 1/4	3,000	3,350	3,850	4,200	5,050	5,875	6,950
1 1/2	2,500	2,800	3,200	3,500	4,200	4,900	5,800
1 3/4	2,150	2,400	2,750	3,000	3,600	4,200	4,975
2	1,875	2,100	2,400	2,625	3,150	3,675	4,350
2 1/4	1,667	1,867	2,133	2,333	2,800	3,275	3,875
2 1/2	1,500	1,680	1,920	2,100	2,525	4,950	3,475
2 3/4	1,364	1,527	1,745	1,909	2,290	2,675	3,150
3	1,250	1,400	1,600	1,750	2,100	2,450	2,900

and successful separations obtained as has been shown earlier. However, if the separation is difficult it may require a 1-m column ½ in. O.D. to separate an equivalent charge of 20 mg. Wolf [3] gave details of the physical and chromatographic properties of a number of 50-cm columns having different diameters; these data are shown in Table 7.4.

Table 7.4 Physical and Chromatographic Properties of Columns 50 cm Long Having Different Diameters

Column	O.D. (in.)	I.D. (cm)	Area (cm^2)	Packing Weight (g)	Flow (ml/min)^a	n^b
A	1/4	0.21	0.036	1.4	0.52	600
B	3/8	0.77	0.475	38.2	6.82	1325
C	1/2	1.09	0.938	74.5	16.3	1550
D	5/8	1.41	1.57	120.4	22.6	1800
E	3/4	1.71	2.52	184.6	32.0	1900
F	1.0	2.36	4.39	338.7	65.0	2350

^{a} Measured at constant carrier velocity of 1.4 cm/sec using inlet pressures of 500-660 psig.
^{b} Taken from the phenanthrene peak ($k' = 0.81$) at a carrier velocity of 1.4 cm/sec. Length of packing bed in column was 0.5 m in all cases.

Columns A, B, and C were packed with Permaphase ODS, 10-40 μ particle diameter, and tested using a solution of polynuclear aromatic hydrocarbons as

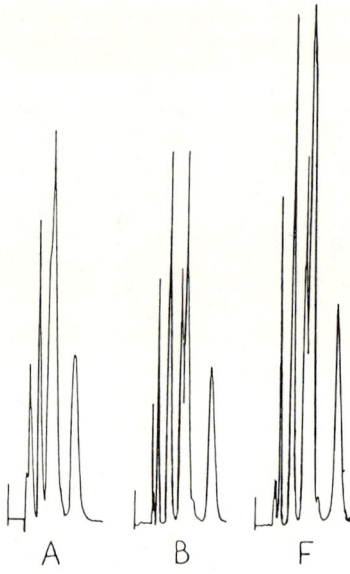

Fig. 7.8. Chromatograms of test mixture on columns A, B, and F with sample sizes of 1, 10, , and 100 µl, respectively.

the sample. The chromatograms Wolfe obtained are shown in Fig. 7.8. It should be emphasized that the sample sizes given on the chromatogram represent sample volumes, *not* sample mass as the packing was a bonded phase which has very limited mass capacity.

Detectors

Unfortunately, there are no detectors that have been specifically designed for preparative liquid chromatography. All detectors presently available have to high a sensitivity, even at their minimum sensitivity setting, to cope with the large sample loads employed. The refractive index detector, being the least sensitive detector, is probably the one most commonly employed in preparative chromatography. However, the multiwavelength UV detector can also be used, by choosing a wavelength for which the solutes of interest have very small extinction coefficients and thus significantly reducing its effective sensitivity. All practical liquid chromatography detectors are concentration-sensitive devices; thus splitting the eluent stream, as in gas chromatography, affords little or no advantage unless the eluent from the column is diluted by another supply of mobile phase just prior to the detector. This possibility, however, is hardly a practical alternative since it further complicates an already highly involved system.

It is hoped that in the not too distant future manufacturers will supply a low-sensitivity detector designed for use with preparative columns.

Fraction Collection

Fraction collecting is best achieved by the use of an appropriate multiport valve. Such a valve can be programmable on a basis of time or can be operated by a peak sensing device; however, a manual override should always be available for nonrepetitive separations. The valve should have at least six or perferably ten collection ports with ports to waste between each fraction port to reduce the volume of solvent that has to be evaporated during recovery. In the author's experience, such a fraction collecting system will be suitable for most applications, but for multicomponent samples, valves with 30 to 40 collection ports may be necessary. If large-diameter columns are being used, the fraction reservoirs may have to be several liters in capacity; such reservoirs will be also necessary where automatic repetitive sampling is employed. Sample recovery is best achieved by evaporating the solvent in a rotary evaporator under reduced pressure and this should be carried out in a well-ventilated fume hood.

Hazards

Preparative liquid chromatography employing large-diameter columns requires the use of large volumes of solvent. It follows that there can be significant fire hazards and toxicity problems, particularly where solvents containing chlorinated hydrocarbons are used as the mobile phase. In carrying out preparative separations careful precautions should therefore be taken to eliminate the possibilities of fire or toxicity risks. It is advisable to operate the entire chromatograph, including its solvent supply and fraction reservoirs, inside a large walk-in fume hood.

Commercial Equipment

The availability of commercial preparative chromatographs capable of using wide-bore columns 4-6 ft in length is rather limited at this time. Many analytical instruments can cope with 3-ft-long columns up to 1 in. I.D., but such columns often strain the capabilities of the instrument. An example of a commercial instrument for preparative liquid chromatography that possesses most of the desirable features for large scale separations is that manufactured by Waters Associates, a photograph is which is shown in Fig. 7.9.

4 STATIONARY PHASES FOR PREPARATIVE LIQUID CHROMATOGRAPHY

The most useful stationary phase for preparative liquid chromatography is silica gel. If the acid nature of silica gel precludes its use, then alumina might

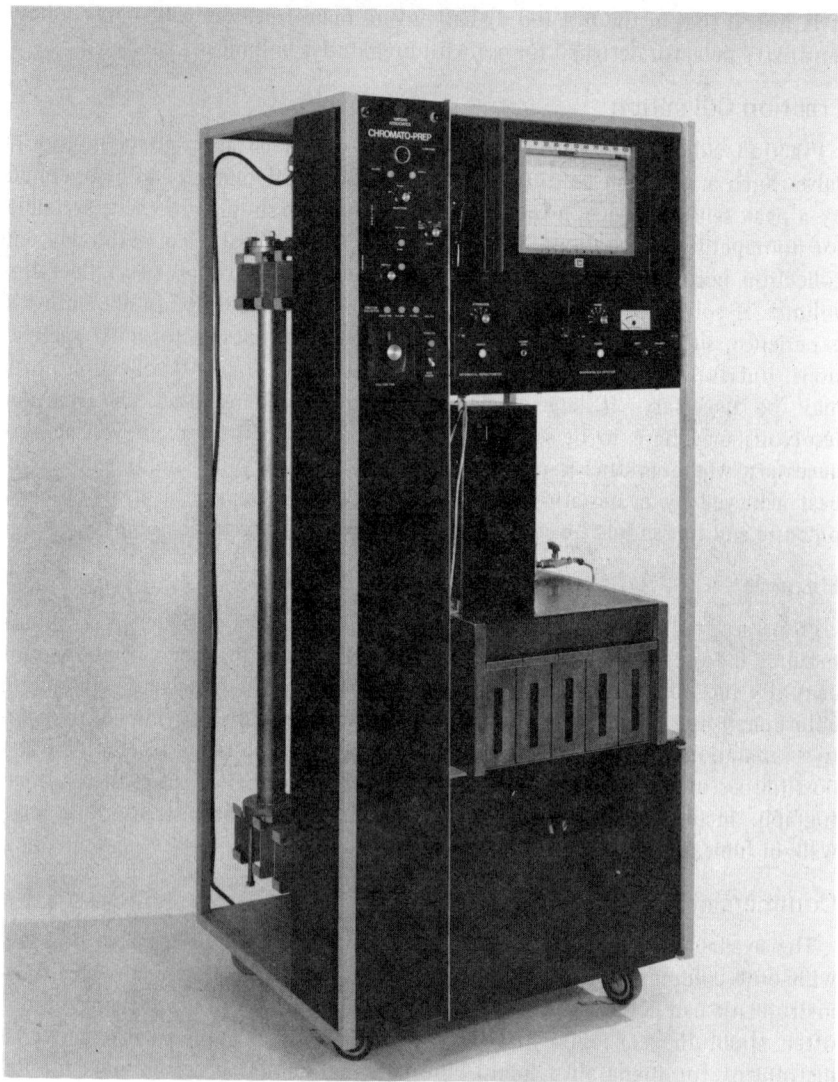

Fig. 7.9. The Waters preparative scale liquid chromatograph.

well be a satisfactory alternative, but the vast majority of preparative separations that have been satisfactorily carried out have utilized silica gel as the stationary phase. Liquid-liquid systems in modern high-pressure columns are rarely employed owing to the difficulties in maintaining phase equilibrium with the large bulk of solvents involved together with the relatively massive apparatus.

The bonded phases presently available do not offer much promise for large-scale separations since their low loading capacities render them unsuitable for preparative work. Bonded phases would only be resorted to under exceptional circumstances where their unique selectivity was necessary or where the quantities of material required were only at the milligram or submilligram level.

An example of the limitations of bonded phases for preparative separation is given in the work of Baker et al. (4) where pyrethrum extract was separated on a Permaphase ODS column 2.3 cm in diameter and 50 cm long. The chromatogram obtained for a 2-mg charge is shown in Fig. 7.10. It is seen that although

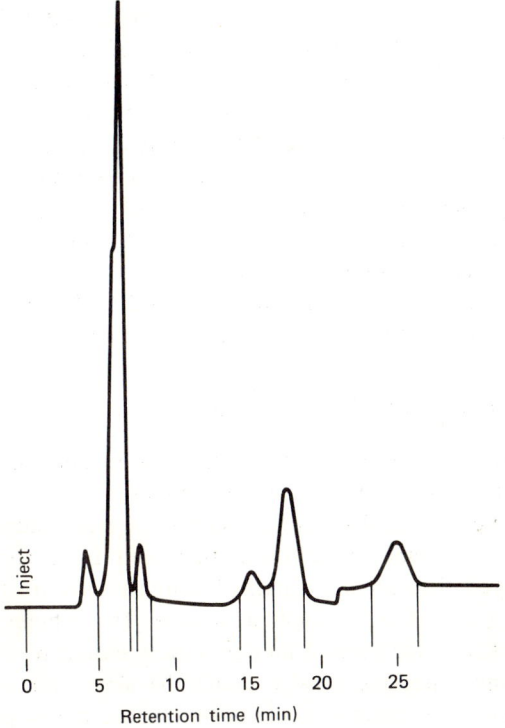

Fig. 7.10. Chromatogram of pyrethrum extract using a larger-diameter preparative column. Sample size, 2 mg; column, 0.5 m × 23 mm I.D. packed with Permaphase ODS; mobile phase, methanol-water (50:50); temperature, 50°; column pressure, 1200 psig; flow rate, 25 ml/min. Vertical lines below the baseline indicate start and finish of fraction collection. AUFS, absorbance units full scale.

the separation is quite satisfactory, the charge of 2 mg is only a fraction of the charge that could be placed on a silica gel column of comparable dimensions. It should be pointed out, however, that the particular separation given may not be possible on a silica gel column, in which case the use of a bonded phase is justified.

Although microparticulate packings having particle diameters of 5 or 10 μ would, without doubt, provide highly efficient columns, the pressure drop across even wide-diameter tubes would exceed the mechanical strength unless very low flow rates were employed. Furthermore, the present cost of such packings would not make preparative columns economically feasible. Silica gels having both greater particle diameters and wider sieve fractions are very much less expensive and can provide columns of adequate efficiency for prepartive work. Biosil A, for example, has a particle diameter range of 20-44 μ, can be dry packed into a column 4 ft. long and 1 in. I.D., and produce about 4000 theoretical plates, which is a very useful efficiency for the majority of preparative scale work. In addition, the cost of the silica gel required to fill such a column would be less than $80.00 at the time of writing this book. For columns of 2-3 in. diameter it is possible to use technical grade silica gel that only costs a few dolloars a pound. The silica gel should be sieved to a 30-50-μ fraction and washed well with concentrated hydrochloric acid to remove inorganic salts, particularly those of iron. The silica gel is then washed free of acid with water, washed with alcohol and dried, and finally activated in an oven at 200°C. The efficiencies obtained from a 4-ft column, 2 in. I.D. may only be 2000-3000 theoretical plates, but such efficiencies would be adequate for many separations.

5 PACKING PREPARATIVE COLUMNS

Although slurry packing preparative columns is feasible and would without doubt provide the highest efficiencies, the procedure would be tedious and difficult in practice, particularly for the larger diameter columns. Dry packing using essentially the same procedure as that used for analytical columns is normally employed. The column is packed using aliquots of silica gel sufficient to pack about 1 in. of the column at a time. The column is trapped both laterally and vertically for about 5 min after each aliquot of silica gel has been added. A maximum packing density should be aimed for as in the case of analytical columns. The column should be terminated at both ends by an appropriate porous frit to contain the packing; the upper frit also acts as a sample diffuser to spread the sample over the complete cross section of the column on injection. In preparative columns, the frit may be of considerable diameter and thus must be thick enough to provide the necessary mechanical strength to cope with the pressure across it; otherwise the frit will collapse. Failing this, the frit should be supported on a coarse filter disk to provide the necessary strength. Preparative scale columns, when packed correctly, should provide between 2000 and 4000 theoretical plates per meter for solutes eluted at a k' value of 1 to 2.

6 PREPARATIVE CHROMATOGRAPHIC PROCEDURE

The Mobile Phase

The correct mobile phase to employ is best determined from the analytical method used for the preliminary examination of the sample. If an analytical separation has not been carried out, or if details of the analytical method are not available, then an appropriate solvent system for a small-scale separation should be developed using the procedures already described. When changing from an analytical scale procedure to a preparative separation it is usually necessary to increase the retention of the solutes and possibly their separation ratios to permit sample overload in the manner previously discussed. Thus in most cases it will be necessary to reduce the polarity of the mobile phase in order to increase the retention of the solutes and provide greater distance between the peaks to accommodate the broadening of the solute bands from column overload. The polarity of the mobile phase can be modified by progressively reducing the proportion of the more polar solvent in the mobile phase, should a mixture of solvents be used, until the necessary separation is obtained. Alternatively, if a single solvent was used as the mobile phase, then a less polar solvent can be chosen from Table 4.8 (Chapter IV) and either used on its own, or used to dilute the original solvent to make it less polar. Solvents used in preparative chromatography have to be eventually evaporated in bulk to recover the separated products, and it is therefore worthwhile to select, where possible, more volatile solvents for the mobile phase to facilitate this recovery procedure. For example, hexane would be a more suitable solvent to chose than heptane since the former has very similar solvent properties but is less volatile. In the same way, if the slight difference in polarity between dichloroethane and dichloromethane can be tolerated, the more volatile dichloromethane would be the better choice for preparative separations.

Sample Injection

The effect of sample volume and sample mass on resolution has already been discussed and, in general, it can be said that as small a volume of charge as possible should be used, which should contain the sample at the maximum concentration permissible to achieve resolution. Large volumes of sample containing the solutes at low concentrations should only be resorted to in cases of poor solute solubility in the mobile phase. If relatively large volumes of sample are to be used, it follows that injection by means of a sample pump is also to be recommended. A practical example of this compromise between sample volume and sample mass in the separation of diastereoisomers is given in Fig. 7.11. The column was 4 ft long, 1 in. I.D. and dry packed with Biosil A to

318   PREPARATIVE LIQUID CHROMATOGRAPHY

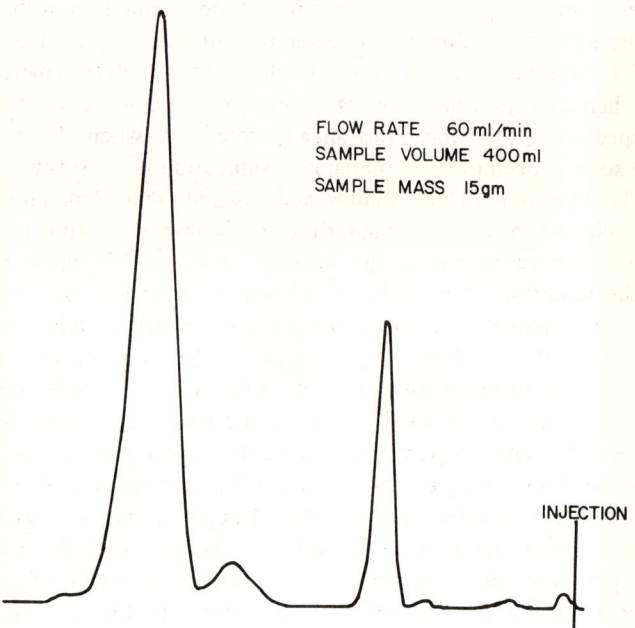

Fig. 7.11. The preparative separation of diastereoisomers.

provide an efficiency of about 4000 theoretical plates. The sample was a reaction product from an organic synthesis, and the two diastereoisomers are the two major peaks in the chromatogram. The sample volume was 200 ml and contained 7.5 g of the material dissolved in the mobile phase. It is seen that an excellent separation is obtained, not merely between the isomers themselves, but from other minor impurities present in the product.

Recycling Development

A method for improving the separation of a particular pair of solutes that are poorly resolved on a preparative column is to use the recycling technique. Recycling in effect increases the number of theoretical plates that can be obtained from the column by repeatedly using the column; it does so at the expense of the scope of the system and its peak capacity. It is another way of trading scope and speed for resolution.

In this method, partially resolved components eluting from the column pass through the detector, and then, instead of being collected or going into fraction collectors, they are directed back into the pump for subsequent passes through the columns and detector. In this fashion, the effective length of the column is increased considerably. The components pass through the detector during each cycle in order to monitor the progress of the separation. In order to avoid remixing the components that are being recycled it is essential to use a very low displacement volume pumping system. Recycle can be very valuable for handling substantial overloads in preparative scale work, thereby increasing yield per injection. This is understood as a consequence of the increased effective length of the column, since doubling the length of a column system in single-pass operation will also increase the loading capability of the column by a factor of about 1.4.

An example of the use of the recycling technique, taken from an application sheet of Waters Associates, is shown in Fig. 7.12. The work was carried out by

Column	3' 8" o.d. × 9' porasil 1
Solvent	1% isopropanol in hexane (V V)
Sample	100 mg IN 2ml 100P
Instrument	Al.C 100 with uv photometer at 06400

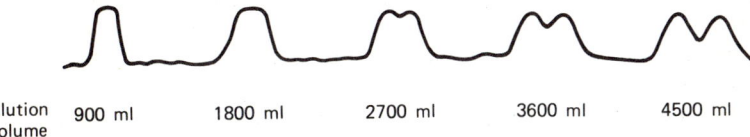

| Elution volume | 900 ml | 1800 ml | 2700 ml | 3600 ml | 4500 ml |

Fig. 7.12. Preparative separation of stereoisomers by recycle liquid chromatography.

Professor Nakanishi of Columbia University and involved the separation of naturally occurring plant hormones.

A 9-ft × 3/8-in. column was employed pack with a fully porous silica and a 100-mg sample injection. The initial elution volume of 900 ml showed no separation, and so recycle was employed; by the fifth pass through the column, complete resolution had been achieved. Using 2 ½ times the loading (250 mg), complete baseline separation was achieved with only one additional pass through the system. This again demonstrates both the resolving capability and the utility of recycle for handling high sample loads.

Gross Overload with Peak Cutting

The sample load that can be placed on a preparative column can be increased to the point where the resolution between the solutes is partially sacrificed, such that although the peaks merge, the major and central portion of the bands are still pure and uncontaminated. This type of overload is often termed gross overload and is an example of the case where high loading is achieved at the expense of that portion of the sample contained in the merged part of the peaks. Providing the contamination of one peak with the other does not extend further than one standard deviation from the maximum of either peak, approximately 75% of the pure solute is still recoverable by appropriate and accurate peak cutting. Under such circumstances the fraction required is cut from the peak between the two points of inflexion, that is, at 60% of the peak height at either side of the peak. Another example of gross overload is where a minor constituent together with a small proportion of the major solute is collected and reinjected onto the column. Because the proportion of what was the major constituent of the mixture is now at a concentration commensurate with that of the original minor constituent, complete separation can now be obtained. To increase the yield the initial gross overload is often repeated several times; the combined contaminated fractions are then injected onto the column and the desired pure minor component collected.

7 THE FUTURE OF PREPARATIVE LIQUID CHROMATOGRAPHY

From the present point of view, preparative liquid chromatography appears capable of being scaled up to a far greater extent than its gas chromatographic counterpart. There is no apparent limit to the column diameter that can be used and, providing the sample is diffused across the entire column cross section on injection, high column efficiencies should be maintained. Furthermore, the efficiency obtainable in large-diameter liquid columns is likely to be much in excess of those obtainable from gas chromatographic columns of similar dimensions. The separation of 100-g quantitites of material is from our present knowledge, predictably possible and separations on the kilogram scale seem eminently feasible. The mechanical problems associated with large-scale columns do not appear serious; solvent pumps with high deliveries at moderate pressures are already available and most of the pipe fittings, valves, and the like can be obtained from manufacturers that supply equipment for general chemical plant operation. The design, for example, of a 10-ft-long 3-ft-I.D. column may present some problems since the heat adsorbed and evolved during the adsorption-desorption processes of chromatographic development will need to be effective dissipated or it could impair the separation obtained. However, there are a limited number of such columns that are being used in industry and have been for many years. The main problem associated with large-scale chromatography, however, is one of economics. The cost of packing a large column with effective absorbent will be high and the life of the adsorbent is not predictable since it will be dependent on the nature of the samples being separated. Considerable volumes of solvent will be required, but the make-up cost of these could be made reasonable, providing an efficienct solvent recovery plant was employed.

In any event, there will be a considerable capital investment involved and the application of large-scale chromatography is likely to be in high-cost product areas such as pharmaceuticals, essential oils, and food flavors.

Chromatographs can be used on a continuous basis by utilizing a modified form of the moving bed process that was introduced by Scott [5] for gas-liquid chromatography. This form of chromatography takes a continuous feed of sample and operates in the following manner. Adsorbent is allowed to fall continuously down the column against an upward flow of mobile phase. The sample enters continuously at the middle of the column at A (Fig. 7.13) and

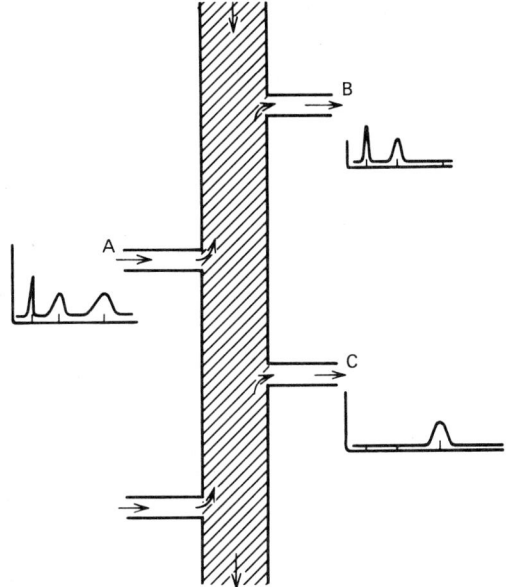

Fig. 7.13. Column arrangement for continuous chromatography.

moves up the column countercurrent to the absorbent and leaves a B. By suitable adjustment of the relative rates of solvent flow and movement of the adsorbent some components of the mixture move up with the solvent and others move down with the adsorbent. If we imagine that the ordinary chromatogram of the mixture is as indicated at the inlet A, the relative speeds of solvent and stationary phase define an imaginary line on this chromatogram, components to the left of which move up and those to the right of which move down. The components that move down are stripped from the adsorbent by arranging that the adsorbent moves into a region of column through which flows a more polar solvent entering the columns at C.

The system as described is hardly practical for liquid chromatography but was used successfully in gas chromatography. Barker et al. [6] used a novel modification of this principle in that they situated the column in circular form around the periphery of a large wheel that had a series of inlet and outlet posts around the wheel perimeter. In this way the column and adsorbent rotated with the wheel, the solvent entering and leaving the column intermittently as the ports passed the solvent supply and take off lines. This device was used by Barker for liquid chromatography separations with moderate success. Another modification described by Barker and Deeble [7] utilized a number of stationary columns programmed by a stationary multiport valve to the same effect. Much development work remains before the continuous form of preparative-scale liquid chromatography can be established as a viable process, but this principle of separation holds exciting prospects for the future.

REFERENCES

1. A. Klinkenberg, *Gas Chromatography 1960*, R. P. W. Scott, Ed., Butterworths, London, 1960.
2. J. J. Destefano and C. Beachell, *J. Chromatogr. Sci.*, **10**, 654 (1972).
3. J. P. Wolf III, *Anal. Chem.*, **45**, 1248 (1973).
4. D. R. Baker et al., *J. Chromatogr.*, **83**, 233 (1973).
5. R. P. W. Scott, *Gas Chromatography 1958*, D. H. Desty, Ed., Butterworths, London, 1958.
6. P. E. Barker, *Preparative Gas Chromatography*, A. Zlatkis and V. Pretorius, Eds., Wiley-Interscience, London, 1971, Chapter 10, p. 325.
7. P. E. Barker and R. E. Deeble, *Anal. Chem.*, **45**, 1121 (1973).

INDEX

Adsorptive capacity, effect on peak capacity, LLC, 93
  LSC, 89
Air dissolved in sample, effect of, 265
Alumina, use as stationary phase, 216
Ancilliary equipment, 182
Apparatus, column detector connections, 137
  columns, 133
  equilibrium systems, 126
  flow programmer, 124
  flow measurement, 185
  gradient elution, 106
  injection systems, 126
  mobile phase supply, 103
  preparative LC, 313
  pulse dampners, 124
  solvent degasing, 105
  solvent pumps, 120
  solvent reservoirs, 104
Analysis qualitative, 268
Analysis quantitative, 269
  procedure, 270
Assessment of stationary phases, 194
  loading capacity, 201
  peak capacity, 197
  permeability, 204
  resolving power, 196
  scope, 198

Balanced density solvent, packing procedure, 247
  preparation of, 248
Base line instability, 267
Basic liquid chromatograph, 9
Bonded phases, 216
  disadvantages of, 221
  preparation of, 217
  properties of, 219
Bubbles elimination of, 266
Bulk property detectors, definition, 137
  sensitivity of, 143
  types of, 144

Carbon as stationary phase, 221

Choice of detector, 181
Choice of mobile phase, 227
Choice of phase system LLC, 225
Choice of stationary phase, 224
Chromatogram data, 259
  differential, 16
  intergral, 16
  nomenclature, 16
  normal, 16
Chromatograph, basic form, 9
Chromatography, classification, 6
  history, 1
  definition, 4
Classification, chromatography, 6
  detectors, 137
Column, conditioning, 249
  design, 133
  detector connections, 137
  efficiency from the plate theory, 33
  how to calculate, 35
  method of measuring, 261
  extra column dispersion, 254
  fractionator, 313
  length, vs flow rate, 251
  vs linear velocity, 252
  vs load, 252
  in preparative LC, 300
  oven, basic form, 16
  oven specification, 136
  overloading, preparative columns, 301, 306
  packing, dry packing, 246
  wet packing, 247
  slurry packing, 247
  preparative columns, 300
  permeability, measurement of, 204
  preparative LC, *see* Preparative LC
  radius, effect on loading in preparative columns, 299
  resolving power, 36
Compression fittings and unions, 185
Computer program for gradient elution, 187
Connecting tubes, dispersion, 76
Contaminated syringe, effect of, 265
Continuous LC, preparative, 320
Critical state chromatography, 8

323

INDEX

Deactivation of silica gel from theory, 210
Dead volume, method of measurement, 260
Degasing systems, 105
Detectors, basic form, 10
  bulk property, 143, 153
  choice of, 181
  classification, specifications, 137
  connection to column, 137
  dielectric constant, 148
  displacement effects, 265
  electrical conductivity, 152
  electrochemical, 176
  fluorometric, 177
  heat of adsorption, 167
  linearity, measurement of, 138
  mass detector, 165
  moving chain, 163
  moving wire, 158
  poleragraphic, 173
  preparative LC, 312
  refractive index, 144
  sensitivity, measurement of, 139
  solute property detectors, 155
  solute transport, 158
  spray injection detector, 179
  UV detector, 155
Development, methods of, displacement development, 15
  elution development, 12
  frontal analysis, 11
Dielectric constant detector, 148
Dissolved air, effect of, 265
Dispersion, in connecting tubes, 76
  radial, 72
Displacement development, 15
Displacement effects, detector, 265
Dry packing, 246

Effective plates, relationship to theoretical plates, 40
  resolving power, 42
Efficiency, from the plate theory, 33
  how to calculate, 35
  method of measuring, 261
Electrical conductivity detector, 152
Electrochemical detector, 176
Elution curve, differential equation, 27
  equation gaussian form, 43
  equation of, 29
  points of inflection, 35
Elution development, 12
Extra column effects, 254

Flow programming, 124
Flow rate, measuring apparatus, 185
  method of measuring, 259

Fluorometer detector, 177
Fraction collector for preparative LC, 313
Frontal analysis, 11

Gas liquid chromatography, applications of, 8
Gas solid chromatography, applications of, 7
Gradient elution, apparatus, 106
  by displacement, 116
  by solvent mixing, 106
  by temperature programming, 116
  Incremental, 111
  other methods, 118
  simple incremental, 239
  theory of, incremental, 227
  solvents for, 232

Hazards, preparative LC, 313
Heat of adsorption detector, 167
HETP, equation of, from the rate theory, empirical equation, 71
  experimental support of, 66
  liquid liquid chromatography, 59
  liquid solid chromatography, 63
History of chromatography, 1

Inflection points on the elution curve, 35
Injection system, basic form, 10
  for preparative LC, 310
  sample valve, 131
  stop flow method, 130
  with solvent flow, 127
Injection of sample, 256
Instability, base line, 267
Ionic exchange media, 222
IR/LC, 294

LC preparative, *see* Preparative LC
LC/IR, 294
LC/MS, direct sampling, 286
  future developments, 293
  methods, 278
  wire transport method, 283
    design of interface, 283
    operation, 286
    performance of, 287
LC/NMR, 294
LC/RAMAN, 294
LC/UV, apparatus, 277
  applications, 277
  method, 275
Length, column, *see* Column, length
Length of column vs flow rate, 250
Liquid chromatography/spectroscopy, 273
Liquid liquid chromatography, applications of, 9
  choice of phase systems, 255

INDEX 325

Liquid solid chromatography, applications, 8
Loading capacity, 201
  definition of, 201
  measurement of, 202
Longitudinal diffusion, 49

Mass detector, 165
Mass of sample, 306
Maximum charge, preparative LC, 301, 306
Micro syringes, 185
Mobile phase, choice of, 227
  preparative LC, 317
  supply, 103
    basic form, 9
Moving chain detector, 163
Moving wire detector, 159
MS/LC, see LC/MS
Multipath process, 49

NMR/LC, 294
Noise, sources of, 265

Overload, column, 301, 306

Packing, dry packing, 246
  slurry packing, 247
  wet packing, 247
Paper chromatography, applications of, 8
Particle diameter, effect of preparative LC, 300
Peak areas, method of measuring, 263
Peak asymmetry, from the plate theory, 31
Peak capacity, 81
  effect of, adsorptive capacity, 89
    detector sensitivity, 86
    sample concentration, 86
  of silica gel, 197
Peak cutting in preparative LC, 319
Peak shape, different forms, 20
Pellicular packing, 215
Plate theory, 25
Points of inflection, 35
Polargraphic detector, 173
Polyamide as a stationary phase, 221
Pratical notes on operation, 265
Preparation of sample, 256
Preparative LC, apparatus for, 309
  choice of mobile phase, 317
  column configuration, 310
  column overload, 301
    effect of charge volume, 301
    effect of sample mass, 306
  continuous LC, 321
  detector, 312
  effect of, column length, 300
    distribution coefficient, 300

effective column radius, 299
factors effecting maximum charge, 298
fraction collecting, 313
future trends, 320
hazards, 313
method of injection, 310
packing density, 300
packing methods, 316
peak cutting, 319
recycling, 318
sample injection, 317
specific surface area, 300
stationary phase, 313
Pulse dampner, 124
Pumps, choice of, 123
  mechanical, 121
  pneumatic, 122
  servicing, 122

Qualitative analysis, 268
Quantitative analysis, 269
  procedure, 270

Radial dispersion, 72
Radius, column, effect of, in preparative LC, 299
RAMAN/LC, 294
Rate theory, 47
Recycling, preparative LC, 318
Recorder, basic requirements, 10
Reduced plate height, 204
Refractive index detector, 144
Resistance to mass transfer, in liquid liquid chromatography, 56
  in liquid solid chromatography, 60
  in mobile phase, 50
  in stationary phase, 51
Resolving power, from effective plates, 42
  from plate theory, 36
Resolution, as a function of retention volume, 38
  as a function k' and, 39
  from plate theory, 36
Retention volume, from plate theory, 29
Retention ratios, method of measuring, 263

Sample, concentration, effect on peak capacity, 86
  injection, method of, 256
    preparative LC, 310
    system, 126
  limits in preparative LC, 301, 306
  mass overload in preparative LC, 306
  preparation of, 256
  solvent, effect of, 265
  volume effect of, 131

on loading limit preparative LC, 301
   volume maximum from the plate theory,
      45
Scope, of silica gel, 198
Sensitivity, detector, measurement by,
      incremental method, 138
   log dilution method, 141
Silica gel, chromatographic properties, 213
   effect of heat, 210
      of OH groups, 210
      of temperature on retention, 211
   formation of, 208
   loading capacity, 201
   peak capacity, 197
   permeability, 204
   preparation of, 208
   resolving power, 196
   scope, 198
   structure of, 208
   surface of, 208
Slurry packing, 247
Solute transport detector, 158
Solvent, equilibrium systems, 126
   properties, table of, 240
   reservoirs, 104
Solvents for incremental gradient elution,
      232

Spectroscopy and LC, 273
Spray impact detector, 179
Stationary phase, assessment of, 194
   choice of, 224
   preparative LC, 313
Support, preparation of, 222
Surface area, loading on prep columns,
      300
Syringe, contamination effects, 265

Temperature, silica gel activation, 210
Theoretical Plates, from the plate theory,
      33
   how to calculate, 35
Thin layer chromatography, applications
      of, 8
Transfer coefficient, 54

Unions, 185
UV Detector, 155
UV/LC, *see* LC/UV

Velocity, mobile phase, measurement of,
      260
Volume of sample, *see* Sample

Wet packing, 247